Practical High-Performance Liquid Chromatography

FIFTH EDITION

Veronika R. Meyer

Swiss Federal Laboratories
for Materials Testing and
Research (EMPA),
St. Gallen,
Switzerland

A John Wiley and Sons, Ltd., Publication

This edition first published 2010
© 2010 John Wiley and Sons, Ltd.

Registered office

John Wiley & Sons Ltd, The Atrium, Southern Gate, Chichester, West Sussex, PO19 8SQ, United Kingdom

For details of our global editorial offices, for customer services and for information about how to
apply for permission to reuse the copyright material in this book please see our website at www.wiley.com.

Library of Congress Cataloging-in-Publication Data

Meyer, Veronika.
        [Praxis der Hochleistungs-Flüssigchromatographie. English]
        Practical high-performance liquid chromatography / Veronika R. Meyer. – 5th ed.
                p. cm.
        Includes bibliographical references and index.
        ISBN 978-0-470-68218-0 (cloth) – ISBN 978-0-470-68217-3 (pbk.)
    1. High performance liquid chromatography.   I. Title.
        QD79.C454M4913 2010
        543'.84–dc22

                                                                      2009052143

A catalogue record for this book is available from the British Library.

ISBN H/bk 978-0470-682180 P/bk 978-0470-682173

Set in 10/12pt, Times Roman by Thomson Digital, Noida
Printed and bound in Great Britain by TJ International, Padstow, Cornwall

Cover photo:
Allmenalp waterfall at Kandersteg, Switzerland (Veronika R. Meyer)

To the memory of Otto Meyer

Alles ist einfacher, als man denken kann,
zugleich verschränkter, als zu begreifen ist.
Goethe, *Maximen*

Everything is simpler than can be imagined,
yet more intricate than can be comprehended.

Contents

Contents ix

Preface to the Fifth Edition

A small jubilee! This book started 30 years ago with the first German edition, with no idea that it could become a success story. Its content became younger with every edition, a fact which is not true concerning the author. In fact, I am sure that the latter cannot be a serious wish. No question: decades of experience are for the benefit of the book.

A new topic is now included: Chapter 20 about quality assurance. Part of it could be found before in chapter 19 but now the subject is presented much broadly and independent of 'Analytical HPLC'. Two chapters in the appendix were updated and expanded by Bruno E. Lendi, namely the ones about the instrument test (now chapter 25) and troubleshooting (now chapter 26). Some new sections were created: 1.7, comparison of HPLC with capillary electrophoresis; 2.11, how to obtain peak capacity; 8.7, van Deemter curves and other coherences; 11.3, hydrophilic interaction chromatography; 17.2, method transfer; 18.4, comprehensive two-dimensional HPLC; 23.3, fast separations at 1000 bar; 23.5, HPLC with superheated water. In addition, many details were improved and numerous references added.

Jump into the HPLC adventure! It can be a pleasure if you know the craft and its theoretical background.

St. Gallen, July 2009   Veronika R. Meyer

Important and Useful Equations for HPLC

This is a synopsis. The equations are explained in Chapters 2 and 8.

Retention factor:

$$k = \frac{t_R - t_0}{t_0}$$

Separation factor, α value:

$$\alpha = \frac{k_2}{k_1}$$

Resolution:

$$R = 2\frac{t_{R2} - t_{R1}}{w_1 + w_2} = 1.18\frac{t_{R2} - t_{R1}}{w_{1/2_1} + w_{1/2_2}}$$

Number of theoretical plates:

$$N = 16\left(\frac{t_R}{w}\right)^2 = 5.54\left(\frac{t_R}{w_{1/2}}\right)^2 = 2\pi\left(\frac{h_P \cdot t_R}{A_P}\right)^2$$

$$N \sim \frac{1}{d_p}$$

Height of a theoretical plate:

$$H = \frac{L_c}{N}$$

Asymmetry, tailing:

$$T = \frac{b_{0.1}}{a_{0.1}} \quad \text{or} \quad T = \frac{w_{0.05}}{2f}$$

Practical High-Performance Liquid Chromatography, Fifth edition   Veronika R. Meyer
© 2010 John Wiley & Sons, Ltd

Linear flow velocity of the mobile phase:

$$u = \frac{L_c}{t_0}$$

Porosity of the column packing:

$$\varepsilon = \frac{V_{\text{column}} - V_{\text{packing material}}}{V_{\text{column}}}$$

Linear flow velocity of the mobile phase if $\varepsilon = 0.65$ (chemically bonded stationary phase):

$$u(\text{mm/s}) = \frac{4F}{d_c^2 \pi \varepsilon} = 33 \frac{F(\text{ml/min})}{d_c^2(\text{mm}^2)}$$

Breakthrough time if $\varepsilon = 0.65$:

$$t_0(\text{s}) = 0.03 \frac{d_c^2(\text{mm}^2) L_c(\text{mm})}{F(\text{ml/min})}$$

Reduced height of a theoretical plate:

$$h = \frac{H}{d_p} = \frac{L}{N d_p}$$

Reduced flow velocity of the mobile phase:

$$v = \frac{u \cdot d_p}{D_m} = 1.3 \times 10^{-2} \frac{d_p(\mu m) F(\text{ml/min})}{\varepsilon D_m(\text{cm}^2/\text{min}) d_c^2(\text{mm}^2)}$$

Reduced flow velocity in normal phase (hexane, analyte with low molar mass, i.e. $D_m \approx 2.5 \times 10^{-3}$ cm²/min) if $\varepsilon = 0.8$:

$$v_{\text{NP}} = 6.4 \frac{d_p(\mu m) F(\text{ml/min})}{d_c^2(\text{mm}^2)}$$

Reduced flow velocity in reversed phase (water/acetonitrile, analyte with low molar mass, i.e. $D_m \approx 6 \times 10^{-4}$ cm²/min) if $\varepsilon = 0.65$:

$$v_{\text{RP}} = 33 \frac{d_p(\mu m) F(\text{ml/min})}{d_c^2(\text{mm}^2)}$$

Note: Optimum velocity at approx. $v = 3$; then $h = 3$ with excellent column packing (analyte with low molar mass, good mass transfer properties).

Reduced flow resistance:

$$\Phi = \frac{\Delta p\, d_p^2 d_c^2 \pi}{4 L_c \eta F} = 4.7 \frac{\Delta p\,(\text{bar})\, d_p^2\,(\mu m^2)\, d_c^2\,(mm^2)}{L_c\,(mm)\, \eta\,(mPas)\, F\,(ml/min)}$$

Note: $\Phi = 1000$ for properly packed and not clogged columns with particulate stationary phase.

$$\Delta p \sim \frac{1}{d_p^2}$$

Total analysis time:

$$t_{\text{tal}} = \frac{L_c d_p}{v D_m}(1 + k_{\text{last}})$$

Total solvent consumption:

$$V_{\text{tal}} = \frac{1}{4} L_c d_c^2 \pi \varepsilon (1 + k_{\text{last}})$$

$$V_{\text{tal}} \sim d_c^2$$

Peak volume:

$$V_{\text{peak}} = \frac{d_c^2 \pi L_c \varepsilon (k + 1)}{\sqrt{N}}$$

A_P	peak area
$a_{0.1}$	width of the leading half of the peak at 10% of height
$b_{0.1}$	width of the trailing half of the peak at 10% of height
d_c	inner diameter of the column
D_m	diffusion coefficient of the analyte in the mobile phase
d_p	particle diameter of the stationary phase
F	flow rate of the mobile phase
f	distance between peak front and peak maximum at 0.05 h
h_P	peak height
k_{last}	retention factor of the last peak
L_c	column length
t_R	retention time

t_0	breakthrough time
V	volume
w	peak width
$w_{1/2}$	peak width at half height
$w_{0.05}$	peak width at 0.05 h
η	viscosity of the mobile phase
Δp	pressure drop

1 Introduction

1.1 HPLC: A POWERFUL SEPARATION METHOD

A powerful separation method must be able to resolve mixtures with a large number of similar analytes. Figure 1.1 shows an example. Eight benzodiazepines can be separated within 70 seconds.

Such a chromatogram provides directly both qualitative and quantitative information: each compound in the mixture has its own elution time (the point at which the signal appears on the screen) under a given set of conditions; and both the area and height of each signal are proportional to the amount of the corresponding substance.

This example shows that *high-performance liquid chromatography* (HPLC) is very efficient, i.e. it yields excellent separations in a short time. The 'inventors' of modern chromatography, Martin and Synge,[1] were aware as far back as 1941 that, in theory, the stationary phase requires *very small particles* and hence a *high pressure* is essential for forcing the mobile phase through the column. As a result, HPLC was sometimes referred to as *high-pressure liquid chromatography*.

1.2 A FIRST HPLC EXPERIMENT

Although this beginner's experiment described here is simple, it is recommended that you ask an experienced chromatographer for assistance.

It is most convenient if a HPLC system with two solvent reservoirs can be used. Use water and acetonitrile; both solvents need to be filtered (filter with $< 1 \, \mu m$ pores) and degassed. Flush the system with pure acetonitrile, then connect a so-called reversed-phase column (octadecyl ODS or C_{18}, but an octyl or C_8 column can be used as well)

---

[1] A.J.P. Martin and R.L.M. Synge, *Biochem. J.,* **35**, 1358 (1941).

Practical High-Performance Liquid Chromatography, Fifth edition   Veronika R. Meyer
© 2010 John Wiley & Sons, Ltd

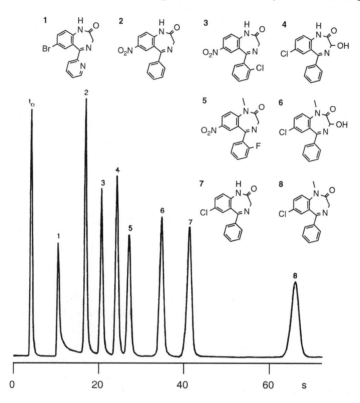

Figure 1.1 HPLC separation of benzodiazepines (T. Welsch, G. Mayr and N. Lammers, *Chromatography*, InCom Sonderband, Düsseldorf, 1997, p. 357). Conditions: samples: 40 ng each; column: 3 cm × 4.6 mm i.d.; stationary phase: ChromSphere UOP C18, 1.5 μm (nonporous); mobile phase: 3.5 ml min^{-1} water–acetonitrile (85:15); temperature: 35 °C; UV detector 254 nm. Peaks: 1 = bromazepam; 2 = nitrazepam; 3 = clonazepam; 4 = oxazepam; 5 = flunitrazepam; 6 = hydroxydiazepam (temazepam); 7 = desmethyldiazepam (nordazepam); 8 = diazepam (valium).

with the correct direction of flow (if indicated) and flush it for *ca.* 10 min with acetonitrile. The flow rate depends on the column diameter: 1–2 ml min^{-1} for 4.6 mm columns, 0.5–1 ml min^{-1} for 3 mm and 0.3–0.5 ml min^{-1} for 2 mm columns. Then switch to water–acetonitrile 8:2 and flush again for 10–20 min. The UV detector is set to 272 nm (although 254 nm will work too). Prepare a coffee (a 'real' one, not decaffeinated), take a small sample before you add milk, sugar or sweetener and filter it (<1 μm). Alternatively you can use tea (again, without additives) or a soft drink with caffeine (preferably without sugar); these beverages must be filtered, too. Inject 10 μl of the sample. A chromatogram similar to the one shown in Figure 1.2 will

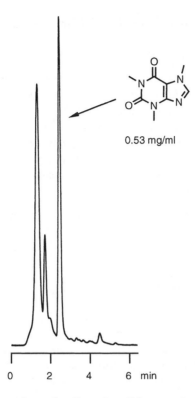

0.53 mg/ml

0   2   4   6 min

Figure 1.2 HPLC separation of coffee. Conditions: column, 15 cm × 2 mm i.d.; stationary phase, YMC 120 ODS-AQ, 3 μm; mobile phase, 0.3 ml min^{-1} water–acetonitrile (8:2); UV detector 272 nm.

appear. The caffeine signal is usually the last large peak. If it is too high, inject less sample and vice versa; the attenuation of the detector can also be adjusted. It is recommended to choose a sample volume which gives a caffeine peak not higher than one absorption unit as displayed on the detector. If the peak is eluted late, e.g. later than 10 min, the amount of acetonitrile in the mobile phase must be increased (try water–acetonitrile 6 : 4). If it is eluted too early and with poor resolution to the peak cluster at the beginning, decrease the acetonitrile content (e.g. 9 : 1).

The caffeine peak can be integrated, thus a quantitative determination of your beverage is possible. Prepare several calibration solutions of caffeine in mobile phase, e.g. in the range 0.1–1.0 mg ml^{-1}, and inject them. For quantitative analysis, peak areas can be used as well as peak heights. The calibration graph should be linear and run through the origin. The caffeine content of the beverage can vary within a large range and the value of 0.53 mg ml^{-1}, as shown in the figure, only represents the author's taste.

After you have finished this work, flush the column again with pure acetonitrile.

1.3 LIQUID CHROMATOGRAPHIC SEPARATION MODES

Adsorption Chromatography

The principle of adsorption chromatography (normal-phase chromatography) is known from classical column and thin-layer chromatography. A relatively polar material with a high specific surface area is used as the stationary phase, silica being the most popular, but alumina and magnesium oxide are also often used. The mobile phase is relatively nonpolar (heptane to tetrahydrofuran). The different extents to which the various types of molecules in the mixture are adsorbed on the stationary phase provide the separation effect. A nonpolar solvent such as hexane elutes more slowly than a medium-polar solvent such as ether.

Rule of thumb: polar compounds are eluted later than nonpolar compounds.

Note: polar means water-soluble, hydrophilic; nonpolar is synonymous with fat-soluble, lipophilic.

Reversed-Phase Chromatography

The reverse of the above applies:

(a) The stationary phase is very nonpolar.
(b) The mobile phase is relatively polar (water to tetrahydrofuran).
(c) A polar solvent such as water elutes more slowly than a less polar solvent such as acetonitrile.

Rule of thumb: nonpolar compounds are eluted later than polar compounds.

Chromatography with Chemically Bonded Phases

The stationary phase is covalently bonded to its support by chemical reaction. A large number of stationary phases can be produced by careful choice of suitable reaction partners. The reversed-phase method described above is the most important special case of chemically bonded-phase chromatography.

Ion-Exchange Chromatography

The stationary phase contains ionic groups (e.g. NR_3^+ or SO_3^-) which interact with the ionic groups of the sample molecules. The method is suitable for separating, e.g. amino acids, ionic metabolic products and organic ions.

Ion-Pair Chromatography

Ion-pair chromatography may also be used for the separation of ionic compounds and overcomes certain problems inherent in the ion-exchange method. Ionic sample molecules are 'masked' by a suitable counter ion. The main advantages are, firstly, that the widely available reversed-phase system can be used, so no ion exchanger is needed, and, secondly, acids, bases and neutral products can be analysed simultaneously.

Ion Chromatography

Ion chromatography was developed as a means of separating the ions of strong acids and bases (e.g. Cl^-, NO_3^-, Na^+, K^+). It is a special case of ion-exchange chromatography but the equipment used is different.

Size-Exclusion Chromatography

This mode can be subdivided into gel permeation chromatography (with organic solvents) and gel filtration chromatography (with aqueous solutions).

Size-exclusion chromatography separates molecules by size, i.e. according to molecular mass. The largest molecules are eluted first and the smallest molecules last. This is the best method to choose when a mixture contains compounds with a molecular mass difference of at least 10%.

Affinity Chromatography

In this case, highly specific biochemical interactions provide the means of separation. The stationary phase contains specific groups of molecules which can only adsorb the sample if certain steric and charge-related conditions are satisfied (cf. interaction between antigens and antibodies). Affinity chromatography can be used to isolate proteins (enzymes as well as structural proteins), lipids, etc., from complex mixtures without involving any great expenditure.

1.4   THE HPLC INSTRUMENT

An HPLC instrument can be a set of individual modules or elements, but it can be designed as a single apparatus as well. The module concept is more flexible in the case of the failure of a single component; moreover, the individual parts need not be from the same manufacturer. If you do not like to do minor repairs by yourself you will prefer a compact instrument. This, however, does not need less bench space than a modular set.

An HPLC instrument has at least the elements which are shown in Figure 1.3: solvent reservoir, transfer line with frit, high-pressure pump, sample injection device, column, detector, and data acquisition, usually together with data evaluation. Although the column is the most important part, it is usually the smallest one. For

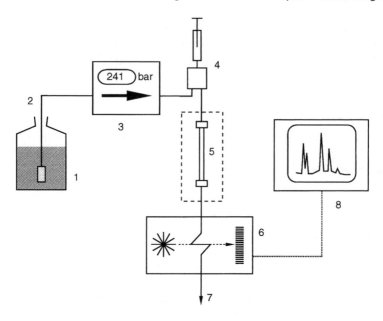

Figure 1.3 Schematic diagram of an HPLC unit. 1 = Solvent reservoir; 2 = transfer line with frit; 3 = pump (with manometer); 4 = sample injection; 5 = column (with thermostat); 6 = detector; 7 = waste; 8 = data acquisition.

temperature-controlled separations it is enclosed in a thermostat. It is quite common to work with more than one solvent, thus a mixer and controller are needed. If the data acquisition is done by a computer it can also be used for the control of the whole system.

1.5 SAFETY IN THE HPLC LABORATORY

Three health risks are inherent in HPLC, these being caused by:

(a) Toxic solvents,
(b) Pulmonary irritation from the stationary phase, and
(c) Dangers resulting from the use of high pressures.

Short- and long-term risks of exposure to solvents and vapours are generally known but too little attention is paid to them. It is good working practice to provide all feed and waste containers with perforated plastic lids, the hole being just large enough to take a PTFE tube for filling or emptying purposes, so that no toxic vapours can escape into the laboratory environment and no impurities can contaminate the highly pure solvent. A good ventilation system should be provided in the solvent handling areas.

The fact that particles of 5 μm and less, as used in HPLC, may pass into the lungs (they are not retained by the bronchial tubes but pass straight through) is less well known and the potential long-term risk to health has not yet been adequately researched. Amorphous silica, as used for stationary phases, is not hazardous[2] but inhalation should be avoided anyway. As a safety precaution, any operation involving possible escape of stationary phase dust (opening phials, weighing etc.) must be carried out in a fume cupboard.

The high-pressure pump does not present too much of a risk. In contrast to gases, liquids are almost incompressible (approximately 1 vol% per 100 bar). Hence, liquids store very little energy, even under high-pressure conditions. A jet of liquid may leak from a faulty fitting but there is no danger of explosion. However, this liquid may cause serious physical damage to the body. A column under pressure which is open at the bottom for emptying purposes must not be interfered with in any way. The description of an accident resulting from this type of action is strongly recommended for reading.[3]

1.6 COMPARISON BETWEEN HIGH-PERFORMANCE LIQUID CHROMATOGRAPHY AND GAS CHROMATOGRAPHY

Like HPLC, gas chromatography[4] (GC) is also a high-performance method, the most important difference between the two being that GC can only cope with substances that are volatile or can be evaporated intact at elevated temperatures or from which volatile derivatives can be reliably obtained. Only about 20% of known organic compounds can be analysed by gas chromatography without prior treatment. For liquid chromatography, the sample must be dissolved in a solvent and, apart from cross-linked, high-molecular-mass substances, all organic and ionic inorganic products satisfy this condition.

The characteristics of the two methods are compared in Table 1.1. In comparison with gas chromatography there are three important differences:

(a) The diffusion coefficient of the sample in the mobile phase is much smaller in HPLC than in GC. (This is a drawback because the diffusion coefficient is the most important factor which determines the speed of chromatographic analysis.)
(b) The viscosity of the mobile phase is higher in HPLC than in GC. (This is a drawback because high viscosity results in small diffusion coefficients and in high flow resistance of the mobile phase.)

---

[2] C.J. Johnston *et al. Toxicol. Sci.*, **56**, 405 (2000).
[3] G. Guiochon, *J. Chromatogr.*, **189**, 108 (1980).
[4] H.M. McNair, J.M. Miller and F.A. Settle, *Basic Gas Chromatography*, Wiley-Interscience, New York, 2009.

TABLE 1.1   Comparison of GC AND HPLC

Problem	GC	HPLC
Difficult separation	Possible	Possible
Speed	Yes	Yes
Automation	Possible	Possible
Adaptation of system to separation problem	By change in stationary phase	By change in stationary and mobile phase
Application restricted by	Lack of volatility, thermal decomposition	Insolubility
Typical number of separation plates	Per column	Per metre
GC with packed columns	2000	1000
GC with capillary columns	50 000	3000
Classical liquid chromatography	100	200
HPLC	5000	50 000

(c) The compressibility of the mobile phase under pressure is negligibly small in HPLC whereas it is not in GC. (This is an advantage because as a result the flow velocity of the mobile phase is constant over the whole length of the column. Therefore optimum chromatographic conditions exist everywhere if the flow velocity is chosen correctly. Moreover, incompressibility means that a liquid under high pressure is not dangerous.)

1.7   COMPARISON BETWEEN HIGH-PERFORMANCE LIQUID CHROMATOGRAPHY AND CAPILLARY ELECTROPHORESIS

Capillary electrophoresis[5] (also termed capillary zone electrophoresis, CZE) is suited for electrically charged analytes and separates them, simply speaking, according to their ratio of charge to size. In addition, the shape of the molecues is another parameter which influences their speed, therefore the separation of isomers or of analytes with identical specific charge is possible. Cations (positively charged molecules) move faster than anions (negatively charged molecules) and appear earlier in the detector. Small, multiply charged cations are the fastest species whereas small, multiply charged anions are the slowest ones.

---

[5] P. Schmitt-Kopplin, *Capillary Electrophoresis*, Humana Press, Totowa, 2008; S. Wren, *Chromatographia Suppl.*, **54**, S-15 (2001).

The separation is performed at high voltage. An electric field of up to 30 kV is applied between the ends of the separation capillary. As a consequence, the buffer solution within the capillary moves towards the negatively charged cathode. The capillaries have a length of 20–100 cm and an inner diameter of 50–250 µm. In contrast to HPLC they are not packed with a stationary phase in the chromatographic sense but in some cases with a gel which allows the separation of the analytes by their size (as in size-exclusion chromatography).

The separation performance can be of much higher order of magnitude than in HPLC (up to 10^7 theoretical plates), making CE an extremely valuable method for peptide mapping or DNA sequencing. However, small molecules such as amino acids or inorganic ions can be separated as well. The absolute sample amounts which can be injected are low due to the small volume of the capillaries. A major drawback is the lower repeatability (precision) compared to quantitative HPLC. Preparative separations are not possible.

Electrokinetic chromatography (see Section 23.6) is a hybrid of HPLC and CE. For this technique the capillaries are packed with a stationary phase and the separation is based on partition phenomena. The mobile phase acts as in CE; it consists of a buffer solution and moves thanks to the applied electrical field.

1.8   UNITS FOR PRESSURE, LENGTH AND VISCOSITY

Pressure Units

The common pressure unit of HPLC is bar, but the SI unit is pascal (Pa): $1\,Pa = 1\,N\,m^{-2}$. The atmosphere (atm or at, respectively) should no longer be used. The unit psi (pounds per square inch) is American and is still in use. Note the difference between psia = psi absolute and psig = psi gauge (manometer), the latter meaning psi in excess of atmospheric pressure.

$$1\,bar = 10^5\,Pa = 10^5\,kg\,m^{-1}\,s^{-2} = 0.987\,atm = 1.02\,at = 14.5\,lb\,in^{-2}\,(psi)$$

Conversion data:

1 MPa = 10 bar (megapascal)
1 atm = 1.013 bar (physical atmosphere)
1 at = 0.981 bar (technical atmosphere, $1\,kp\,cm^{-2}$)
1 psi = 0.0689 bar

Rule of thumb:

$1000\,psi \approx 70\,bar$, $100\,bar = 1450\,psi$

Length Units

English units are often used in HPLC to indicate tube or capillary diameters, the unit being the inch (in or ″). Smaller units are not expressed in tenths but as 1/2, 1/4, 1/8, or 1/16 in, or multiples of these.

Outer diameters:

$1'' = 25.40$ mm	$1/2'' = 12.70$ mm	$3/8'' = 9.525$ mm	$1/4'' = 6.35$ mm
$3/16'' = 4.76$ mm	$1/8'' = 3.175$ mm	$1/16'' = 1.59$ mm	

Inner diameter of capillaries:

$0.04'' = 1.0$ mm $0.02'' = 0.51$ mm $0.01'' = 0.25$ mm $0.007'' = 0.18$ mm

Viscosity Units

The SI unit of the dynamic viscosity is the pascal second: $1\,\mathrm{Pa\,s} = 1\,\mathrm{kg\,m^{-1}\,s^{-1}}$. Solvents have viscosities around $1 \cdot 10^{-3}\,\mathrm{Pa\,s} = 1\,\mathrm{mPa\,s}$. The old unit was the centipoise (cP): $1\,\mathrm{mPa\,s} = 1\,\mathrm{cP}$.

1.9 SCIENTIFIC JOURNALS

Journal of Chromatography A (all topics of chromatography) ISSN 0021–9673.

Journal of Chromatography B (biomedical sciences and applications) ISSN 1570-0232.

Until volume 651 (1993) this was one journal with some volumes dedicated to biomedical applications. Afterwards the journal was split and continued with separate volumes having the same number but not the same letter (e.g. 652A and 652B). Elsevier Science, P.O. Box 211, NL-1000 AE Amsterdam, The Netherlands.

Journal of Chromatographic Science, ISSN 0021–9665, Preston Publications, 6600 W. Touhy Avenue, Niles, IL 60714–4588, USA.

Chromatographia, ISSN 0009–5893, Vieweg Publishing, P.O. Box 5829, D-65048, Wiesbaden, Germany.

Journal of Separation Science (until 2001 *Journal of High Resolution Chromatography*), ISSN 1615–9306, Wiley-VCH, P.O. Box 10 11 61, D-69451 Weinheim, Germany.

Journal of Liquid Chromatography & Related Technologies, ISSN 1082–6076, Marcel Dekker, 270 Madison Avenue, New York, NY 10016–0602, USA.

LC GC Europe (free in Europe, formerly *LC GC International*), ISSN 1471–6577, Advanstar Communications, Advanstar House, Park West, Sealand Road, Chester CH1 4RN, UK.

LC GC North America (free in the USA, formerly *LC GC Magazine*), ISSN 0888–9090, Advanstar Communications, 859 Willamette Street, Eugene, OR 97401, USA.

LC GC Asia Pacific (free in the Asia Pacific region), Advanstar Communications, 101 Pacific Plaza, 1/F, 410 Des Voeux Road West, Hong Kong, People's Republic of China.

Biomedical Chromatography, ISSN 0269–3879, John Wiley & Sons, Ltd, 1 Oldlands Way, Bognor Regis PO22 9SA, UK.

International Journal of Bio-Chromatography, ISSN 1068-0659, Gordon and Breach, P.O. Box 32160, Newark, NJ 07102, USA.

Separation Science and Technology, ISSN 0149–6395, Taylor & Francis, Mortimer House, 37–41 Mortimer Street, London, W1T 3JH, UK.

Chromatography Abstracts, ISSN 0268–6287, Royal Society of Chemistry, Thomas Graham House, Cambridge CB4 0WF, UK.

The *Journal of Microcolumn Separations* (John Wiley & Sons, Ltd, ISSN 1040–7865) merged with the *Journal of Separation Science* after issue 8 of volume 13 (2001).

1.10 RECOMMENDED BOOKS

J.W. Dolan and L.R. Snyder, *Troubleshooting LC Systems*, Aster, Chester, 1989.

M.W. Dong, *Modern HPLC for Practicing Scientists*, John Wiley & Sons, Ltd, Chichester, 2006.

N. Dyson, *Chromatographic Integration Methods*, Royal Society of Chemistry, London, 2nd ed., 1998.

S. Kromidas, ed., *HPLC Made to Measure*, Wiley-VCH, Weinheim, 2006.

H.J. Kuss and S. Kromidas, *Quantification in LC and GC*, Wiley-VCH, Weinheim, 2009.

V.R. Meyer, *Pitfalls and Errors of HPLC in Pictures*, Wiley-VCH, Weinheim, 2nd ed., 2006.

S.C. Moldoveanu and V. David, *Sample Preparation in Chromatography*, Elsevier, Amsterdam, 2002.

U.D. Neue, *HPLC Columns – Theory, Technology, and Practice*, John Wiley & Sons, Inc., New York, 1997.

H. Posch and B. Trathnigg, *HPLC of Polymers*, Springer, Berlin, Heidelberg, 1998.

P.C. Sadek, *Troubleshooting LC Systems*, John Wiley & Sons, Inc., New York, 1999.

L.R. Snyder, J.J. Kirkland, and J.W. Dolan, *Introduction to Modern Liquid Chromatography*, John Wiley & Sons, Ltd, Chichester, 3rd ed., 2010.

L.R. Snyder and J.W. Dolan, *High-Performance Gradient Elution*, Wiley-Interscience, Hoboken, 2007.

L.R. Snyder, J.J. Kirkland and J.L. Glajch, *Practical HPLC Method Development*, Wiley-Interscience, New York, 2nd ed., 1997.

General textbooks on chromatography:

E. Heftmann, ed., *Chromatography, Part A: Fundamentals and Techniques, Part B: Applications*, Elsevier, Amsterdam, 6th ed., 2004.

J.M. Miller, *Chromatography – Concepts and Contrasts*, John Wiley & Sons, Ltd, Chichester, 2nd ed., 2009.

C.F. Poole, *The Essence of Chromatography*, Elsevier, Amsterdam, 2002.

K. Robards, P.E. Jackson and P.R. Haddad, *Principles and Practice of Modern Chromatographic Methods*, Academic, London, San Diego, 1995.

2 Theoretical Principles

2.1 THE CHROMATOGRAPHIC PROCESS

Definition

Chromatography is a separation process in which the sample mixture is distributed between two phases in the chromatographic bed (column or plane). One phase is stationary whilst the other passes through the chromatographic bed. The *stationary phase* is either a solid, porous, surface-active material in small-particle form or a thin film of liquid coated on a solid support or column wall. The *mobile phase* is a gas or liquid. If a gas is used, the process is known as gas chromatography; the mobile phase is always liquid in all types of liquid chromatography, including the thin-layer variety.

Experiment: Separation of Test Dyes

A 'classical' 20 cm long chromatography column with a tap (or a glass tube tapered at the bottom, ca. 2 cm in diameter, with tubing and spring clip) is filled with a suspension of silica in toluene. After settling, about 50–100 µl of dye solution (e.g. test dye mixture II N made by Camag, Muttenz, Switzerland) is brought onto the bed by means of a microlitre syringe, and toluene is added as eluent.

Observations

The various dyes move at different rates through the column. The six-zone separation is as follows: Fat Red 7B, Sudan Yellow, Sudan Black (two components), Fat Orange, and Artisil Blue 2 RP. Compounds that tend to reside in the mobile phase move more quickly than those that prefer the stationary phase.

Practical High-Performance Liquid Chromatography, Fifth edition   Veronika R. Meyer
© 2010 John Wiley & Sons, Ltd

Phase preference can be expressed by the *distribution coefficient, K*:

$$K_X = \frac{c_{stat}}{c_{mob}}$$

where c_{stat} is the concentration (actual activity) of compound X in the stationary phase and c_{mob} is the concentration of X in the mobile phase, or the *retention factor, k* (formerly termed *capacity factor k'*):

$$k_X = \frac{n_{stat}}{n_{mob}}$$

where n_{stat} is the number of moles of X in the stationary phase and n_{mob} is the number of moles of compound X in the mobile phase. The stationary and mobile phases must obviously be in intimate contact with each other in order to ensure a distribution balance.

The various components present must have different distribution coefficients and hence different capacity factors in the chromatographic system if the mixture is to be separated.

Graphical Representation of the Separation Process

(a) A mixture of two components, ▲ and ●, is applied to the chromatographic bed (Figure 2.1a).

(b) The ▲ component resides for preference in the stationary phase and the ● component more in the mobile phase (Figure 2.1b). Here $k_{\blacktriangle} = 5/2 = 2.5$ and $k_{\bullet} = 2/5 = 0.4$.

(c) A new equilibrium follows the addition of fresh eluent: sample molecules in the mobile phase are partly adsorbed by the 'naked' stationary phase surface, in accordance with their distribution coefficients, whereas those molecules that have previously been adsorbed appear again in the mobile phase (Figure 2.1c).

(d) After repeating this process many times, the two components are finally separated. The ● component prefers the mobile phase and migrates more quickly than the ▲ component, which tends to 'stick' in the stationary phase (Figure 2.1d).

As the diagrams show, here the new balance is found along a section corresponding to about $3\frac{1}{2}$ particle diameters of the stationary phase. Hence, this distance represents a *theoretical plate*. The longer is the chromatographic bed, the more theoretical plates it contains and the better the degree of separation of a mixture. This effect is partly compensated by *band broadening*. As experiments show, substance zones become increasingly broader the greater the distance along the column and the longer the retention time.

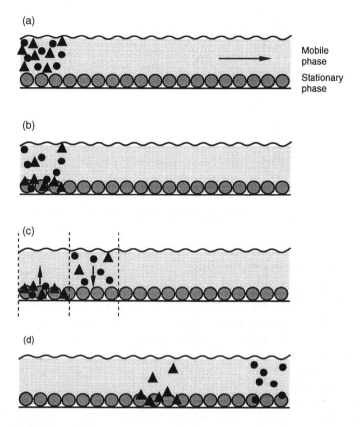

Figure 2.1   Representation of a chromatographic separation.

2.2   BAND BROADENING

There are many reasons for band broadening and it is important that these are understood and the phenomenon kept to a minimum so that the number of theoretical plates in the column is high.

First Cause: Eddy Diffusion

The column is packed with small stationary phase particles. The mobile phase passes through and transports the sample molecules with it (Figure 2.2). Some molecules are 'fortunate' and leave the column before most of the others, after having travelled by chance in roughly a straight line through the chromatographic bed. Other sample molecules leave later, having undergone several diversions along the way.

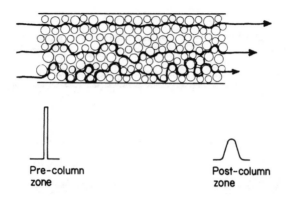

Pre-column   Post-column
zone   zone

Figure 2.2   Eddy diffusion in a chromatographic column.

Second Cause: Flow Distribution

The mobile phase passes in a laminar flow between the stationary phase particles (Figure 2.3). The flow is faster in the 'channel' centre than it is near a particle. The arrows in Figure 2.3 represent mobile phase velocity vectors (the longer the arrow, the greater the local flow velocity). Eddy diffusion and flow distribution may be reduced by packing the column with evenly sized particles.

The *first principle* on which a good column is based is that *the packing should be composed of particles with as narrow a size distribution as possible.* The ratio between the largest and the smallest particle diameters should not exceed 2.0, 1.5 being even better (example: smallest particle size 5.0 µm, largest particle size 7.5 µm).

The broadening due to eddy diffusion and flow distribution is little affected, if at all, by the mobile phase flow velocity.

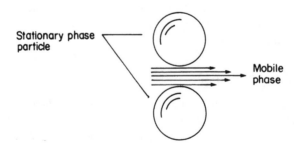

Figure 2.3   Flow distribution in a chromatographic bed.

Figure 2.4 Band broadening by longitudinal diffusion. Left: Sample zone immediately after injection. It will spread out in all three axes of space (arrow directions). Right: Sample zone at a later moment. It is larger now due to diffusion and it has also been transported by the flowing mobile phase.

Third Cause: Sample Molecule Diffusion in the Mobile Phase

Sample molecules spread out in the solvent without any external influence (just as a sugar lump dissolves slowly in water even without being stirred). This is longitudinal diffusion (Figure 2.4) and has a disadvantageous effect on plate height only if:

(a) small stationary phase particles,
(b) too low a mobile phase velocity in relation to the particle diameter, and
(c) a relatively large sample diffusion coefficient

coincide in the chromatographic system.

The second principle is that the mobile phase flow velocity should be selected so that longitudinal diffusion has no adverse effect. This applies when $u > 2D_m/d_p$, where u is the linear flow velocity of the mobile phase, D_m the diffusion coefficient of the sample in the mobile phase and d_p the particle diameter. Further details can be found in Section 8.5.

Fourth Cause: Mass Transfer between Mobile, 'Stagnant Mobile', and Stationary Phases

Figure 2.5 shows the pore structure of a stationary phase particle: the channels are both narrow and wide, some pass right through the whole particle and others are closed off.

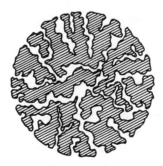

Figure 2.5 Pore structure of a stationary phase particle.

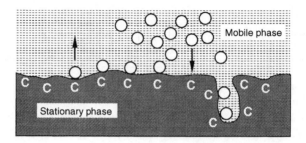

Figure 2.6 Mass transfer between mobile and stationary phase. The stationary phase has 'adsorptive' centres C (in a broad sense) which attract the molecules around them. Molecules adsorb to the centres (middle) and desorb (left). The access to centres within the pores is more difficult and therefore slower (right).

The pores are filled with mobile phase which does not move (it stagnates). A sample molecule entering a pore ceases to be transported by the solvent flux and changes its position by means of diffusion only. However, two possibilities present themselves:

(a) The molecule diffuses back to the mobile flux phase. This process takes time, during which molecules that have not been retained in the pores move on slightly further. The resulting band broadening is smaller the shorter are the pores, i.e. the smaller are the stationary phase particles. In addition, the diffusion rate of the sample molecules in a solvent is larger under lower viscosity conditions (i.e. they diffuse faster in and out of the pores) than it is in a more viscous medium.

(b) The molecule interacts with the stationary phase itself (adsorbent or liquid film) and is adsorbed. For a while, it remains 'stuck' to the stationary phase and then passes on once more. Again, this mass transfer takes a fair amount of time (Figure 2.6).

In both cases, band broadening increases with increasing mobile phase flow velocity: the sample molecules remaining in the moving solvent become further removed from the stagnant molecules the faster is the solvent flux (but less time for solute elution is necessary).

The *third principle* is that *small particles or those with a thin, porous surface layer should be used as the stationary phase.*

The *fourth principle* is that *low-viscosity solvents should be used.*

The *fifth principle* is that *high analysis speed is achieved at the expense of resolution and vice versa.* However, this effect is much less pronounced with smaller than with larger particles.

The theoretical plate height, H, can be expressed as a function of mobile phase flow velocity, u (Figure 2.7).[1] The H/u curve is also called the van Deemter curve. The optimum flow rate u_{opt} depends on the properties of the analyte.

---

[1] J.J. van Deemter, F.J. Zuiderweg and A. Klinkenberg, *Chem. Eng. Sci.*, **5**, 271 (1956).

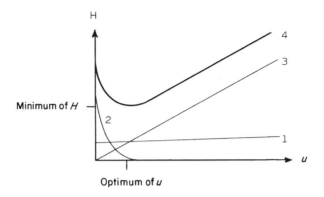

Figure 2.7   Van Deemter curve *(H/u* curve). 1 = eddy diffusion and flow distribution component of band broadening; 2 = longitudinal diffusion component: flow rates at which this diffusion is not a factor of any significance should be used in liquid chromatography; 3 = mass-transfer component: the slope of the line is greater for 50 µm than it is for 5 µm particles; 4 = the resultant van Deemter *H/u* curve.[1]

2.3   THE CHROMATOGRAM AND ITS PURPORT

The eluted compounds are transported by the mobile phase to the detector and recorded as Gaussian (bell-shaped) curves.[2] The signals are known as peaks (Figure 2.8) and the whole entity is the chromatogram.

The peaks give qualitative and quantitative information on the mixture in question:

(a) *Qualitative*:   the retention time of a component is always constant under identical chromatographic conditions. The retention time is the period that elapses between sample injection and the recording of the signal maximum. The column dimensions, type of stationary phase, mobile phase composition and flow velocity, sample size and temperature provide the chromatographic conditions. Hence, a peak can be identified by injecting the relevant substance and then comparing retention times.

(b) *Quantitative*:   both the area and height of a peak are proportional to the amount of a compound injected. A calibration graph can be derived from peak areas or heights obtained for various solutions of precisely known concentration and a peak-size comparison can then be used to determine the concentration of an unknown sample.

---

[2] If the MS is used as a detector, the peaks may show a misshapen shape due to the low data rate. The quantification will be impended.

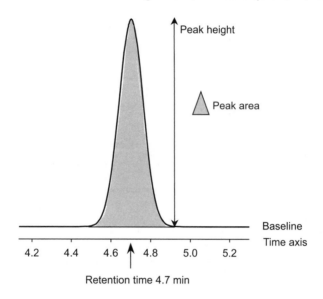

Figure 2.8 Shape of a peak.

The chromatogram can be used to provide information on separation efficiency (Figure 2.9). Here w is the peak width at the baseline,[3] t_0 is the *dead time* or retention time of an unretained solute, i.e. the time required by the mobile phase to pass through the column (also called the *breakthrough time*). Hence the linear flow velocity, u, can be calculated as

$$u = \frac{L}{t_0}$$

where L is the column length. A nonretained compound, i.e. one that is not retained by the stationary phase, appears at the end of the column at t_0. t_R is the *retention time*;[4] this is the period between sample injection and recording of the peak maximum. Two compounds can be separated if they have different retention times. t'_R is the *net retention time* or *adjusted retention time*. Figure 2.9 shows that $t_R = t_0 + t'_R$. t_0 is identical for all eluted substances and represents the mobile-phase residence time. t'_R is the stationary phase residence time and is different for each separated compound. The longer a compound remains in the stationary phase, the later it becomes eluted.

Retention time is a function of mobile phase flow velocity and column length. If the mobile phase is flowing slowly or if the column is long, then t_0 is large and hence so is t_R; t_R is therefore not suitable for characterizing a compound.

---

[3] $w = 4\sigma$, where σ is the standard deviation of a Gaussian peak.
[4] Retention volume $V_R = F t_R$ (F = volume flow rate in ml min^{-1}). Void volume $V_0 = F t_0$.

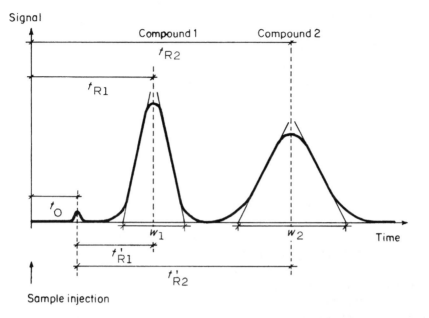

Figure 2.9   The chromatogram and its characteristic features.

Therefore the *retention factor* or *k value* (formerly known as the *capacity factor*, *k'*) is preferred:

$$k = \frac{t'_R}{t_0} = \frac{t_R - t_0}{t_0}$$

k is independent of the column length and mobile phase flow rate and represents the molar ratio of the compound in the stationary and the mobile phase, as mentioned earlier (Section 2.1).

Problem 1

Calculate the k values of compounds 1 and 2 in Figure 2.9.

Solution

$t_0 = 12.5$ mm; $t_{R1} = 33.1$ mm; $t_{R2} = 70.5$ mm.

$$k_1 = \frac{t_{R1} - t_0}{t_0} = \frac{33.1 - 12.5}{12.5} = 1.6$$

$$k_2 = \frac{t_{R2} - t_0}{t_0} = \frac{70.5 - 12.5}{12.5} = 4.6$$

Retention factors between 1 and 10 are preferred. If the k values are too low, then the degree of separation may be inadequate (if the compounds pass too rapidly through the column, no stationary phase interaction occurs and hence no chromatography). High k values are accompanied by long analysis times.

The k value is connected with the distribution coefficient described in Section 2.1 in the following way:[5]

$$k = K \frac{V_S}{V_M}$$

where V_S is the volume of stationary phase and V_M the volume of mobile phase in the column.

The retention factor is directly proportional to the volume occupied by the stationary phase and more especially to its specific area $(m^2 g^{-1})$ in the case of adsorbents. A column packed with porous-layer beads produces lower k values and hence shorter analysis times than a column containing completely porous particles if the other conditions remain constant. Silica with narrow pores produces larger k values than a wide-pore material.

Two components in a mixture cannot be separated unless they have different k values, the means of assessment being provided by the *separation factor*, α, formerly known as the *relative retention*

$$\alpha = \frac{k_2}{k_1} = \frac{t_{R2} - t_0}{t_{R1} - t_0} = \frac{K_2}{K_1}$$

with $k_2 > k_1$. If $\alpha = 1$, then no separation takes place as the retention times are identical. The separation factor is a measure of the chromatographic system's potential for separating two compounds, i.e. its *selectivity*. Selection of the stationary and mobile phases can affect the value of α.

Problem 2

Calculate the α value of compounds 1 and 2.

Solution

Problem 1 showed that $k_1 = 1.6$ and $k_2 = 4.6$; hence $\alpha = 4.6/1.6 = 2.9$.

---

[5] This is only valid within the so-called linear range where k is independent of sample load but no longer under conditions of mass overload.

The *resolution, R,* of two neighbouring peaks is defined by the ratio of the distance between the two peak maxima, i.e. the distance between the two retention times, t_R, and the arithmetic mean of the two peak widths, w:[6]

$$R = 2\frac{t_{R2}-t_{R1}}{w_1 + w_2} = 1.18\frac{t_{R2}-t_{R1}}{w_{1/2_1} + w_{1/2_2}}$$

where $w_{1/2}$ is the peak width at half-height.

The peaks are not completely separated with a resolution of 1, but two peaks can be seen. The inflection tangents touch each other at the baseline. For quantitative analysis a resolution of 1.0 is too low in most cases. It is necessary to obtain baseline resolution, e.g. $R = 1.5$. If one of the peaks is markedly smaller than its neighbour even higher resolution is needed. See paragraph 19.5.

Problem 3

Calculate the resolution of compounds 1 and 2.

Solution

$t_{R1} = 33.1$ mm, $t_{R2} = 70.5$ mm; $w_1 = 17$ mm, $w_2 = 29$ mm.

$$R = \frac{2(70.5-33.1)}{17+29} = 1.6$$

Manual Determination of Peak Width at the Baseline:

Inflection point tangents[7] are drawn to each side of the Gaussian curve; this generally causes few problems. However, the recorder line width must be considered and this is best done by adding it to the signal width on one side but not the other (Figure 2.10). The base width is the distance along the baseline between the two inflection tangents.

---

[6] These equations are less suited for peak pairs of highly unequal area and for asymmetric peaks. In such cases the peak separation index (PSI) is a better criterion: $PSI = 1-\frac{b}{a}$.

[7] Inflection point = position at which the curvature or slope changes sign; $\int$ = positive curvature; $\int$ = negative curvature.

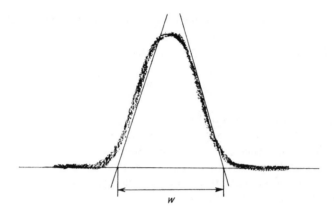

Figure 2.10 Construction of inflection tangents.

Finally, the chromatogram can be used to calculate the *number of theoretical plates, N,* in the column:

$$N = 16\left(\frac{t_R}{w}\right)^2$$

$$N = 5.54\left(\frac{t_R}{w_{1/2}}\right)^2$$

where $w_{1/2}$ is the peak width at half-height;

$$N = 2\pi\left(\frac{h_P t_R}{A}\right)^2$$

where h_P is the peak height and A the peak area.

All three equations yield correct results only if the peak has a Gaussian shape. This is hardly ever the case with real-life chromatograms.[8] Correct values for asymmetric peaks are obtained by the momentum method.[9] Approximately correct values are obtained by the equation:[10]

$$N = 41.7\frac{(t_R/w_{0.1})^2}{T + 1.25}$$

where $w_{0.1}$ is the peak width at 10% of the peak height and T is tailing $b_{0.1}/a_{0.1}$ (Figure 2.25).

---

[8] B.A. Bidlingmeyer and F.V. Warren, *Anal. Chem.*, **56**, 1583A (1984).
[9] Definition: $m_n = \int_0^x t^n f(t)\mathrm{d}t$. The second moment with $n = 2$ corresponds to the variance σ of the peak. With $w = 4\sigma$ it is possible to calculate N. Literature: N. Dyson, *Chromatographic Integration Methods*, Royal Society of Chemistry, London, 2nd ed., 1998, p. 23.
[10] J.P. Foley and J.G. Dorsey, *Anal. Chem.*, **55**, 730 (1983).

The plate number calculated from a nonretained peak is a measure of the column packing efficiency, whereas in the case of peaks eluted later, mass-transfer processes also contribute to the plate number. As a general rule, N is higher for retained compounds because their relative band broadening by extra-column volumes (see Section 2.6) is lower than for the early eluted peaks.

Problem 4

How many theoretical plates emerge from calculations based on the last peak in Figure 2.9?

Solution

$$t_{R2} = 70.5 \text{ mm}, \ w_2 = 29 \text{ mm}.$$

$$N = 16 \left(\frac{70.5}{29} \right)^2 = 94$$

The *height of a theoretical plate, H*, is readily calculated provided the length of the column is known:

$$H = \frac{L}{N}$$

where H is the distance over which chromatographic equilibrium is achieved (see Figure 2.1) and is referred to as the *height equivalent to a theoretical plate* (HETP).

Problem 5

Fat Red 7B, 1-[(p-butylphenyl)azo]-2-naphthol, 1-[(p-methoxyphenyl)azo]-2-naphthol, 1-[(m-methoxyphenyl)azo]-2-naphthol and 1-[o-methoxyphenyl)-azo]-2-naphthol red test dyes were chromatographed on a Merck low-pressure column. Silica gel 60 was the stationary phase (40–63 μm) and 50% water-saturated dichloromethane was used as the mobile phase at a flow rate of 1 ml min^{-1}.

Using the chromatogram shown in Figure 2.11, calculate:

(a) retention factors for peaks 1–5;
(b) separation factors for the best and worst resolved pair of peaks;
(c) resolution of these two pairs of peaks;
(d) plate number for each of peaks 1–5.

Solution

(a) $k_1 = 0.51$; $k_2 = 1.15$; $k_3 = 1.51$; $k_4 = 1.88$; $k_5 = 2.50$.
(b) $\alpha_{12} = 2.2$; $\alpha_{34} = 1.2$.

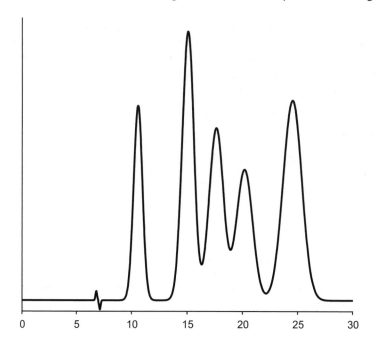

Figure 2.11   Chromatogram of red test dyes.

(c) $R_{12} = 2.4$; $R_{34} = 0.79$.
(d) $N_1 = 720$; $N_2 = 850$; $N_3 = 760$; $N_4 = 770$; $N_5 = 850$.

(Your results may differ depending on measuring precision and the formulae used.)

2.4   GRAPHICAL REPRESENTATION OF PEAK PAIRS WITH DIFFERENT DEGREE OF RESOLUTION[11]

With some experience it is not difficult to get a feeling concerning the meaning of resolution R. The graphical representations in Figs. 2.13–2.18 are a help. They show peak pairs with different resolutions and peak size ratios of $1:1, 2:1, 4:1, 8:1, 16:1, 32:1, 64:1$, and $128:1$. These figures can be mounted near the data system so that a comparison with real chromatograms can be made at any time. In reality one will note some deviations from these idealistic drawings: real peaks are often not symmetrical

---

[11] L.R. Snyder, J. *Chromatogr. Sci.*, **10**, 200 (1972).

(i.e. they are tailed, see Section 2.7), and it is rare that neighbouring peaks are of really identical width.

The figures allow to estimate the resolution of peak pairs with adequate accuracy. For semi-quantitative discussions it is not necessary to calculate R by one of the formulas given above.

Problem 6

Estimate the resolutions of the peak pairs indicated by an arrow in Figure 2.12.

Solution

Peak pair	1	2	3	4	5	6
Size ratio	1:1	2:1	1:1	1:2	1:2	1:8
Resolution	0.8	0.6	> 1.25	1.25	0.7	0.8

The drawings incorporate also points and arrows. The points show the true peak height (and the true retention time as well). In cases of poor resolution it is impossible to set this point intuitively to the true position which is often below the sum curve. The arrows show the positions at which both peaks are separated into fractions of equal purity by preparative chromatography. The number above each arrow indicates the percentage purity level attained. These numbers, however, are only true if the ratio between the amount of material and the signal (peak height as

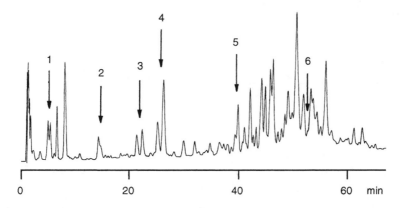

Figure 2.12 Separation of peptides obtained from pepsin-digested lactalbumin. Gradient separation with water–acetonitrile (0.1% trifluoroacetic acid) on a butyl phase, detection at UV 210 nm.

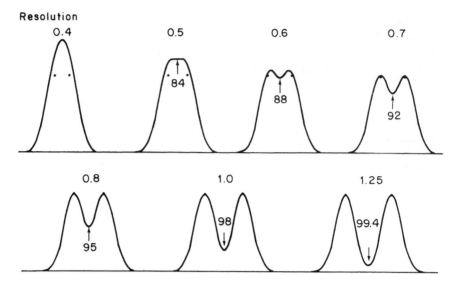

Figure 2.13   Resolution of neighbouring peaks, peak-size ratio 1 : 1. Reproduced with permission from L.R. Synder, *J. Chromatogr. Sci.*,**10**, 200 (1972).

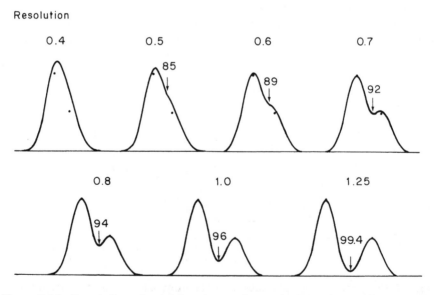

Figure 2.14   Resolution of neighbouring peaks, peak-size ratio 2 : 1. Reproduced with permission from L.R. Synder, *J. Chromatogr. Sci.*,**10**, 200 (1972).

Resolution

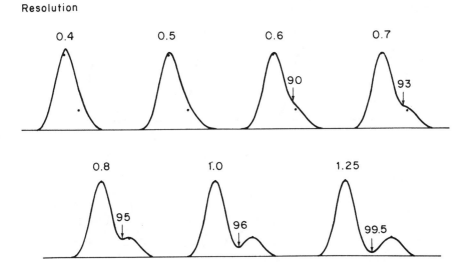

Figure 2.15   Resolution of neighbouring peaks, peak-size ratio 4 : 1. Reproduced with permission from L.R. Synder, *J. Chromatogr. Sci.*, **10**, 200 (1972).

Resolution

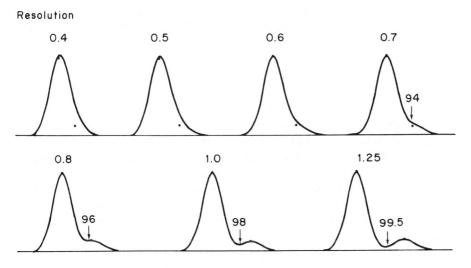

Figure 2.16   Resolution of neighbouring peaks, peak-size ratio 8 : 1. Reproduced with permission from L.R. Synder, *J. Chromatogr. Sci.*, **10**, 200 (1972).

Resolution

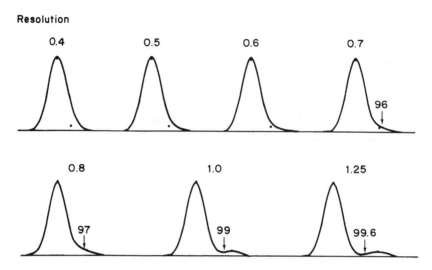

Figure 2.17 Resolution of neighbouring peaks, peak-size ratio 16:1. Reproduced with permission from L.R. Synder, *J. Chromatogr. Sci.*, **10**, 200 (1972).

well as peak area) can be assumed to be equal for both components. This can be approximately the case for homologues; for other peak pairs, such an assumption can be totally wrong because even small deviations in molecular structure can lead to a different detector response.

Graphical representations of this type for any peak area ratios and resolutions can be easily obtained by a spreadsheet calculation (such as Lotus, Excel, etc.).[12] The following equation describes the shape of Gaussian peaks:

$$f(t) = \frac{A_p}{\sigma\sqrt{2\pi}} e^{-(t-t_R)^2/2\sigma^2}$$

where t is the time axis, t_R is the retention time (time of the peak maximum), $f(t)$ is the signal (peak height) as a function of time, A_p is the peak area and σ is the standard deviation of the Gauss function which can be taken as 1. As peak areas, the numbers of the ratio are used, e.g. 2 and 1 for drawings such as the ones shown in Figure 2.14. The desired resolution is obtained by the known relationship $R = 2\Delta t_R/(w_1 + w_2)$ with $w = 4\sigma$. If $\sigma = 1$ then $w_1 + w_2 = 8$. t_R of the first peak is taken at any number whatever, then the second peak has $t_R + \Delta t_R$. For asymmetric peaks the mathematical description is rather complicated.

---

[12] V.R. Meyer, *LC GC Int.*, **7**, 590 (1994).

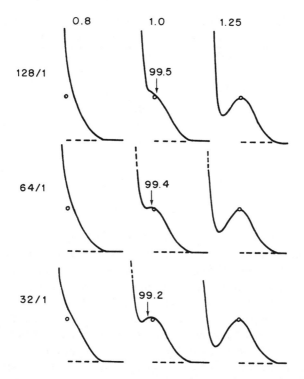

Figure 2.18 Resolution of neighbouring peaks, peak-size ratios 32:1, 64:1, and 128:1. Reproduced with permission from L.R. Synder, *J. Chromatogr. Sci.*, **10**, 200 (1972).

2.5 FACTORS AFFECTING RESOLUTION

The resolution, R, of two peaks is dependent on the separation factor, α, the number of theoretical plates, N, and the retention factor, k:

$$R = \frac{1}{4}(\alpha-1)\sqrt{N}\left(\frac{k_1}{1+\bar{k}}\right) = \frac{1}{4}\frac{\alpha-1}{\alpha}\sqrt{N}\frac{k_2}{1+\bar{k}}$$

$$\bar{k} = \frac{k_1+k_2}{2}$$

where the assumption $N_1 = N_2$ is valid, i.e. isocratic separation.[13]

---

[13] J.P. Foley, *Analyst*, **116**, 1275 (1991). If $N_1 \neq N_2$: $R = 1/4[(\alpha-1)/\alpha](N_1N_2)^{0.25}\,\bar{k}/(1+\bar{k})$. If $k_1 \approx k_2$: $R = 1/4(\alpha-1)\sqrt{N}k/(1+k)$.

TABLE 2.1 Effect of the number of theoretical plates on resolution for different separation factors, $k_1 = 10$

Separation factor α	Theoretical plates	
	$R = 1.0$	$R = 1.5$
1.005	780 000	1 750 000
1.01	195 000	440 000
1.05	8100	18 000
1.10	2100	4800
1.25	320	860
1.50	120	260
2.0	40	90

α is a measure of the chromatographic selectivity, the value of which is high when the components in question interact with the mobile and/or stationary phase at different levels of intensity. The type of interaction forces engendered (e.g. dispersion forces, dipole–dipole interactions, hydrogen bonding, π–π interactions, pH value in ion-exchange chromatography) is a highly significant factor.

N characterizes the column efficiency. The number of theoretical plates increases as a function of better packing, longer column length and optimum mobile phase flow rate conditions. A column with a high number of plates can also separate mixtures in which the components have similar separation factors, α. If α is small, then the required degree of resolution can only be achieved by incorporating more plates, as shown in Table 2.1.

Figure 2.19 shows the effect of separation factor and plate number on the separation of two neighbouring peaks. If the relative retention is high, satisfactory resolution can be obtained even if the number of theoretical plates is not great (a). The column is poor, yet the system is selective. A high separation factor and a large number of theoretical plates give a resolution that is beyond the optimum value and the analysis takes an unnecessarily long period of time (b). The resolution drops too much when the number of theoretical plates is the same as in (a), i.e. small, and the separation

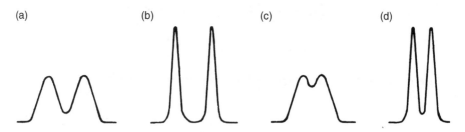

Figure 2.19 Relative retention, plate number and resolution.

factor is low (c). A column with a large number of theoretical plates is required to guarantee a satisfactory resolution when the separation factor is low (d).

The retention factor, k, is dependent only on the 'strength' of the eluent (with a constant volume ratio between the mobile and stationary phase). A mobile phase is strong if the components are eluted quickly and weak if the elution process is slow.

Improving the Resolution

If the chromatogram is inadequately resolved, then various means can be employed to improve the situation:

(a) As already mentioned in Section 2.3, k should be between 1 and 10 (and not less than 5 if separation is difficult). If other k values are calculated, the strength of the mobile phase needs to be adjusted until a favourable k is obtained (see discussion on individual chromatographic methods).

(b) If the separation is still poor, then the number of theoretical plates may be increased by: (i) buying or preparing a better column, (ii) using a longer column or two arranged in series (only recommended if both columns are good; one good column combined with a bad column leads to unsatisfactory results), or (iii) optimizing the mobile phase flow rate. Figure 2.7 shows that the plate height and hence the number of theoretical plates are a function of flow rate. However, in order to find the optimum it is necessary to vary the volume flow rate over a wide region. If the diffusion coefficients of the sample molecules in the mobile phase are known it is possible to estimate by means of the reduced flow velocity v, presented in Sections 8.5 and 8.6, if the separation is done near the van Deemter optimum. Resolution is proportional only to the square root of the number of theoretical plates. Doubling the number of plates improves the resolution by a factor of $\sqrt{2} = 1.4$. At the same time, the analysis time increases. More theoretical plates may be the answer if the separating column is patently inferior, i.e. supplies only 3000 plates instead of the expected 6000, in which case a new column must be installed.

(c) The most effective but often the most difficult means of improving the resolution is to increase the separation factor, α. If necessary, a different stationary phase could be considered (e.g. alumina instead of silica or a reversed-phase instead of a normal-phase system). However, a change in mobile phase should be the first step. The new solvent should have a similar strength to the initial solvent (as the k values have already been optimized), but should produce a different interaction. For example, diethyl ether and dichloromethane produce the same solvent strength in adsorption chromatography yet differ in their proton acceptor and dipole properties.

Figure 2.20 shows how the chromatogram changes on modifying k, α or N and leaving other factors unchanged, for an initial resolution of 0.8.

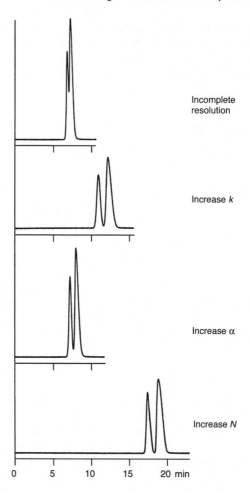

Figure 2.20 Three possibilities to improve resolution. Problem: separation of acetophenone (first peak) and veratrole. Original conditions: column, 5 cm × 2 mm i.d.; stationary phase, YMC-Pack ODS-AQ, 3 µm; mobile phase, 0.3 ml min^{-1} water–acetonitrile (80 : 20); UV detector 254 nm. k was increased by using water–acetonitrile (85 : 15), all other conditions being unaltered. α was increased by using water–methanol (70 : 30). N was increased by using a column of 15 cm length under original conditions.

Problem 7

A preparative column with 900 and an analytical column with 6400 theoretical plates were used. Preliminary tests showed that the retention factors of the two compounds in a mixture were approximately the same on both columns. A resolution of 1.25 was required for preparative separation. Calculate the optimum resolution required for the analytical column enabling subsequent transfer of separation to the preparative column.

Solution

The resolution equation can be rewritten as:

$$\frac{4R}{\sqrt{N}} = (\alpha - 1)\frac{k_1}{1 + \bar{k}}$$

The k values are identical on both columns and hence the α are also identical. Consequently,

$$\frac{4R_{an}}{\sqrt{N_{an}}} = \frac{4R_{pr}}{\sqrt{N_{pr}}}$$

$$R_{an} = \frac{R_{pr}\sqrt{N_{an}}}{\sqrt{N_{pr}}} = \frac{1.25\sqrt{6400}}{\sqrt{900}} = \frac{1.25 \times 80}{30} = 3.3$$

The solution to this problem corresponds to the stage between (b) and (a) in Figure 2.19.

Problem 8

The literature describes an HPLC separation in which $k_A = 0.75$, $k_B = 1.54$, $k_C = 2.38$, $k_D = 3.84$. Can this separation be carried out in a low-pressure column having only 300 plates, with the same phase system and a minimum resolution of 1?

Solution

$$\alpha_{AB} = \frac{1.54}{0.75} = 2.05; \quad \alpha_{BC} = \frac{2.38}{1.54} = 1.55; \quad \alpha_{CD} = \frac{3.84}{2.38} = 1.61$$

The worst resolved peak pair is BC; hence the critical separation factor is 1.55.
Data: $\alpha = 1.55$; $k_1 = k_B = 1.54$; $\bar{k} = (1.54 + 2.38)/2 = 1.96$; $R = 1$; $N_{min} = ?$

$$\sqrt{N_{min}} = \frac{4R}{\alpha - 1}\left(\frac{1 + \bar{k}}{k_1}\right)$$

$$\sqrt{N_{min}} = \frac{4}{1.55 - 1}\left(\frac{1 + 1.96}{1.54}\right) = 14$$

$$N_{min} = 196$$

Hence separation is possible.

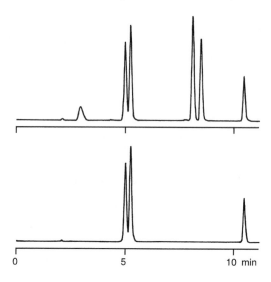

Figure 2.21　Reduced chromatogram. Top: separation of a mixture of explosives with peaks t_0, acetone, octogen, hexogen, tetryl, trinitrotoluene, and nitropenta. Bottom: reduced chromatogram which shows only the signals of t_0, the critical peak pair of octogen and hexogen, and the last peak of nitropenta. Conditions: column, 25 cm × 4.6 mm i.d.; stationary phase, Grom-Sil 80 ODS-7 PH, 4 µm; mobile phase, 1 ml min⁻¹, gradient from 50–70% acetonitrile in water within 8 min; UV detector 220 nm.

The chromatogram of a complex mixture can be represented in 'reduced' form: the critical peak pair[14] determines the necessary separation performance (all other peak pairs which are close to each other are resolved better), whereas the last peak defines the k range which needs to be considered for the solution of the separation problem. Figure 2.21 gives an example of a reduced chromatogram.

2.6　EXTRA-COLUMN VOLUMES (DEAD VOLUMES)[15]

All volumes within an HPLC instrument which affect the separation are termed extra-column volumes. (The term dead volumes should not be used because sometimes the retention volume at t_0 is also called dead volume.) This concerns the volume of the injector, the capillary between injector and column, the capillary between column and detector, the volume of the detector cell, the volume of the connecting fittings and,

[14] G.K. Webster *et al.*, *J Chromatogr. Sci.*, **43**, 67 (2005).
[15] R.P.W. Scott, *J. Liquid Chromatogr.*, **25**, 2567 (2002).

depending on the instrument design, also additional parts such as heat exchange capillaries (often hidden within the detector) or switching valves. Volumes of the system which are located before the injector or behind the detector are not extra-column volumes.

Extra-column volumes should be kept as low as possible because they affect the separation. They are one cause of band broadening and of tailing. Their contribution to band broadening is additive.[16] If each part adds 5% to peak width the decrease in separation performance is not negligible at all. Whereas it can be difficult to modify the instrument with regard to injector and detector, it is possible to use adequate tubing to connect the parts. 'Adequate' means capillaries with an inner diameter of 0.17 or 0.25 mm, see Section 4.4. Fittings must be mounted properly, see Section 4.5.

The requirements for the instrument design increase (or the allowed volumes decrease) with decreasing retention volume of the peak of interest, i.e.:

(a) with decreasing column length,
(b) with decreasing column diameter,
(c) with decreasing retention factor k,
(d) with increasing efficiency of the column,
(e) with decreasing diffusion coefficient of the sample in the mobile phase (reversed-phase separations are somewhat more critical than normal-phase separations).

If early eluting peaks show theoretical plate numbers markedly lower than the late eluting ones, too large extra-column volumes can be the reason.

2.7 TAILING

Closer study of a chromatographic peak shows that the Gaussian form is usually not completely symmetrical, the rear being spread out to a greater or lesser extent, forming a 'tail': hence the expression 'tailing'. In exceptional cases, the opposite phenomenon may arise; if the front is flatter than the back then it is referred to as 'fronting' or 'leading'. Peak shapes are illustrated in Figure 2.22.

Slight deviations from the Gaussian form are insignificant and in fact may almost be regarded as normal. However, a greater level of asymmetry means that the chromatographic system is not the optimum one. Tailing reduces the column plate number, which in turn impairs the resolution, so the cause must be found and eliminated. With

---

[16] Variances σ^2 from injection, capillaries, fittings, column (i.e. the separation process) and detector are additive, therefore the width $w = 4\sigma$ of a Gauss peak is:

$$w = 4\sqrt{\sigma_{\text{inj}}^2 + \sigma_{\text{cap}}^2 + \sigma_{\text{fit}}^2 + \sigma_{\text{col}}^2 + \sigma_{\text{det}}^2}$$

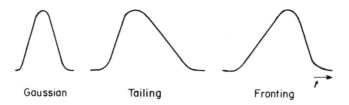

Figure 2.22   Peak shapes.

tailed peaks it is more difficult for the integrator to determine the peak delimination. Any of the following may produce these adverse conditions.

Poorly Packed or Deteriorated Column

This gives rise to shoulders or even double peaks.

Extra-Column Volumes

These cause not only general band broadening but also tailing, and should be investigated when:

(a) The tailing of peaks that are eluted early is greater than tailing of those eluted later.
(b) Tailing is greater when the flow rate of the mobile phase is faster rather than slower.

Remedy: shorter line between the injection valve and chromatographic bed and from the end of column to the detector, narrower capillaries, smaller detector cell.

Column Overload

The column is not capable of separating any quantity of substances. If too much material is injected, then the retention factor and the peak width cease to be independent of sample size (see Figure 2.23). The peaks become wider and asymmetric and the retention time changes at the same time. Tailing is accompanied by decreased retention and fronting by an increase in k.

By definition, the column has not quite reached the overload level if the change in the k value is less than 10% with respect to infinitely small amounts of sample. Most phases are not overloaded if the sample mass is less than 10 μg per gram of stationary phase. For the determination of column efficiency by injecting test compounds and calculating the number of theoretical plates, it is best to use sample sizes of 1 μg per gram of stationary phase; it can be assumed that the column contains 1 g of phase (in reality from 0.5 to 5.0 g, depending on the column dimensions).

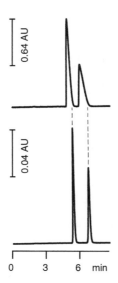

Figure 2.23   Mass overload. Top: 2 mg each of acetophenone (first peak) and veratrole; bottom: 2 µg each. Column, 25 cm × 3.2 mm i.d.; stationary phase, LiChrosorb SI 60 5 µm; mobile phase, hexane-diethyl ether 9 : 1, 1 ml min^{-1}; UV detector, 290 nm, with preparative or analytical cell, respectively.

'Chemical Tailing', Incompatibility of Sample with Stationary and/or Mobile Phase

Tailing may also arise even if the unavoidable extra-column volumes are kept to a minimum and the sample size is below $1 \mu g\, g^{-1}$. In this instance, the chosen chromatographic system of mobile and stationary phase is unsuitable and does not fit the sample; this is known as chemical or thermodynamic tailing. For examples see also Figures 9.9 and 10.10.

Some possible causes are:

(a) Poor solubility of the sample in the mobile phase.
(b) Mixed interaction mechanism. In reversed-phase chromatography the residual silanol groups which did not react with the alkyl reagent (Section 7.5) can interact with basic analytes, see Figure 2.24. Stationary phases should be synthesized from highly pure silica and silanol groups should no longer be accessible.
(c) Ionic analytes in cases where a stationary phase is used which cannot handle ions.
(d) Too active stationary phase in adsorption chromatography (see Section 9.4).

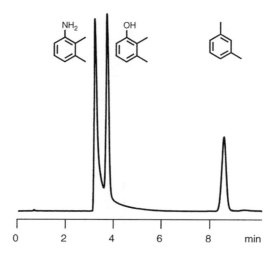

Figure 2.24 Chemical tailing of a basic analyte. Conditions: sample, 2,3-xylidine, 2,3-dimethylphenol, and *m*-xylene; column, 25 cm × 4 mm i.d.; stationary phase, LiChrospher 60 RP-Select B, 5 μm; mobile phase, 1.5 ml min^{-1} water–methanol (35 : 65); UV detector, 260 nm.

Remedy: change the mobile and/or stationary phase,[17] change the pH or use a different method (e.g. ion-pair chromatography). A change in pH may also be required in adsorption chromatography: addition of acetic or formic acid if a sample is acidic, or pyridine, triethylamine or ammonia if it is basic.

Time Constant Too High

Too high a time constant of the detector (Section 6.1) results in broad and asymmetric peaks of decreased height.

Remedy: choose an appropriate time constant.

The *peak asymmetry*, *T*, can be described in various ways, see Figure 2.25. Left: A line is laid at 10% of the peak height, thereby defining the sections $a_{0.1}$ and $b_{0.1}$. The first one describes the distance from the peak front to the maximum, the second one the distance from maximum to peak end:

$$T = \frac{b_{0.1}}{a_{0.1}}$$

---

[17] J.W. Dolan, *LC GC Int.*, **2** (7), 18 (1989).

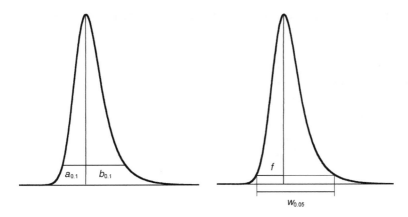

Figure 2.25   Two possibilities to describe peak asymmetry.

Right: the definition according to United States Pharmacopoeia (USP) uses a line at 5% of the peak height. The first section is defined as f and the peak width at 5% height as $w_{0.05}$:

$$T = \frac{w_{0.05}}{2f}$$

In both cases $T > 1.0$ when peaks are tailed and $T < 1.0$ with fronting peaks. A peak with $T = 1.0$ is symmetrical.

Problem 9

Calculate the tailing of the peak shown in Figure 2.25 with two methods.

Solution

$T \approx 1.6$ (left) or 1.4 (right, USP equation).

Asymmetric peaks can be described by an exponentially modified Gauss function (EMG).[18] The deviation from Gauss shape needs the expansion of the function by a time constant term τ.

---

[18] M.S. Jeansonne and J.P. Foley, J. *Chromatogr. Sci.*, **29**, 258 (1991).

2.8 PEAK CAPACITY AND STATISTICAL RESOLUTION PROBABILITY[19]

The efficiency of a separating system is best demonstrated by its peak capacity. This shows how many components can be separated in theory within a certain k range as peaks of resolution 1. The number of theoretical plates, N, of the column used must be known, as the isocratic peak capacity, n, is proportional to its square root:[20]

$$n = 1 + \frac{\sqrt{N}}{4} \ln(1 + k_{max}) \quad \text{under isocratic conditions}$$

where $k_{max} = $ maximum k value.

Problem 10

A column with 10 000 theoretical plates has a breakthrough time of 1 min. Calculate the number of peaks of resolution 1 that can possibly be separated over a 5 min period.

Solution

$t_0 = 1$ min; $t_{R\,max} = 5$ min.

$$k_{max} = \frac{5-1}{1} = 4$$

$$n = 1 + \frac{\sqrt{10\,000}}{4} \ln(1+4) = 1 + (25 \times 1.61)$$

$$n = 41$$

The above-mentioned equation for n in fact is only valid for isocratic separations and if the peaks are symmetric; the peak capacity is larger with gradient separations.[21] Tailing decreases the peak capacity of a column. In real separations the theoretical plate number is not constant over the full k range. However, it is even more important to realize that a hypothetical parameter is discussed here. It is necessary to deal with peaks that are *statistically* distributed over the accessible time range. The theory of probabilities allows us to proceed from ideal to near-real separations. Unfortunately, the results are discouraging.

---

[19] V.R. Meyer and T. Welsch, *LC GC Int.*, **9**, 670 (1996); A. Felinger and M.C. Pietrogrande, *Anal. Chem.*, **73**, 619 A (2001).
[20] E. Grushka, *Anal. Chem.*, **42**, 1142 (1970).
[21] U.D. Neue, *J. Chromatogr. A*, **1079**, 153 (2005).

What is the Probability that a Certain Component of the Sample Mixture will be Eluted as a Single Peak and not Overlapped by Other Components?

$$P \approx e^{-2m/n}$$

where:

P = probability related to a single component
m = number of components in the sample mixture

Note that m is never known in real samples. One always has to presume the presence of unknown and unwanted compounds which will be eluted at any time, perhaps even together with the peak of interest. Nevertheless, it is possible to calculate the following problem with the equation given above.

Problem 11

The sample mixture consists of ten compounds. Calculate the probability that a given peak will be eluted as a single peak on the column with peak capacity 41.

Solution

$$m = 10$$
$$n = 41$$
$$P \approx e^{-2 \times 10/41} = 0.61$$

The probability of sufficient resolution to the neighbouring peaks is only little more than 60% for each peak. We can expect that six of the compounds present will be resolved; the other four will be eluted with inadequate or missing resolution. Quantitative analysis can be impeded or almost impossible and fractions obtained by preparative chromatography can be impure, perhaps without any sign of warning to the user. The situation is even worse if the peaks are of unequal size, which in fact is the rule.

It is possible to calculate the necessary peak capacity for a 95% probability by converting the equation. For the ten-component mixture this is 390, which corresponds to a plate number of almost one million ($k_{max} = 4$) or to a retention factor of 5.7 millions ($N = 10\ 000$).

What is the Probability that all Components of a Mixture will be Separated?

$$P' = \left(1 - \frac{m-1}{n-1}\right)^{m-2}$$

P' = probability related to all components

Problem 12

Calculate the probability that all components of the ten-compound mixture will be separated on the column with a peak capacity of 41.

Solution

$$m = 10$$
$$n = 41$$
$$P' = \left(1 - \frac{9}{40}\right)^8 = 0.13$$

P' is only 13% and is markedly lower than P because now, in contrast to the above-discussed problem, *all* peaks need to be resolved. To reach this goal with a 95% probability it would be necessary to perform the separation on a column with a peak capacity of 1400, i.e. with 12 million theoretical plates ($k_{max} = 4$)![22]

Computer generated chromatograms, obtained with random numbers, are shown in Figure 2.26. They represent four possible separation patterns in a case with $m = 10$ and $n = 41$, i.e. $N = 10\,000$ and a retention time window between 1 and 5 min. In one

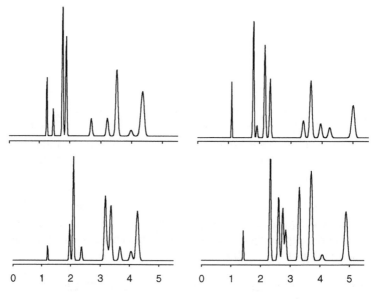

Figure 2.26 Simulated chromatograms, generated with random numbers, representing a separation at peak capacity 41 and a sample with ten compounds.

---

[22] Separation with peak capacity 900: P. Sandra and G. Vanhoenacker, *J. Sep. Sci.*, **30**, 241 (2007).

chromatogram all ten peaks are visible. In the others, only nine peaks can be seen, sometimes with poor resolution. In two cases it is not known where the tenth peak is and in one separation its position can be supposed. These graphs look like real chromatograms, and real chromatograms often follow the same rules of random elution patterns.

These considerations do not mean that the separation methods available today are inadequate. Often the situation is better than presented here because the peaks are not statistically distributed but are eluted as given by their physical properties. Homologues will be separated by increasing hydrocarbon chain length on a reversed phase, as an example. Nevertheless, even the highest separation performance does not help against troublesome surprises in qualitative or quantitative analysis. The knowledge of the equations presented here is an aid in the realistic judgement of separations obtained from complex mixtures.

Remedy: highly specific detection, derivatization (see Section 19.8), coupling with spectroscopic methods (Section 6.10), optimized gradient elution (Sections 18.2 and 18.5) or column switching (multidimensional separation, Sections 18.3 and 18.4).

2.9   EFFECTS OF TEMPERATURE IN HPLC[23]

It is not possible to set generally valid rules about the influence of temperature on HPLC separations. At increased temperature the performance of a column often increases because of the decrease of mobile phase viscosity which improves mass transfer; however, it is also possible that performance decreases. The separation factor can increase or decrease. An advantage is the shortening of analysis time due to the possibility to use higher flow rates of the mobile phase due to the increase in diffusion coefficients. If the eluent or the sample solution are viscous it is even necessary to work at higher temperatures: less pressure is needed to pump the mobile phase or it is possible only under these circumstances to inject the sample, respectively.

For the optimization of a separation the influence of temperature should always be checked; perhaps it is possible that the analysis can be done better (i.e. faster or more accurate) at a temperature higher or lower than usual. It is even possible to handle high temperatures, up to 200 °C, although some changes in instrumentation are necessary.[24] Figure 2.27 shows the separation of cyclosporins at 80 °C which is used for routine analysis. Especially for the separation of high molecular mass compounds, including proteins, this temperature range is of interest.

---

[23] C. Zhu, D.M. Goodall and S.A.C. Wren, *LC GC Eur.*, **17**, 530 (2004) or *LC GC North Am.*, **23**, 54 (2005); G. Vanhoenacker and P. Sandra, *J. Sep. Sci.*, **29**, 1822 (2006).
[24] C.V. McNeff *et al.*, *J. Sep. Sci.*, **30**, 1672 (2007); S. Heinisch and J.L. Rocca, *J. Chomatogr. A*, **1261**, 642 (2009).

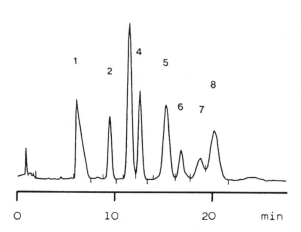

Figure 2.27 Separation at higher temperature. Reproduced with permission from N.M. Djordjevic *et al.*, *J. Chromatogr.*, **550**, 27 (1991). Conditions: sample, cyclosporins (CS); column, 25 cm × 4.6 mm i.d.; stationary phase, Ultrasphere ODS 5 μm; mobile phase, 2 ml min^{-1} acetonitrile–water–*tert*-butylmethyl ether (50 : 45 : 5); temperature, 80 °C; UV detector, 210 + 230 nm. Peaks: 1 = iso-CSA; 2 = CS C; 3 = CS B; 4 = CS L; 5 = CS A; 6 = dihydro-CS A; 7 = CS T; 8 = CS G.

An increase in temperature has the following disadvantages:

(a) Solvent or sample components are more likely to decompose.
(b) The vapour pressure of the solvent rises, thus increasing the risk of bubbles in the detector, which in turn produce an uneven baseline, ghost peaks or even complete light absorption.
(c) All chromatographic equilibria are temperature dependent, in particular all ion equilibria in aqueous mobile phases (reversed-phase and ion-exchange chromatography!). Likewise, the equilibrium between water dissolved in the mobile phase and adsorbed water in adsorption chromatography is temperature dependent and is more difficult to control at elevated temperature. In most cases these equilibria are crucial for the chromatographic properties of the column. Therefore, the reproducibility may be poor if thermostating is inadequate.[25]

---

[25] An extreme example of this has been given by M. Vecchi, G. Englert, R. Maurer and V. Meduna, *Helv. Chim. Acta*, **64**, 2746 (1981). It describes the separation of β-carotene isomers on alumina with a defined water content (hexane as mobile phase). The separation alters significantly with a change in temperature of 1 °C!

(d) The solubility of silica is greatly increased in all mobile phases as the temperature rises. Thermostated scavenger columns containing silica are recommended to be installed prior to the injection valve and also when working with chemically bonded phases based on silica.

Thermostat control is no problem in the more expensive compact units with an incorporated column oven. However, a glass or metal jacket fitted around the column, which is then connected to a thermostat bath, is an effective self-help measure. The mobile phase must also be thermostatically controlled by the same bath prior to entry into the column, by means of suitable equipment, e.g. a long spiral. Pumps with thermostatically controlled heads are also available as well as detectors with thermostated cell.

The maximum temperature is around $120\,^{\circ}\mathrm{C}$ for silica columns and should not exceed $80\,^{\circ}\mathrm{C}$ for chemically bonded phases.

The fact that the mobile phase is heated by flow resistance as it passes through the column should not be overlooked. A rule of thumb is a $0.1\,^{\circ}\mathrm{C}$ temperature increase per 1 bar pressure drop or $10\,^{\circ}\mathrm{C}$ per 100 bar ($0.025\,^{\circ}\mathrm{C}$ per bar or $2.5\,^{\circ}\mathrm{C}$ per 100 bar for water).[26] It may be advisable to cool the column and work with low volume flow rates when very low-boiling mobile phases such as pentane (boiling point $36\,^{\circ}\mathrm{C}$) are involved.

2.10   THE LIMITS OF HPLC

Martin *et al.* published a paper on the theoretical limits of HPLC which is well worth reading.[27] They used relatively simple mathematics to calculate pressure-optimized columns for which the length L, particle size d_p and flow rate u of the mobile phase were selected such that a minimum pressure Δp is required to solve a separation problem. It has been shown that these optimized columns are operated at their van Deemter curve minima. Some astonishing facts have emerged from the study, provided that the chromatography is performed on well packed columns (reduced plate height $h = 2$–3; see Section 8.5).

'Normal' Separation Problem: 5000 Theoretical Plates are Required and Analysis Extends over a 5 min Period

A pressure-optimized column has $L = 10\,\mathrm{cm}$ and $d_p = 6.3\,\mu\mathrm{m}$. The recommended pressure is 11.8 bar (for a mobile phase viscosity of 0.4 mPa, which is a typical value in

---

[26] The relationship is given by $\Delta T = -\Delta p / C_v$, where p is pressure and C_v the heat capacity of the liquid at constant volume.
[27] M. Martin, C. Eon and G. Guiochon, *J. Chromatogr.*, **99**, 357 (1974).

adsorption chromatography; the viscosity may be up to five times greater in reversed-phase chromatography, increasing the pressure required to around 50 bar). This pressure is much lower than usual in most separations.

Simple Separation Problem: $N = 1000$, $t_R = 1$ min

A pressure-optimized column has $L = 2.2$ cm, $d_p = 6.9$ μm and $\Delta p = 2.3$ bar! Shorter columns are preferred for simple problems with a separation factor of ca. 1.2. Both time and solvent are saved by using columns 3–5 cm in length (also available commercially), but the injection and extra-column volumes and the detector time constant must be kept small in order not to deteriorate the separation performance.

An example of an eight-component separation on a 5 cm column is given in Figure 2.28.

Difficult Separation Problem: $N = 10^5$, $t_R = 30$ min

A pressure-optimized column has $L = 119$ cm, $d_p = 3.8$ μm and $\Delta p = 780$ bar. These conditions can be achieved, as the separation of polycyclic aromatic compounds using 68 000 theoretical plates shows.[28]

The optimized pressure concept can be extended. Figures 2.29 and 2.30 show nomograms for well packed columns operated at their van Deemter curve minima.[29] The nomograms show the interrelationship between column length, pressure, number of theoretical plates, particle size of the stationary phase and breakthrough time. Two of these five parameters may be selected at random, the other three being geared to the optimum flow rate.

Figure 2.29 is based on low-viscosity mobile phases typical of adsorption chromatography (viscosity $\eta = 0.44$ mPa s, e.g. dichloromethane). Figure 2.30 relates to higher viscosity mobile phases which are generally aqueous in nature, as found in reversed-phase and ion-exchange chromatography ($\eta = 1.2$ mPa s; e.g. methanol–water, 8:2).

Strictly, the nomograms apply only for columns with a very specific van Deemter curve and samples with defined diffusion coefficients. However, they can always be used as a means of establishing whether or not a specific system is suitable for solving a specific separation problem. Also, they provide a view of the HPLC limits for those who understand how to interpret them.

Problem 13

Calculate the number of theoretical plates that can be achieved with a pressure of 10 bar and 10 μm silica under optimum conditions (low viscosity system).

---

[28] M. Verzele and C. Dewaele, *J. High Resolut. Chromatogr. Commun.*, **5**, 245 (1982).
[29] I. Halász and G. Görlitz, *Angew. Chem.*, **94**, 50 (1982).

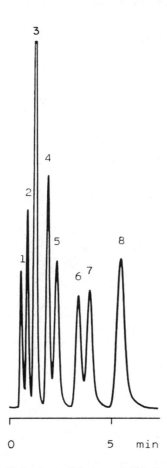

Figure 2.28 Separation of phenothiazine derivatives on a short column (Prolabo). Conditions: column, 5 cm × 4 mm i.d.; stationary phase, silica, 6.2 µm; mobile phase, 0.8 ml min^{-1} diisopropyl ether–methanol (1:1) containing 2.6% water and 0.2% triethylamine; pressure, 15 bar; UV detector, 254 nm; number of theoretical plates, 850 (final peak). Peaks: 1 = 3-chlorophenothiazine; 2 = chlorophenethazine; 3 = chloropromazine; 4 = promazine; 5 = 5,5-dioxy-chlorpromazine; 6 = oxy-chlorpromazine; 7 = 2-chloro-10-(3-methylaminopropyl)-phenothiazine; 8 = N-oxychlorpromazine.

Solution

These conditions are marked with a circle in Figure 2.29 and the required parameters can be read off: $N = 10\,000$. Other conditions are as follows: $L = 30$ cm, $t_0 = 300$ s, optimum flow rate, $u_{opt} = 1$ mm s^{-1}.

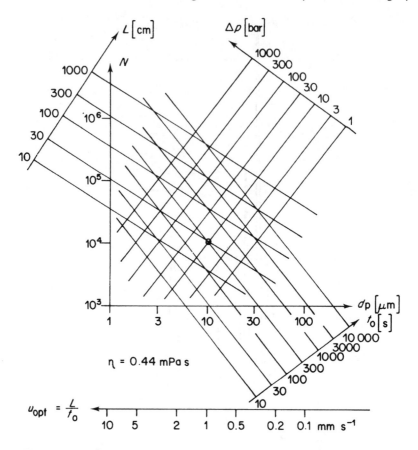

Figure 2.29   Optimum parameters for low-viscosity mobile phases.

Problem 14

The pump supplies 1000 bar. Calculate the optimum conditions for achieving 100 000 theoretical plates (high viscosity system).

Solution

The result is marked with a circle in Figure 2.30. $d_p = 3.5\,\mu$m, $L = 100$ cm, $t_0 = 700$ s, $u_{opt} = 1.4$ mm s^{-1}.

In such separation problems the required plate number N is given by the separation factor α_{min} of the 'critical peak pair', i.e. those two components of the mixture that are most difficult to separate. Consequently the optimizing strategy can also be developed on α_{min} alone.[30]

[30] R.P.W. Scott, *J. Chromatogr.*, **468**, 99 (1989).

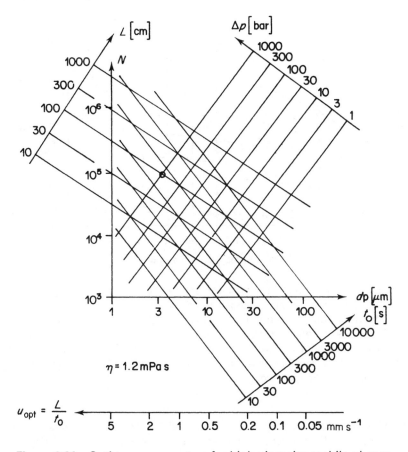

Figure 2.30 Optimum parameters for high-viscosity mobile phases.

A comprehensive presentation of the "limits of HPLC" was published by G. Guichon as "The limits of the separation power of unidimensional column liquid chromatography".[31]

2.11 HOW TO OBTAIN PEAK CAPACITY

From a practical point of view, peak capacity (Section 2.8) is more important than the number of theoretical plates. Peak capacity is needed in order to solve analytical problems and to separate as many peaks as possible. A glance at the Halász diagrams (Figures 2.29 and 2.30) makes clear that, to some degree, the analyst is free to work

---

[31] G. Guiochon, *J. Chomatogr. A*, **1126**, 6 (2006).

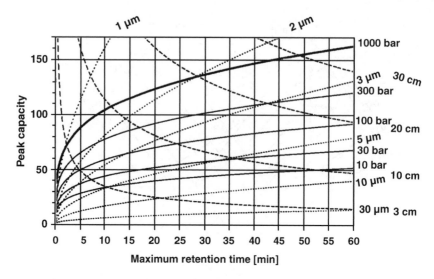

Figure 2.31 Peak capacity as a function of analytical run time. The graph is valid for isocratic reversed-phase systems which are run at their van Deemter optimum. The maximum retention factor k is 20, i.e. the maximum retention time is t_0 21, then the separation ends. The figure is only valid for small analytes with a diffusion coefficient of approx. $1 \cdot 10^{-9}\ m^2\ s^{-1}$ and not for macromolecules. Dotted lines represent the particle diameter, dashed lines the column length, and solid lines the pressure, respectively.

fast, with short columns and small particle diameter or slow, with longer columns and coarser packing; the difference lies in the pressure needed. Figure 2.31 is a transformation of Figure 2.30, now with the peak capacity and the analysis time as the interesting parameters.[32] The separation system is in reversed-phase mode as well, the eluent viscosity is 1.2 mPa s, the total porosity is 0.70 (Section 8.3) and the possible columns are run in their van Deemter optimum. The graph is valid for isocratic separations. The maximum pressure is 1000 bar. On the very right it can be seen that, e.g., a peak capacity of 90 can be obtained within 1 h, using a column of 20 cm length with 4 µm packing; the pressure needed is 100 bar. (90 means more than the doubled peak capacity of 41 which was assumed in Problems 10 and 11.) A peak capacity of 50 is possible within 5 min, using a 3 cm column with approx. 2.5 µm packing at 100 bar as well – or with not more than 30 bar, a 10 cm column, a 5 µm phase, and a separation time of almost 40 min.

[32] V.R. Meyer, *J. Chromatogr. A*, **1187**, 138 (2008).

Separations in the normal-phase mode allow to get higher peak capacities at identical pressure or identical capacitites as in this figure at lower pressure due to the lower viscosity of the eluent: at 20 °C, hexane has a viscosity of 0.33 mPa s but water has 1 mPa s (Section 5.1). We can also compare the time needed instead of the pressure. Normal-phase separations are faster than the ones on reversed phases if the goal is identical capacity at identical pressure.

Gradient separations allow to reach much higher peak capacities than shown here with isocratic runs.

3 Pumps

3.1 GENERAL REQUIREMENTS

An HPLC pump unifies two different features: It must be a sturdy device capable of generating high pressure up to 350 or even 500 bar; conjointly the user demands high flow accuracy and precision at any flow rate which is chosen. The range of flow usually goes from $0.1 \, ml \, min^{-1}$ to 5 or $10 \, ml \, min^{-1}$ and can be set at any value in steps of $0.1 \, ml \, min^{-1}$. The delivered flow rate must be independent of the back pressure, even if this changes during a separation, which is usually the case with gradient elution. Moreover, the flow should be pulseless, especially when a refractive index, conductivity or electrochemical detector is used.

Other, more practical, requirements refer to the ease of operation. The pump should be ready to use with only a few simple manipulations, including the priming of the pump head (i.e. the flushing and filling of the check valves with new eluent). Its internal volume must be low in order to enable a quick solvent change, although a certain extra volume is necessary for pulse dampening. Any handling for maintenance and repair needs to be simple and easy to perform.

It should be mentioned here again that high pressure per se is not what one is looking for in HPLC. Unfortunately, it comes from the fact that the mobile phase is a liquid with its rather high viscosity which needs to be pressed through a densely packed bed of very fine particles. Small particles have short diffusion paths and therefore yield a high number of theoretical plates per unit length.

3.2 THE SHORT-STROKE PISTON PUMP

The usual type of HPLC pump is the so-called short-stroke piston pump. Its general design, although simplified, is shown in Figure 3.1. The mobile phase is driven by a

Practical High-Performance Liquid Chromatography, Fifth edition   Veronika R. Meyer
© 2010 John Wiley & Sons, Ltd

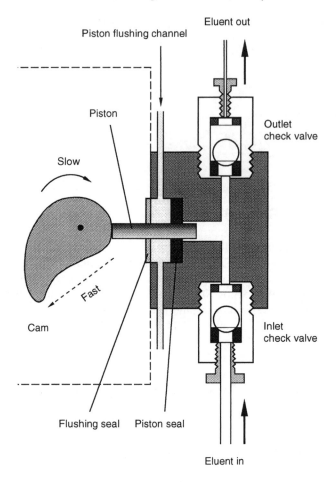

Figure 3.1   Principle of a short-stroke piston pump.

piston whereas check valves open or close the path of the liquid towards the appropriate direction. The dark grey part which carries the check valves is the pump head.

The pumping action comes from a stepper motor which drives a rotating disk or cam. The cam then guides the piston to its forward and backward direction. With each stroke, the piston conveys a small amount of liquid, usually in the 100 µl range. The check valves are built in such an asymmetric way that they close when pressure comes from the top and open when the pressure at their bottom is higher than at the top. The

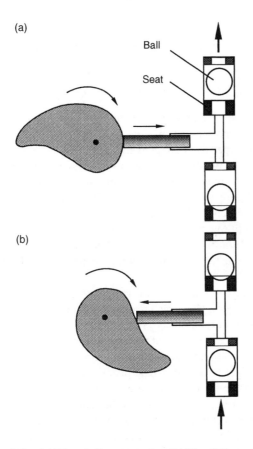

Figure 3.2 (a) The delivery stroke. (b) The filling stroke.

schematic drawings of Figure 3.2 clarifiy this principle. Electrically driven valves are also used because they are advantageous for low flow rates.

With a circle-shaped cam and a uniformly working stepper motor the outlet of the pump would be discontinuous and not smooth:

In order to obtain as smooth a flow as possible, the cam has an irregular shape and is rotated according to a well designed time program depending on its actual position. During the delivery stroke of the piston the cam turns slowly; its shape guarantees a constant flow. For the filling stroke the cam turns rapidly which results in an interruption of the flow which is kept as short as possible:

The cam can be replaced by an electronically controlled linear spindle. Optimum flow constancy is then obtained by a perfectly matching software. Such systems, however, are more expensive than conventional ones.

The flow can be smoothed by using two pistons which work parallel but in opposite directions. Another design is the use of tandem pistons arranged in series, whereby the first one conveys slightly more than the second. This allows to compensate for mobile phase compressibility and helps to suppress cavitation, i.e. the formation of vapour bubbles during the filling stroke. Some pumps use double-ball check valves with a second seat and ball on top of the first one. The electronics of the pumps are also designed to compensate for the compressibility of the eluent (which fortunately is low, e.g. 1% per 100 bar).

The piston is made from sapphire or ceramics, therefore it is fragile and will break when dropped to the floor. However, during normal pump operation, piston breaking is extremely rare. If buffer solution is left in the interior of the pump head when it is not in use, the solvent may evaporate and crystals will grow within the pump head. If the pump is then swiched on again, the surface of the piston will be scratched and solvent delivery will no longer be accurate. The piston seal is from an inert polymer, e.g. a mixture of teflon and graphite, and has often an embedded spring which is only visible from one side. Therefore the orientation of the seal is important when it needs to be replaced. The seal is designed to withstand the high pressure for which the pump is built but it shows a minute degree of leakage. The leaking mobile phase is used to lubricate the piston and to keep the seal elastic. Most pumps today have a flushing channel where a liquid, usually water, can circulate and which flushes the piston free from buffer salts or abrasives from the piston seal. The flushing channel itself needs a seal, a piece of teflon which is not under pressure.

The valve balls are from ruby, the seats from sapphire. In its seat, the ball seals only through the thin ring where the two parts are in contact. If a particle of dust or a buffer crystal is trapped there, the valve will be leaky. This is one of the reasons why mobile phases must rigorously be kept particle-free.

3.3 MAINTENANCE AND REPAIR

A pump is a precision instrument and must be treated with as much care as an analytical balance. All recommendations given in the manual should be followed. Even under best circumstances some maintenance will be necessary from time to time. Also for such operations the manual must be consulted in advance.

Occasionally, the cam needs some lubrication. Follow the instructions.

For maintenance at or in the pump head it is necessary to work cleanly and with the appropriate tools. Enough bench space, wipes, cleaning solvent and time is needed and a beaker or bowl should be placed below the pump head which will prevent pistons from dropping and valve balls from disappearing somewhere on the floor.

Piston seals become worn and need to be replaced, e.g. once a year (they cannot be repaired). For this purpose the pump head is removed from the pump body; then the seals become visible. They are removed with the special tool which was delivered together with the pump. Removing the seals with a screwdriver can scratch the pump head which causes leaking! The new seals are brought in place with another special tool. It is necessary to follow the instructions exactly. One must not mix up the two different sides of a seal. A seal which is placed the wrong way must be removed and thrown away; it cannot be used once again. The new seals need to be broken in with a low solvent flow and without a column connected. Water should be avoided to break in a new seal. Well suited solvents are isopropanol and methanol.

Blocked or leaky check valves[1] can probably be fixed by sonicating them. Perhaps a number of solvents need to be tried, e.g. water with detergent, tetrahydrofuran or dichloromethane. Some valves fall apart when they are removed from the pump head (beware of loss of the ball!), some can be opened and some are a unit which should not be opened. With the first two designs the ball can be replaced, maybe also the seat, when these parts are scratched and do not seal well any longer. The other types of check valves must be replaced totally if they do not work properly and if a treatment with ultrasonics does not help.

Two recommendations are of the utmost importance:

(a) A pump must not run dry!
(b) Do not switch off a pump when it is filled with buffer solution! Always replace the buffer by water; if you think it is advantageous, then replace the water by pumping an organic solvent. In the case of a longer break it is best to store the pump with a liquid of at least 10% organic solvent, thus preventing the growth of algae.

3.4 OTHER PUMP DESIGNS

In a special type of short-stroke piston pump the piston does not convey the mobile phase directly. The piston moves within an oil-filled channel, and oil and eluent are separated from each other by an elastic steel membrane. By this design the piston is not in contact with the mobile phase which may be aggressive; on the other hand, the eluent cannot be contaminated by abrasives from the piston seal.

---

[1] J.W. Dolan, *LC GC Eur.*, **19**, 140 (2006) and **21**, 514 (2008) or *LC GC North Am.*, **24**, 132 (2006) and **26**, 532 (2008).

For very low solvent flows a pump type can be used which does not convey the liquid by rapid strokes but which acts like an oversized syringe. A slow-moving piston pushes the eluent directly from its reservoir. This flow is pulsation-free. The reservoir has a limited volume, e.g. 10 ml, therefore the chromatography must be stopped from time to time and the pump refilled if a single-piston design is used. The operation is more convenient with a two-piston pump where one of the pistons is delivering the eluent while the other one simultaneously fills its reservoir. As soon as the first piston has reached its end-point the solvent delivery is taken over by the second one. Gradients are not possible with syringe pumps.

For flow rates in the microliter per minute range, special two-piston pumps are available. With such instruments it is even possible to run rather precise high-pressure gradients if the total flow is $50 \, \mu l \, min^{-1}$ or higher.

Pneumatic amplifier pumps are used for column packing and for applications where a pressure of over 500 bar is needed. With this design, a relatively low gas pressure activates the larger cross-sectional area of a piston which at its other end is in contact with the eluent over a much smaller area. The force is the same on both sides of the piston but the pressure is higher on the smaller area, according to the ratio of the two areas. The pressure amplification can be as high as 70-fold; this means that a gas pressure of 10 bar can generate an eluent pressure of 700 bar.

For preparative separations there are pumps on the market which convey up to $300 \, ml \, min^{-1}$ even at a pressure of 300 bar.

4 Preparation of Equipment up to Sample Injection

4.1 SELECTION OF THE MOBILE PHASE

The mobile phase must obviously be chosen for its chromatographic properties: it must interact with a suitable stationary phase to separate a mixture as fast and as efficiently as possible. As a general rule, a range of solvents is potentially able to solve any particular problem, so selection must be based on different criteria:

(a) *Viscosity*: a low-viscosity solvent produces a lower pressure drop than a solvent with higher viscosity for a specific flow rate. It also allows faster chromatography as mass transfer takes place faster. The (dynamic) viscosity η of a solvent is expressed in millipascal seconds (mPa s; formerly in centipoise (cP); the values remain the same).

(b) *UV transparency*:[1] if a UV detector is used, the mobile phase must be completely transparent at the required wavelength, e.g. ethyl acetate is unsuitable for detection at 254 nm as it does not become sufficiently optically transparent until 275 nm ($< 10\%$ absorption). The UV transparency of buffer salts, ion-pair reagents and other additives must also be considered.

(c) *Refractive index*: only important if a refractive index detector is used. The difference between the refractive indices of the solvent and the sample should be as great as possible when working at the detection limits.

(d) *Boiling point*: a lower boiling point of the mobile phase is required if the eluate is to be recovered and further processed; this means less stress during evaporation of the mobile phase for heat-sensitive compounds. But, solvents

---

[1] C. Seaver and P. Sadek, *LC GC Int.*, **7**, 631 (1994).

Practical High-Performance Liquid Chromatography, Fifth edition   Veronika R. Meyer
© 2010 John Wiley & Sons, Ltd

with a high vapour pressure at the operating temperature tend to produce vapour bubbles in the detector.

(e) *Purity*: this criterion has a different meaning depending on intended use: (i) absence of compounds that would interfere with the chosen mode of detection, (ii) absence of compounds that disturb gradient elution (see Figure 18.6), (iii) absence of nonvolatile residues in the case of preparative separations (see Problem 43 in Section 21.4). Hexane for HPLC does not necessarily need to be pure *n*-hexane but may also contain branched isomers because the elution properties are not influenced by this (but it must not contain benzene, even in traces, if used for UV detection).

(f) *Inert with respect to sample compounds*: the mobile phase must not react at all with the sample mixture (peroxides!). If extremely oxidation-sensitive samples are involved, 0.05% of the antioxidant 2,6-di-*tert*-butyl-*p*-cresol (BHT) may need to be added to the solvent. BHT is readily removed from the eluent by vaporization, but it absorbs in the UV region below 285 nm.

(g) *Corrosion resistance*:[2] light promotes the release of HCl from chlorinated solvents. The ever-present traces of water combine with this to produce hydrochloric acid, which attacks stainless steel. Corrosion is intensified by the presence of polar solvents; mixtures such as tetrahydrofuran–carbon tetrachloride or methanol–carbon tetrachloride are particularly reactive. All iron complex-forming compounds, e.g. chloride, bromide, iodide, acetate, citrate or formate ions (buffer solutions!) have a corrosive effect. Lithium salt buffers are also very corrosive at low pH.[3] Steel may even corrode with methanol or acetonitrile.[4] Perhaps a passivation of the apparatus with nitric acid[5] or a change in instrument design[6] is necessary. In any case a wash with a halogen- and ion-free solvent after a chlorinated mobile phase or a salt solution has been used is recommended.

(h) *Toxicity*: here the onus is on each individual laboratory to avoid toxic products as far as possible. Chlorinated solvents may release the highly poisonous phosgene gas. Toluene should always replace benzene (carcinogenic) wherever possible.

(i) *Price!*

As a general rule the mobile phase should not be detector-active, i.e. it should not have a property which is used for detection (exception: indirect detection, see Section 6.9). Otherwise it is very possible that unwanted baseline effects and extra peaks will show up in the chromatogram. However, this recommendation cannot be followed in the case of bulk-property detectors such as the refractive index detector.

Solvent properties are listed in Section 5.1.

---

[2] K.E. Collins *et al.*, *LC GC Eur.*, **13**, 464 and 642 (2000) or *LC GC Mag.*, **18**, 600 and 688 (2000).
[3] P.R. Haddad and R.C.L. Foley, *J. Chromatogr.*, **407**, 133 (1987).
[4] R.A. Mowery, *J. Chromatogr. Sci.*, **23**, 22 (1985).
[5] R. Shoup and M. Bogdan, *LC GC Int.*, **2** (10), 16 (1989) or *LC GC Mag.*,**7**, 742 (1989).
[6] M.V. Pickering, *LC GC Int.*, **1** (6), 32 (1988).

4.2   PREPARATION OF THE MOBILE PHASE

Careful selection of the mobile phase should be followed by just as careful preparation.

Usually the best choice is to use solvents and reagents which are sold as 'HPLC grade' or a similar quality. This is even true for water if the quality produced in the laboratory by a water-purifying system is not good enough. HPLC solvents guarantee the best possible UV transparency and the absence of contaminants which alter the elution strength or give rise to extra peaks in gradient separation. If such a quality is not available, the solvents of poorer purity need to be purified by fractional distillation or by adsorption chromatography.[7]

For mixed mobile phases or buffers the mode of preparation should be described in detail. The eluent properties may differ depending on the order of the necessary steps such as the dissolution of various buffer salts, the pH adjustment or the addition of nonionic additives. When mixing water and methanol (to some extent also other water-miscible solvents), a volume contraction effect occurs: the volume of the mixture is smaller than the individual volumes. Therefore it is necessary to measure the two solvents separately before mixing them.

If an eluent is not used directly from its original bottle it is necessary to filter it through a 0.5 or 0.8 μm filter immediately before use. There is always a danger of the presence of particulate matter which would damage the pump valves, block the capillaries and frits or plug the column. Take care to use a filter material which is inert with regard to the solvent used.

For various reasons the eluents used for HPLC should be degassed, especially the polar ones which can dissolve fairly high amounts of air (water, buffer solutions, all other aqueous mixtures, water-miscible organic solvents),[8] otherwise some problems may occur: gas bubbles will show up in the detector where the back pressure is low, the baseline will be noisy or spiked with extra peaks, analytical precision may be affected, and peaks may be small because dissolved oxygen absorbs at low UV wavelengths and quenches fluorescence. Gradient separations are particularly delicate as gas is released when air-containing solvents are mixed.

A mobile phase can be degassed offline by putting the reservoir flask into an ultrasonic bath and sonicating it under vacuum for one minute. This works for several hours and is sufficient if the demands on analytical precision are low. With ion-pair reagents problems may arise due to foaming. Online degassing is more convenient (and more expensive), either by continuous helium sparging (helium is much less soluble than other gases) or by the use of a degassing module. With such instruments

---

[7] Detailed instructions are given in: *Purification of Solvents with ICN Adsorbents*, ICN Biomedicals, P.O. Box 1360, D-37269 Eschwege, Germany.

[8] J.W. Dolan, *LC GC Eur.*, **12**, 692 (1999) or *LC GC Mag.*, **17**, 908 (1999).

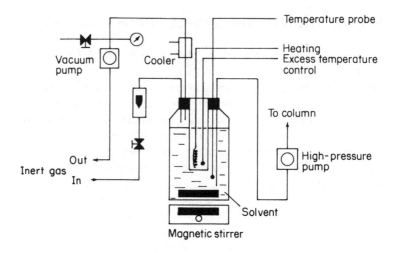

Figure 4.1 Solvent preparation. Reproduced by permission of Hewlett-Packard.

the mobile phase flows through a membrane which allows the gases to penetrate into the surrounding vacuum. They have a rather high internal volume which is a drawback when the eluent needs to be changed.

Solvents and eluents must not be stored in plastic bottles because plasticizers and other low-molecular compounds may diffuse into the liquid. Eluent reservoirs should always be capped, either gas-tight or not depending on whether the solvents are stored or in use. Figure 4.1 shows a solvent supply vessel with all possible accessories: heating (for thermostating), reflux condenser, magnetic stirrer, vacuum pump, inert gas line and excess temperature control.

The level in the supply vessel must be permanently controlled. The pump must never be allowed to run dry, otherwise severe mechanical damage could ensue. The pump feed line must have as large a diameter as possible (ca. 2 mm).

4.3 GRADIENT SYSTEMS

The composition of the mobile phase may need to be changed as the substances migrate through the column when complex sample mixtures are involved. Changes must be accompanied by an increase in mobile phase elution strength so that peaks which would otherwise be eluted late or not at all are accelerated. Depending on construction, two different gradient systems can be distinguished, called low- or

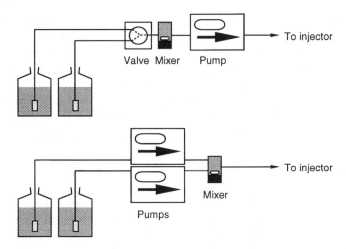

Figure 4.2 Low-pressure gradient system (top) and high-pressure system (bottom).

high-pressure gradients, depending on whether the various solvents are mixed prior or after the high-pressure pump.

For example, a *low-pressure gradient system* (Figure 4.2, top) may consist of two or more solvent reservoirs connected to a proportioning valve and a mixing chamber (volume < 1 ml) fitted with a magnetic stirrer. The reservoirs contain solvents that differ in elution strength. Valve ports are opened for different periods of time, thus changing the mixing chamber composition. A pump conveys the mixed eluent to injector and column. With such a design the solvents are mixed on the low-pressure side of the pump, therefore the term low-pressure gradient system is used. It is relatively cheap although the proportioning valve needs to be controlled very precisely. Its dwell volume (see below) may be large and problems rising from air bubbles are not uncommon.

High-pressure gradient systems (Figure 4.2, bottom) require two or three high-pressure pumps for conveying different solvents (which must, of course, be inter-mixable). Initially the weak eluent pump conveys most (if not all) and the strong eluent pump very little (or even nothing). The flow rate of each pump is linearly or exponentially modified either continuously or intermittently. Electronic control must ensure that the total volume always remains constant. Here the solvent mixing occurs on the high-pressure side of the pumps, therefore the design is known as a high-pressure gradient system. Because one pump per solvent is necessary it is expensive. Its dwell volume is small.

The *dwell volume*[9] is the volume within the HPLC system from the point where the solvents are mixed to the entrance of the column. In a low-pressure system it consists of the volumes of the proportioning valve, the mixer, the pump head, the injector and the connecting capillaries. In a high-pressure system it is smaller because only the mixer, injector and the capillaries add their volumes.

A large dwell volume is rather disadvantageous. It delays and smoothes the gradient and requires a long equilibration time before the next separation. Large dwell volumes are not suited for low flow rates.

4.4 CAPILLARY TUBING

Capillaries with an outer diameter of 1/16 in (1.6 mm) are generally used to connect the various components of the chromatograph. For the inner diameter there is a choice of 0.18, 0.25, 0.50 and 1.0 mm and others (i.e. 0.007, 0.01, 0.02 or 0.04 in). For all connections within the system without any contact to the samples, the 1 mm i.d. capillaries are preferred. They will (usually) not clog and have the lowest possible flow resistance. Their internal volume is 0.8 ml m^{-1}. For all critical connections with regard to extra-column volumes, i.e. between injector and column as well as between column and detector, it is necessary to use 0.18 or 0.25 mm i.d. capillaries.[10] The extra-column volume is by far lowest with the 0.18 mm type (23 µl m^{-1}) but special precautions such as careful filtration of mobile phase and samples are necessary to avoid clogging. The 0.25 mm capillaries (50 µl m^{-1}) are less critical but they cannot be recommended for the use with columns which generate very narrow peaks.

General aspects of extra-column volumes in the HPLC instrument have been discussed in Section 2.6. Various formulas have been proposed which allow to calculate the maximum allowed capillary length with regard to band broadening; an adaption after M. Martin *et al.* is as follows:[11]

$$l_{cap}(cm) = 4 \times 10^{10} \frac{F(ml\ min^{-1})t_R^2(min)D_m(m^2s^{-1})}{d_{cap}^4(mm)N}$$

where F is the volume flow rate, t_R is the retention time of the peak of interest, D_m is the diffusion coefficient of the solute molecule in the mobile phase, d_{cap} is the inner diameter of the capillary, and N is the theoretical plate number of the peak. The equation gives the length which broadens the peak by not more than 5% of its width. For a rough estimation of how long a capillary may be (the equation is not intended to give more than this), a mean diffusion coefficient for small molecules is assumed to be

---

[9] L.R. Snyder and J.W. Dolan, *LC GC Int.*, **3** (10), 28 (1990).

[10] i>,/i>, **19**, 644 (2006) or *LC GC North Am.*, **24**, 1078 (2006).

[11] M. Martin, C. Eon and G. Guiochon, *J. Chromatogr.*, **108**, 229 (1975).

$1 \times 10^{-9}\,\mathrm{m^2\,s^{-1}}$. Therefore the equation can be written in a simpler form:

$$l_{cap}(cm) = 40\frac{F(\mathrm{ml}\cdot\mathrm{min}^{-1})t_R^2(\mathrm{min})}{d_{cap}^4(\mathrm{mm})N}$$

Note that band broadening increases with the fourth power of the capillary inner diameter! Therefore it is allowed to use quite long capillaries if they are thin enough. For analytical separations it is not advisable to use 0.5 mm capillaries for critical connections.

Problem 15

With a flow rate of 1 ml min^{-1}, a peak has a retention time of 2.6 min on a column with 7900 theoretical plates. What is the maximum length for 5% peak broadening if either a 0.25 mm i.d. or a 0.18 mm i.d. capillary is used?

Solution

$$l_{cap} = 40\frac{1 \times 2.6^2}{0.25^4 \times 7900}\,\mathrm{cm} = 9\,\mathrm{cm} \text{ for the 0.25 mm capillary.}$$

Analogously, 33 cm for the 0.18 mm capillary.

The pressure drop Δp generated by capillaries is low compared to the flow resistance which comes from the column. It can be calculated by the Hagen–Poiseuille equation (for laminar flow) which, if expressed in the usual HPLC units, appears as:

$$\Delta p(\mathrm{bar}) = 6.8 \times 10^{-3}\frac{F(\mathrm{ml\,min}^{-1})l_{cap}(\mathrm{m})\eta(\mathrm{mPas})}{d_{cap}^4(\mathrm{mm})}$$

where F is the volume flow rate and η is the viscosity of the eluent. For the calculation of Δp in kPa the factor is 0.68.

Problem 16

Calculate the pressure drop generated by a flow of 2 ml min^{-1} of water through a capillary of 0.8 m length and 0.25 mm inner diameter.

Solution

Water has a viscosity of 1 mPas (Section 5.1).

$$\Delta p = 6.8 \times 10^{-3}\frac{2 \times 0.8 \times 1}{0.25^4}\,\mathrm{bar} = 2.8\,\mathrm{bar}$$

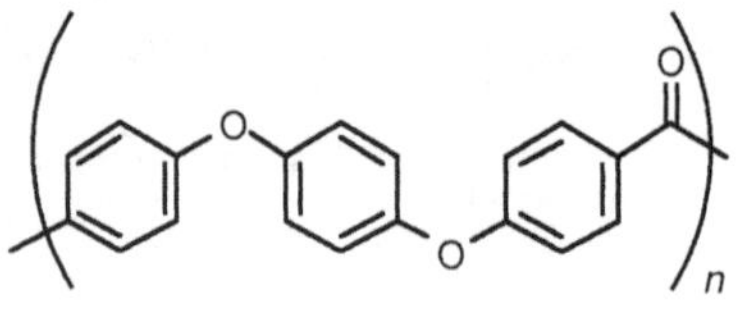

Figure 4.3 Chemical formula of peek.

For materials one can choose between steel, teflon and peek. Steel is pressure-resistant and generally also corrosion-resistant; for exceptions see Section 4.1. Biocompatibility cannot be taken for granted but the adsorption of proteins can occur. Teflon is only slightly resistant against pressure and temperature, so it cannot be used for all applications. Peek (polyether ether ketone, see Figure 4.3) can replace steel if its limited chemical and pressure resistance is considered: peek is not stable in dichloromethane, tetrahydrofuran, dimethyl sulfoxide, concentrated nitric and concentrated sulfuric acid. The material is biocompatible. Its upper pressure limit is approx. 200 bar, although some suppliers claim 350 bar. Peek capillaries are colour-coded: 0.18 mm yellow, 0.25 mm blue, 0.5 mm orange, 1.0 mm grey. Ultrapek and Carbon Peek have a slightly different chemical composition from peek, which leads to better stability against pressure and chemicals.

It is recommended to use all plastic capillaries only together with plastic (finger-tight) fittings. With steel fittings there is a danger that they will be tightened too strongly and that the capillary will be damaged.

Teflon and peek capillaries are easily cut with a razor blade but cutting instruments are also available. Steel is difficult to cut because thin-walled tubes are prone to deformation and thick-walled ones to clogging or to a reduction of their inner diameter. It is best to use one of the special instruments which are available from the vendors of capillaries; follow the instructions. Without such a device at hand, cut a score over the whole circumference of the capillary by means of a small file. It is then grasped with two smooth-jawed pliers or with a vice and one plier very close to the score and bent back and forth until it breaks. If a freshly cut capillary is built into the instrument for the first time, attach it with the new end downstream from the pump and do not connect another instrument but allow the effluent, with possible dust, to go directly to waste.

4.5   FITTINGS[12]

All the small items which are needed to connect the various parts of an HPLC system are called fittings. Usually they guide a capillary to the column or to an instrument. A

---

[12] J.W. Batts, *All About Fittings,* Scivex, Upchurch Scientific Division, P.O. Box 1529, Oak Harbor, WA 98277, USA; i>,/i>, **19**, 644 (2006) or *LC GC North Am.*, **24**, 1078 (2006).

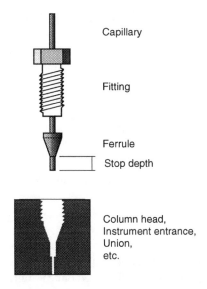

Capillary

Fitting

Ferrule
Stop depth

Column head,
Instrument entrance,
Union,
etc.

Figure 4.4   High-pressure connection.

possible design is shown in Figure 4.4 but other set-ups are also in use. As in the drawing, the term 'fitting' is often used for the part which is movable on the capillary; here this is a male nut. It can be made from steel or plastic (Kel-F or ketone resin).

Steel fittings usually have a separate ferrule. Some designs even use two ferrules, a thin ring and a conical, longer part. Once installed, the steel ferrules are pinched permanently to the capillary which becomes slightly compressed. They cannot be removed again. Although steel fittings can also be mounted on teflon or peek tubing this is not recommended because plastic can be compressed too much if the installation is not made very carefully. In any case wrenches are necessary for installation.

Plastic fittings come with separate or integrated ferrule, i.e. nut and ferrule are one single piece. Capillary compression is less pronounced with plastic fittings, so they can be removed again. They are recommended for all types of plastic tubing but can also be used on steel capillaries. No tools are necessary for installation, they are 'fingertight'.

Installation must be done according to the guidelines coming with the product. Overtightening and botching must be avoided (but it is a common phenomenon). It is best to tighten the screw just slightly and to test if the connection is pressure-resistant and does not leak. If necessary the nut is tightened slightly more. Overtightened fittings can damage the capillary and the whole set-up; they have a short lifetime and cannot be reused many times. The threads and conical areas must be kept scrupulously free of silica and other particles.

Many different types are available from numerous manufacturers. Different brands must not be mixed! They often differ in the angle of the ferrule and in their stop depth, from 0.1700 in = 4.32 mm (Rheodyne) to 0.0800 in = 2.03 mm (Valco). Usually the thread is English not metric but the length of the thread is also not identical with all brands. If the stop depth is less than it should be there is an extra-column volume in the system; if it is too long the capillary will be damaged or the column frit will be pushed into the chromatographic bed.

Note that the designation 1/16 in fitting means that it fits onto capillaries of this outer diameter. The nut width can be 5/16 in or the like.

4.6 SAMPLE INJECTORS

Sample feed is one of the critical aspects of HPLC. Even the best column produces a poor separation result if injection is not carried out carefully. In theory, an infinitely small volume of sample mixture should be placed in the centre of the column head, care being taken to prevent any air from entering at the same time.

There are various possibilities for sample injection:

(a) with syringe and septum injector;
(b) with a loop valve;
(c) with an automated injection system (autosampler).

Although injection through a septum into the column head is the most direct mode of sample application yielding the lowest possible band broadening, it is not suitable for HPLC. It is only applicable at pressures lower than 100 bar. There is always a danger of needle clogging by septum particles; moreover the septum needs to be chemically resistant against the various mobile phases used.

Figure 4.5 shows clearly how a *loop valve*[13] operates. The loop is filled with sample solution and the internal channelled rotor seal is then turned to bring the loop within the eluent flux. These valves are known as six-port valves as they have six connections.

Loops can be used in two different ways:

(a) *Complete filling.* As the loop is filled, the sample cannot displace the existing solvent like a plug but tends to mix with it. As a result, the loop must be filled with five times the volume of sample (e.g. 100 µl of sample solution for a 20 µl loop) so that it contains no more than 1% of residual solvent for quantitative external

---

[13] J.W. Dolan, *LC Mag.*, **3**, 1050 (1985).

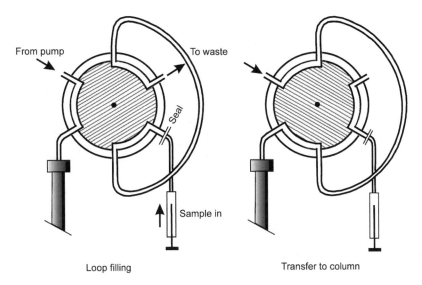

Loop filling   Transfer to column

Figure 4.5   Loop valve. The hatched area is the rotor seal.

standard analysis (obviously, a smaller amount can be used if the quantitative analysis is on an internal standard basis).

(b) *Partial filling.* In cases where it is important that no sample is lost, the loop should be only half-filled, following the manufacturer's guidelines. The loss is due to partial mixing of sample and liquid which already is in the loop; the reason for this is the parabolic stream profile within the capillary. To prevent this, one can try to inject with 'leading bubble': a small air bubble which is in front of the solution to be injected.[14] In any case the mobile phase should flow through the loop in opposite direction than was the sample loading direction, as shown in Figure 4.5. By this mode of operation any band broadening during the transfer to the column is suppressed, even if the sample fills only a small fraction of the loop volume.

Loops are available in sizes ranging from 5 to 2000 µl. They are easily attached to the valve and quickly changed. Smaller amounts of sample, as required for 3 µm and microcolumns, are best injected using a special valve with an internal loop of capacity between 0.5 and 5 µl.

The rotor seal is usually made from vespel which has the best mechanical properties and can be used in the range of pH 0–10. If the mobile phase has a

---

[14] M.C. Harvey and S.D. Stearns, *J. Chromatogr. Sci.*, **20**, 487 (1982).

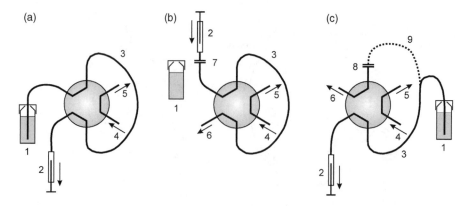

Figure 4.6   Filling the loop by three different autosampler designs: a, pull-loop; b, push-loop; c, integral-loop injection. 1 = sample vial, 2 = syringe with stepper motor, 3 = loop, 4 = from pump, 5 = to column, 6 = to waste, 7 = low-pressure seal, 8 = high-pressure seal, 9 = position of the movable loop when the valve is switched and the sample transferred to the column.

higher pH a tefzel seal is needed which is pH-resistant up to pH 14. However, this material is less suited to withstand high pressures and its channels become narrower under such conditions. Tefzel surfaces are less smooth than the ones obtained by the manfacturing of vespel. Therefore it is recommended to use tefzel rotor seals only when necessary.

The heart of an *autosampler*[15] is the six-port valve as well. Its loop can be filled partially or totally as in manual operation. Flushing steps are necessary between the individual samples but they are part of the automation routine. The different liquids (sample, mobile phase, flush solvent) can be separated from each other by the aspiration of air bubbles. Different working principles (and variations thereof) are in use.

(a) *Pull-loop injection* (Figure 4.6a).   The sample is pulled by a syringe with stepper motor from the vial through the loop. If a series of samples needs to be analyzed either the vials or the syringe at position 1 must be movable. This design is technically simple but additional sample is needed because the capillary between vial and loop is filled, too, this latter sample liquid being then discarded and lost. Pulling a liquid can be prone to cavitation, i.e. bubble formation due to reduced pressure, and therefore to inaccuracy.

---

[15] J.W. Dolan, *LC GC Eur.*, **14**, 276 (2001) or *LC GC North Am.*, **19**, 386 (2001).

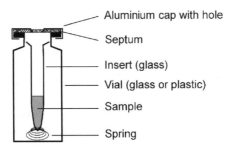

Aluminium cap with hole

Septum

Insert (glass)

Vial (glass or plastic)

Sample

Spring

Figure 4.7   Vial with conical insert and spring for small sample volumes.

(b) *Push-loop injection* (Figure 4.6b). This design imitates manual injection since the sample is sucked into the syringe and then pushed into the loop through a low-pressure needle seal. The syringe needs to be moved, perhaps also the vials, depending on the set-up of the autosampler. This is a flexible design with regard to the variability of the injection volume and little extra sample is lost.

(c) *Integral-loop injection* (Figure 4.6c). The loop capillary itself switches between the sample vial and a high-pressure seal of the six-port valve. No sample is lost and the problem of carry-over (see below) hardly exists. The high-pressure seal can be the weak part of this design.

Autosamplers may not utilize the last droplet of sample liquid within a vial because the needle must be kept at some distance from the bottom. A possible way out is the use of vials with conical inner shape or of conical inserts mounted on a small spring within the vial (Figure 4.7).

The injector is part of the dwell volume of the HPLC system (see Section 4.3). This must be kept in mind if the loop volume is changed drastically, e.g. from 50 µl to 1 ml. It may then be necessary to adjust the gradient profile of a complex separation.

The flush solvent is usually the mobile phase but without buffer additives. It must have excellent dissolving properties to clean off the sample traces on the needle and capillaries.

Important features of all injection systems are their precision and carryover[16] (the appearance of traces of the preceding sample in the chromatogram). They are determined following the routine described in Section 25.6. Precision should be better than 0.3% relative standard deviation. Carryover should be less than 0.05%. These low values are only possible with well maintained instruments where the rotor seal and needle are replaced in regular intervals; in addition, the sample needs to have

---

[16] J.W. Dolan, *LC GC Eur.*, **19**, 522 (2006) or *LC GC North Am.*, **24**, 754 (2006).

good solubility properties without the tendency to adsorb on seals or on the inner surface of capillaries.

4.7 SAMPLE SOLUTION AND SAMPLE VOLUME

Sometimes the *sample preparation* is a difficult problem, especially in clinical and environmental chemistry.[17] General procedures are filtration (perhaps by means of a dedicated membrane which retains compounds selectively), solid phase extraction with disposable cartridges[18] (also with dedicated selectivity), protein precipitation and desalting. A special case is sample preparation for biopolymer analysis.

The *sample solution* should pass through a preliminary filtration step if necessary, to ensure that it contains no solids. If possible, it is always best to dissolve the sample in the mobile phase solvent (extreme insolubility in the mobile phase may give rise to tailing and column blockage, so a different eluent should be substituted). If the breakthrough time is recorded as a refractive index change, then the sample solvent should be weaker than the mobile phase but have as similar a chemical composition as possible, e.g. pentane for elution with hexane (normal phase) or water for elution with mixtures of methanol and water (reversed phase). The plate number of the column may even be higher if the sample is injected in a much weaker solvent than the mobile phase itself, as the components are concentrated at the column inlet.[19] If the sample solvent is markedly stronger than the mobile phase this can lead to considerable band broadening or even to strange peak shapes.[20] This is shown in Figure 4.8 with the case of a reversed-phase separation; the mobile phase contains only 8% of acetonitrile. If the sample solution is higher in acetonitrile (which means higher elution strength in reversed-phase chromatography), the effect is first band broadening and, as concentration increases, a heavily distorted peak shape.

The *amount of sample* that may be injected has already been discussed in Section 2.7 and the volume of solvent required for dissolution may vary. The sample volume should be kept as small as possible, thereby not increasing the inevitable band broadening. On the other hand, it may be preferable to dissolve the sample in a relatively large volume to prevent mass overload at the column inlet. Obviously there

---

[17] *Guide to Sample Preparation*, supplement to *LC GC Mag.* (1998), or *LC GC Eur.* (2000), Advanstar Communications; S.C. Moldoveanu and V. David, *Sample Preparation in Chromatography*, Elsevier, Amsterdam (2002); S.C. Moldoveanu, *J. Chromatogr. Sci.*, **42**, 1 (2004); Y. Chen *et al.*, *J. Chromatogr. A*, **1184**, 191 (2008).

[18] M.C. Hennion, *J. Chromatogr. A*, **856**, 3 (1999); C.R. Poole, *Trends Anal. Chem.*, **22**, 362 (2003).

[19] See Figures 8 and 9 in P.J. Naish, D.P. Goulder and C.V. Perkins, *Chromatographia*, **20**, 335 (1985).

[20] S. Keunchkarian *et al.*, *J. Chromatogr. A*, **1119,** 20 (2006). Exceptions: E. Loeser, S. Babiak and P. Drumm, *J. Chromatogr. A*, **1216**, 3409 (2009).

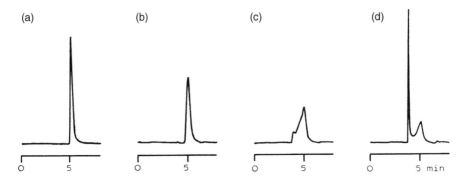

Figure 4.8 Influence of sample solvent on peak shape [reproduced with permission from N. E. Hoffmann, S. L. Pan and A. M. Rustum, *J. Chromatogr.*, **465**, 189 (1989)]. Conditions: sample, phenylalanine; mobile phase, buffer pH 3.5-acetonitrile (92:8); stationary phase, reversed phase; UV detector, 210 nm. The sample is dissolved in buffer with: (a) 0%, (b) 30%, (c) 50%, (d) 70% acetonitrile.

is an upper limit to the injection volume (which is knowingly exceeded during preparative work) because the peak width at the column outlet is at least as great as the input volume (See Section 19.2). Hence, the injection volume should be smaller than the degree of band broadening occurring in the column (measured as the elution volume) to prevent any severe drop in plate number.

5 Solvent Properties

5.1 TABLE OF ORGANIC SOLVENTS

Table 5.1 lists a large number of solvents in order of their polarity. Many of them are not suitable as mobile phases for reasons which are discussed in Section 4.1. The table makes clear why a certain compound is not a good choice, e.g. because the UV absorption or the viscosity is too high. Some solvents are used as additives in low concentration for certain applications: with a small amount of an amine the mobile phase becomes basic, with an acid one gets an acidic eluent.

The table lists the following properties of the solvents:

Strength: $\varepsilon°$ is a parameter which defines the elution strength of the solvent when it is used as a mobile phase on silica. It represents the adsorption energy of an eluent molecule per unit area of the adsorbent. A list according to strength is referred to as an *eluotropic series*. [In many tables the $\varepsilon°$ values refer to alumina as adsorbent which gives higher numbers: $\varepsilon°(Al_2O_3) = 1.3\ \varepsilon°\ (SiO_2)$.]

Viscosity: η is given in mPa s at 20 °C. Solvents with a higher viscosity than water with $\eta = 1.0$ are less suited for HPLC because the pressure drop will be high.

Refractive index: As the abbreviation shows, n_D^{20} is given at 20 °C.

UV cutoff: This is the wavelength at which the absorbance of the pure solvent is 1.0, measured in a 1 cm cell with air as the reference (10 % transmittance). The values listed here are only valid for highly pure solvents. Less pure solvents have higher UV cutoffs.

Boiling point: Too low a boiling point is less convenient. There is a danger of vapour bubbles in the HPLC system. A solvent loss can occur during degassing.

Dipole character: π^* is a measure of the ability of the solvent to interact with a solute by dipolar and polarization forces.

Practical High-Performance Liquid Chromatography, Fifth edition   Veronika R. Meyer
© 2010 John Wiley & Sons, Ltd

TABLE 5.1   Eluotropic series with solvent properties.

Solvent	Strength $\varepsilon°$	Viscosity η(mPa s)	Refractive index n_D^{20}	UV cutoff (nm)	Boiling point (°C)	Dipole π^*	Acidity α	Basicity β
Fluoroalkane FC-78	−0.19	0.4	1.267	210	50			
n-Pentane	0.00	0.23	1.3575	195	36			
n-Hexane	0.00	0.33	1.3749	190	69			
Isooctane	0.01	0.50	1.3914	200	99			
Cyclohexane[a]	0.03	1.00	1.4262	200	81			
Cyclopentane	0.04	0.47	1.4064	200	49			
Carbon tetrachloride	0.14	0.97	1.4652	265	77			
p-Xylene	0.20	0.62	1.4958	290	138	0.81	0.00	0.19
Diisopropyl ether	0.22	0.37	1.3681	220	68	0.36	0.00	0.64
Toluene	0.22	0.59	1.4969	285	111	0.83	0.00	0.17
Chlorobenzene	0.23	0.80	1.5248	290	132	0.91	0.00	0.09
Benzene	0.25	0.65	1.5011	280	80	0.86	0.00	0.14
Diethyl ether	0.29	0.24	1.3524	205	34.5	0.36	0.00	0.64
Dichloromethane	0.30	0.44	1.4242	230	40	0.73	0.27	0.00
Chloroform	0.31	0.57	1.4457	245	61	0.57	0.43	0.00
1,2-Dichloroethane	0.38	0.79	1.4448	230	83	1.00	0.00	0.00
Triethylamine	0.42	0.38	1.4010	230	89	0.16	0.00	0.84
Acetone[b]	0.43	0.32	1.3587	330	56	0.56	0.06	0.38
Dioxane	0.43	1.54	1.4224	220	101	0.60	0.00	0.40
Methyl acetate	0.46	0.37	1.3614	260	56	0.55	0.05	0.40
Tetrahydrofuran	0.48	0.46	1.4072	220	66	0.51	0.00	0.49
tert. Butylmethyl ether	0.48	0.35	1.3689	220	53	0.36	0.00	0.64
Ethyl acetate	0.48	0.45	1.3724	260	77	0.55	0.00	0.45
Dimethyl sulfoxide	0.48	2.24	1.4783	270	189	0.57	0.00	0.43
Nitromethane	0.49	0.67	1.3819	380	101	0.64	0.17	0.19
Acetonitrile	0.50	0.37	1.3441	190	82	0.60	0.15	0.25
Pyridine	0.55	0.94	1.5102	305	115	0.58	0.00	0.42
Isopropanol	0.60	2.3	1.3772	210	82	0.22	0.35	0.43
Ethanol	0.68	1.20	1.3614	210	78	0.25	0.39	0.36
Methanol	0.73	0.60	1.3284	205	65	0.28	0.43	0.29
Acetic acid	High	1.26	1.3719	260	118	0.31	0.54	0.15
Water	Higher	1.00	1.3330	<190	100	0.39	0.43	0.18
Salt solutions, buffers	Highest							

[a] Depending on temperature, cyclohexane becomes solid at > 300 bar: B.E. Lendi and V.R. Meyer, *LC GC Eur.*, **19**, 476 (2006).

[b] If sufficiently diluted, acetone has a 'UV window' at 210 nm. Water/acetone 9 : 1 shows an absorbance of approx. 0.4 at this wavelength.

Acidity: α is a measure of the ability of the solvent to act as a hydrogen bond donor towards a basic (acceptor) solute.

Basicity: β is a measure of the ability of the solvent to act as a hydrogen bond acceptor towards an acidic (donor) solute.

Note: π^*, α and β are normalized in such a way that their sum gives 1.00 and therefore are only relative numbers. These so-called *solvatochromic parameters*[1] are useful for the characterization of the selectivity properties of a solvent, see Section 5.2. Solvents of low polarity, from fluoroalkanes to carbon tetrachloride, do not interact with solutes by dipoles or hydrogen bonds, therefore no solvatochromic parameters can be listed for them.

5.2 SOLVENT SELECTIVITY

The most powerful means to influence the separation is by changing the selectivity properties of the phase system. This can be done by the use of another method (e.g. normal vs reversed phase), the use of another stationary phase (e.g. octadecyl vs phenyl silica) or the use of another mobile phase. In the latter case it will be best to choose solvents with large differences in their selectivity properties which are listed in Table 5.1 as solvatochromic parameters.

If these parameters are used for the construction of a diagram as shown in Figure 5.1, a solvent selectivity triangle is obtained which clearly shows the differences between the individual solvents with regard to their dipolar (π^*), acidic (α) and basic (β) properties.[2] The largest differences in the elution pattern can be expected if solvents are chosen which are as far apart from each other as possible. Because mixtures of two solvents, A and B, are used in most cases, only such solvents can be chosen which are miscible with each other. The usual A solvent in normal-phase separations is hexane, in reversed-phase separations it is water. Therefore the possible B solvents are limited in number. With regard to selectivity, it makes no real sense to try a normal-phase separation with diethyl ether as well as with *tert.* butyl methyl ether because all aliphatic ethers are located at the same spot in the selectivity triangle. Likewise it is not necessary to try several aliphatic alcohols for reversed-phase separations.

Benzene and its derivatives are rarely used because they do not allow UV detection.

5.3 MISCIBILITY

Figure 5.2 gives an overview of the miscibility properties of the common HPLC solvents at ambient temperature. Miscibility depends on temperature. Moreover,

---

[1] L.R. Snyder, P.W. Carr and S.C. Rutan, *J. Chromatogr. A*, **656**, 537 (1993).
[2] See footnote 1.

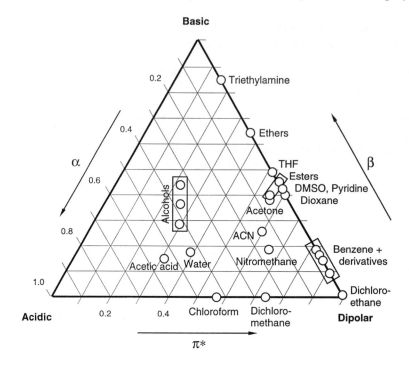

Figure 5.1 Selectivity triangle.

many solvent mixtures are partially soluble at certain volume proportions. This is not represented in the figure because it is always best to use solvents which are fully miscible with each other over the full range from pure A to pure B.

Solvents which are fully miscible with all other solvents (from hexane to water) are acetone, glacial acetic acid, dioxane, absolute ethanol, isopropanol, and tetrahydrofuran.

5.4 BUFFERS[3]

Buffers are required in ion-exchange chromatography and frequently also in reversed-phase chromatography. If ionic or ionizable compounds need to be separated, it is often, although not always, an imperative necessity to run the chromatography at a

[3] R.J. Beynon and J.S. Easterby, *Buffer Solutions – The Basics*, IRL Press, Oxford (1996); J.W. Dolan, *A Guide to HPLC and LC-MS Buffer Selection*, Advanced Chromatography Technologies, Aberdeen, free at www.ace-hplc.com.

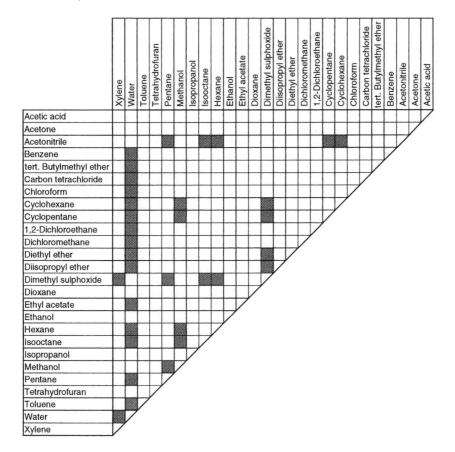

Figure 5.2 Miscibility graph. Nonmiscible solvent pairs are marked in grey.

well defined pH. The analytes can be forced either into the nondissociated or the ionized form, depending on the mode of separation. The chosen pH of the buffer must be two units apart from the pK_a of the compounds of interest in order to get the vast majority of the molecules in one single form.

Acidic analytes: Buffer at 2 pH units lower than pK_a gives nondissociated species. Buffer at 2 pH units higher than pK_a gives ionized species (anions).

Basic analytes: Buffer at 2 pH units higher than pK_a gives nondissociated species. Buffer at 2 pH units lower than pK_a gives ionized species (cations).

Concerning the ionic strength of a buffer, it is a good start to prepare a 25 mM solution. Too low an ionic strength has no effect, i.e. the buffer capacity is low.

TABLE 5.2   pK_a values of common buffers

Ion	pK_a
Acetate	4.8
Ammonium	9.2
Borate	9.2
Citrate	3.1, 4.7, 5.4
Diethylammonium	10.5
Formate	3.7
Glycinium	2.3, 9.8
1-Methylpiperidinium	10.3
Perchlorate	−9.0
Phosphate	2.1, 7.2, 12.3
Pyrrolidinium	11.3
Triethylammonium	10.7
Trifluoroacetate[a]	0.2
Tris[b]	8.3

[a] Trifluoroacetate TFA is a common, volatile buffer additive for the separation of proteins and peptides. Depending on quality and batch it can give rise to ghost peaks. General properties of TFA: Y. Chen et al., *J. Chromatogr. A*, **1043**, 9 (2004).
[b] Tris-(hydroxymethyl)aminomethane.

High ionic strengths (e.g. 100 mM) can lead to solubility problems with organic solvents.[4] It is recommended to check the total miscibility of the chosen buffer with the chosen organic solvent over the full range of mixing ratios. Buffer solutions must be filtered *after* preparation! They are prone to bacterial growth, therefore it can be advisable to add 0.1% sodium azide. A buffer solution must not remain within the HPLC system and column when not in use. It must be replaced by water and, for reversed-phase columns, afterwards by 100% organic solvent. If this procedure is too tedious (e.g. at the end of the day), it is necessary to keep running the pump at a low flow rate.

Buffers are effective within ±1 pH unit of their pK_a. Table 5.2 lists some common buffer ions together with their pK_a values.

The pH value of a buffer is clearly defined in water and can be calculated from the composition of salt and acid or salt and base, respectively. If an organic solvent is also present the ion relationships are rather complicated and the pH changes. However, there is no need to deal with such effects. When the mobile phase is prepared it is necessary to adjust the pH of the *aqueous* solution correctly and to add the organic part later; the resulting apparent pH value will not be determined afterwards.

A buffer solution is often prepared from a commercially available concentrate; if the dilution is performed correctly the wanted pH value will result which needs not

---

[4] A.P. Schellinger and P.W. Carr, *LC GC North Am.*, **22**, 544 (2004).

TABLE 5.3   Volatile buffer systems.[8]

pH	Composition	Counter-ion
2.0	Formic acid	H^+
2.3–3.5	Pyridine/formic acid	$HCOO^-$
3.0–6.0	Pyridine/acetic acid	CH_3COO^-
6.8–8.8	Trimethylamine/HCl	Cl^-
7.0–12.0	Trimethylamine/CO_2	CO_3^-
7.9	Ammonium bicarbonate	HCO_3^-
8.0–9.5	Ammonium carbonate/ammonia	CO_3^-
8.5–10.0	Ammonia/acetic acid	CH_3COO^-
8.5–10.5	Ethanolamine/HCl	Cl^-

necessarily be controlled. If it is not possible or wanted to buy buffer concentrates it is usually best to weigh in the required chemicals (whose amount is calculated or can be found in tables) and to dilute them to the correct volume[5].

Volatile buffers are needed for light scattering detection, the coupling with mass spectroscopy, and for preparative separations; they are listed in Table 5.3. The volatility of acids, bases and their salts was investigated by Petritis et al.[6] Buffer additives for ESI-MS were published by García.[7]

5.5   SHELF LIFE OF MOBILE PHASES

It is a good practice to prepare only as much mobile phase as will be used within short time. The shelf life of aqueous solutions without an organic solvent is very limited if rigorous quality standards need to be followed.[9]

Water from water purification system	3 days
Aqueous solutions (without buffer)	3 days
Buffer solutions	3 days
Aqueous solutions with <15% organic solvent	1 month
Aqueous solutions with >15% organic solvent	3 months
Organic solvents	3 months

---

[5] G.W. Tindall and J.W. Dolan, *LC GC Eur.*, **16**, 64 (2003) or *LC GC North Am.*, **21**, 28 (2003).

[6] K. Petritis et al., *LC GC Eur.*, **15**, 98 (2002).

[7] M.C. García, *J. Chromatogr. B*, **825**, 111 (2005).

[8] Reproduced with permission from D. Patel, *Liquid Chromatography Essential Data*, John Wiley & Sons, Ltd, Chichester, 1997, p. 89.

[9] Data after B. Renger, Byk Gulden, Konstanz, Germany, personal communication, 1998.

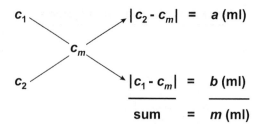

Figure 5.3   Mixing cross. c_1, content (%) of mixture or solvent 1, c_2, content of mixture or solvent 2, c_m, content of the new mixture, a, b, m, volumes of the mixtures or solvents, with $m = a + b$.

5.6   THE MIXING CROSS

The mixing cross is a useful aid for the calculation of solvent mixtures. Figure 5.3 shows how to use it. The contents c (usually in vol%) with regard to a certain solvent of the three mixtures involved (or two mixtures and a pure solvent) are noted as well as the content differences between the initial mixtures and the needed one, thus representing a cross with the new mixture in the middle. The negative sign that occurs in one of the two subtractions is ignored. The obtained differences represent the volume fractions to be mixed which finally must be converted into the needed volume parts.

Problem 17

One liter of water/acetonitrile $65:35$ is needed. There is a leftover of a $80:20$ mixture.

Solution

Mixture 1 is 20% in acetonitrile. What needs to be added is pure acetonitrile, i.e. $c_2 = 100\%$.

old mixture: 20   65 parts =   812 ml
                               35
acetonitrile: 100   15 parts =   188 ml
new mixture: 80 parts = 1000 ml

The volume fractions are 65 parts of mixture 1 and 15 parts of solvent 2, giving a total of 80 parts which must represent 1000 ml. Therefore 65 parts correspond to 812 ml and 15 parts to 188 ml. (The control of the result is simple: 812 ml ×

0.2 = 162 ml, plus 188 ml = 350 ml of acetonitrile in 1 l of new mixture.) The problem can also be solved on the base of the water content:

old mixture: 80.   ↗ 65
                      >65<
acetonitrile: 0 ↗   ↘ 15

Other problems to practise with:

(a) 600 ml of hexane/THF 60 : 40 need to be diluted to a ratio of 85 : 15. Solution: add 1 l of hexane.
(b) Prepare 1.5 l of water/acetonitrile 50 : 50 from mixtures of water/acetonitrile 90 : 10 and 30 : 70. Solution: mix 0.5 l of the 90 : 10 mixture and 1 l of the 30 : 70 mixture.
(c) A mixture of water/dioxane 70 : 30 is needed, what is at hand are 800 ml of a 45 : 55 mixture. Solution: add 667 ml of water.

Be aware of the fact that old mixtures do not necessarily still have their original contents of solvents 1 and 2, therefore a new mixture may also differ from the ratio that is wanted. Due to the effect of volume contraction, it is not allowed to prepare water/methanol mixtures in a measuring flask by filling the flask to the mark.

The mixing cross does not work with buffer mixtures when a certain pH of the new buffer is needed!

6 Detectors

6.1 GENERAL

The detector should be able to recognize when a substance zone is eluted from the column. Therefore, it has to monitor the change in mobile phase composition in some way, convert this into an electric signal and then convey the latter to the display where it is shown as a deviation from the baseline. The ideal detector should:

(a) either be equally sensitive to all eluted peaks or record only those of interest;
(b) not be affected by changes in temperature or in mobile phase composition (as happens, for example, in gradient elution);
(c) be able to monitor small amounts of compound (trace analysis);
(d) not contribute to band broadening, hence the cell volumes should be small;
(e) react quickly to pick up correctly narrow peaks which pass rapidly through the cell;
(f) be easy to manipulate, robust and cheap.

The above specifications are rather utopic in nature and some preclude the presence of others. Hence, detectors are better considered in terms of the following characteristics.

Concentration or Mass Sensitivity

Concentration-sensitive detectors produce a signal, S, which is proportional to the concentration, c, of the sample in the eluate:

$$S \propto c(\text{g ml}^{-1})$$

Mass-sensitive detectors produce a signal which is proportional to the mass flux, i.e. to the number, n, of sample molecules or ions per unit time, Δt, in

---

Practical High-Performance Liquid Chromatography, Fifth edition   Veronika R. Meyer
© 2010 John Wiley & Sons, Ltd

the eluate:

$$S \propto n/\Delta t(\mathrm{g\ s}^{-1})$$

The type of detector used is determined by switching off the pump at the peak maximum; the signal retains its actual value in concentration-sensitive detectors but drops to its initial level (baseline) in mass-sensitive detectors.

With the exception of mass-sensitive electrochemical, light-scattering, conductivity and photoconductivity detectors, all those described in Sections 6.2–6.7 are concentration-sensitive.

Selectivity

Nonselective detectors react to the bulk property of the solution passing through. A refractive index detector monitors the refractive index of the eluate. The pure mobile phase has a specific refractive index which changes when any compound is eluted. The detector senses this difference and records all peaks: hence the term nonselective or bulk-property detector. This is why the refractive index as well as the conductivity detector is not suited for gradient elution.

In contrast, a UV detector identifies only those compounds that absorb a minimum amount of UV light at a chosen wavelength and is, therefore, selective.

Noise[1]

Increased amplification of the detector signal shows that the baseline is irregular even when no peak is eluted. High-frequency (short-term) noise (Figure 6.1a) may be due to inadequate grounding of the detector and/or data system. However, a minimum noise level persists despite good grounding and the use of the most modern equipment. High-frequency noise oscillates with slower, low-frequency (long-term) noise (Figure 6.1b).

In general, the mobile phase is the more important factor (impurities, bubbles, stationary phase particles, change in flow rate), but rapid fluctuations of the ambient

(a)   (b)

Figure 6.1   (a) High- and (b) low-frequency noise.

---

[1] J.W. Dolan, *LC GC Eur.*, **14**, 530 (2001) and **15**, 142 (2002) or *LC GC North Am.*, **19**, 688 (2001) and **20**, 114 (2002).

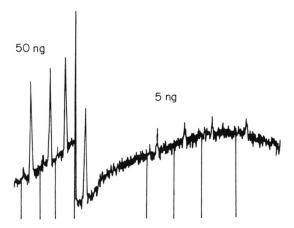

Figure 6.2 Smallest detectable signal. S/N is approximately 10 with the first four injections, with the later ones only approximately 2. Conditions: column, 50 cm × 3 mm i.d.; stationary phase, Perisorb A; mobile phase, heptane/ethyl acetate (95 : 5); UV detector, 254 nm. Reproduced with permission from F. Eisenbeiss and H. Sieper, *J. Chromatogr.*, **83**, 439 (1973).

temperature may also play a part (if the detector is placed in a draughty location). Power supply disturbances such as thermostat switching pulses may also have an effect, in which case a low-frequency filter is recommended. Sometimes a home-made 'Faraday cage' from aluminium foil may help.

Detection Limit

The smallest detectable signal cannot be less than double the height of the largest noise peak, e.g. 5 ng of dithianone (a pesticide) in Figure 6.2. If the amount injected is even less, then the signal ceases to be distinguishable from noise. For qualitative analysis the signal-to-noise ratio (S/N) should not be lower than 3–5; for quantitative analysis it must be higher than 10.

Statements such as 'detector X can pick up as little as 1 ng of benzene' should be regarded with caution. This amount can be affected by many limiting factors and a statement such as 'detection limit 1 ng ml^{-1}' is more correct for concentration-sensitive detectors.

Drift

This describes the deviation of the baseline and appears in the data acquisition as a slope (Figure 6.3). Drift is normal when the detector is switched on. However,

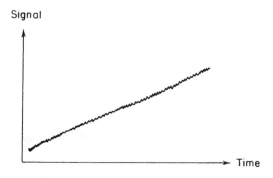

Figure 6.3   Drift.

as soon as the electronic system and the lamps have reached their normal operating temperature, the baseline will be stable unless drift is produced by one of the following:

(a) gradient elution (mobile phase refractive index changes, which can be picked up by a UV detector);
(b) old mobile phase is not yet fully displaced by new solvent after an eluent change;
(c) change in mobile phase composition (evaporation, too intense helium degassing);
(d) change in ambient temperature.

Linear Range

The ideal detector gives a signal with an area proportional to the amount injected, irrespective of whether it is large or small. Obviously, this linearity is not infinite but should embrace as wide a range as possible. The slope of the line drawn in Figure 6.4 is known as the *response* and represents the extent to which the detector reacts to a specific change in concentration.

The linear range is greater in UV than in refractive index detectors, its general range being $1 : 10\,000$, i.e. an upper limit of $5 \times 10^{-4}\,\mathrm{g\,ml^{-1}}$ corresponds to a lower linear range limit of $5 \times 10^{-8}\,\mathrm{g\,ml^{-1}}$, as an example.

Time Constant

The time constant, τ, is a measure of how quickly a detector can record a peak. As the detector works together with the data acquisition, the time constant of this combination is an important factor. The time constant can be defined as the minimum time required by a system to reach 63% of its full-scale value ($1 - 1/e = 0.63$; 98% is also often specified). It should not be greater than 0.3 s for detecting very narrow HPLC peaks, i.e. those that are rapidly eluted at a volume of less than 100 µl, and should not

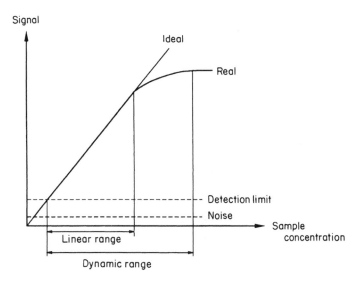

Figure 6.4   Linear detector range.

exceed 0.1 s for high-speed HPLC. Too small a time constant increases noise level, but if it is too large the peaks become wide and tailed (Figure 6.5). Better detectors offer a choice of time constant.

Cell Volume

The volume of the detector cell should have a negligible influence on peak broadening and should, therefore, amount to less than 10% of the elution volume of the narrowest (first) peak. A cell volume of 8 µl is standard. Too small a cell volume impairs the detection limit (as a certain amount of analyte is needed to produce any signal at all). The figure must be well below 8 µl for microHPLC. The cell must have no dead corners which may prevent the peak being fully removed by subsequent eluent.

Most of the original pressure has gone from the eluate by the time it leaves the column. The risk of bubble formation is greatest at this point and can never be fully eliminated whatever measures are taken. For this reason, a long steel capillary or a PTFE capillary tube should be fitted to the detector outlet so that the cell is always kept under a slight excess pressure (1–2 bar), considering the cell pressure stability limits (the same applies for detectors arranged in series). There are UV detectors with cells that can withstand pressures of up to 150 bar. *Note:* if a small amount of substance of much higher polarity is added to the mobile phase (e.g. 1% ethanol in hexane), then a film of it may coat the cell windows. The elution of a relatively polar peak may remove this coating, resulting in the appearance of ghost peaks.

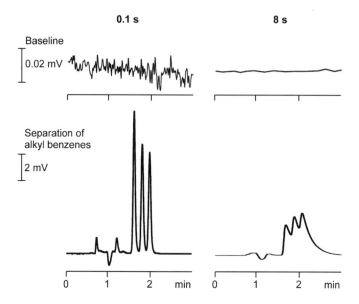

Figure 6.5 Influence of the time constant (reproduced by permission of D. Stauffer, Hoffmann-La Roche, Basel). Conditions: sample, ethyl, butyl and *tert*-butyl benzene; column, 12.5 cm × 4 mm i.d.; stationary phase, LiChrospher RP-18, 5 μm; mobile phase, 1 ml min^{-1} water/methanol (95 : 5); UV detector, 250 nm.

6.2   UV DETECTORS[2]

This is the most commonly used type of detector as it can be rather sensitive, has a wide linear range, is relatively unaffected by temperature fluctuations and is also suitable for gradient elution. It records compounds that absorb ultraviolet or visible light. Absorption takes place at a wavelength above 200 nm, provided that the molecule has at least:

(a) a double bond adjacent to an atom with a lone electron pair, $X = Y - Zl$ (e.g. vinyl ether);
(b) bromine, iodine or sulfur;
(c) a carbonyl group, $C=O$; a nitro group NO_2;
(d) two conjugated double bonds, $X=X-X=X$;
(e) an aromatic ring;
(f) thus including inorganic ions: Br^-, I^-, NO_3^-, NO_2^-.

---

[2] B.E. Lendi and V.R. Meyer, *LC GC Eur.*, **18**, 156 (2005). For diode-array detectors, see Section 6.10.

These groups do not absorb to the same extent or at the same wavelength. The absorption intensity and the wavelength of maximum absorption are also affected by neighbouring atom groups in the molecule. The molar absorptivity, ε, which is found in tables in many handbooks for a large number of compounds, is a measure of the light absorption intensity. E.g., aromatic molecules have higher and ketones (with the functional group C=O) comparatively smaller molar absorptivities.

The degree of absorption resulting from passage of the light beam through the cell is a function of the molar absorptivity ε, the molar concentration, c, of the compound and length of the cell, d. The product of ε, c and d is known as the absorbance, A:

$$A = \varepsilon c d$$

A is a dimensionless number and an absorbance of 1 is normally expressed as 1 absorption unit (AU). The expression '1 a.u.f.s.' or '1 absorption unit full scale', now almost obsolete, means that the recorder is registering an absorbance of 1 at full deflection.

A UV detector measures the absorbance of the eluate. The mobile phase should be selected for optical transparency at the detector lamp wavelength, i.e. its absorbance should be zero or at least be adjustable to zero electronically. If this is the case the detector signal itself is also zero, and the integrator is set to the required position and produces the baseline. According to the above equation, peak height is a function of the molar absorptivity and the concentration of a substance passing through the detector cell. Compounds with a higher molar absorptivity produce larger peaks than those with a small molar absorptivity when identical amounts of a compound are injected.

As the signal amplitude is a function of light path, d, the flow cells in UV detectors should be as long as possible, 10 mm being by no means unusual. However, they must also be thin in order to keep cell volume small. Light travels in a longitudinal direction through the cells. Figure 6.6 shows one such type of design.

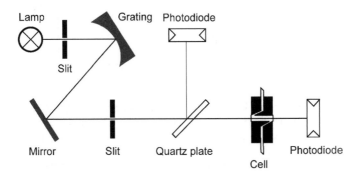

Figure 6.6 Principle of UV detector (simplified).

The requested wavelength is positioned in the optical path by means of a grating. A quartz plate is positioned in front of the detector cell which reflects a small fraction of the light onto the reference photodiode. The main part of the light throws through the plate and the cell on the measuring photodiode. The reference diode allows the electronical elimination of the minor light intensity fluctuations which always may occur. The detector cell is shaped in such an optimized geometry that any refractive index fluctuation or drift (always present with gradient elution) leads to minimum baseline noise or drift.

Two types of lamps are in use.

Deuterium lamps emit a continuous UV spectrum (up to about 340 nm, stray light up to 600 nm). The wavelength at which detection actually occurs can be adapted to the absorption maximum of the compound in question but a suitable choice also permits the removal of interfering peaks, thus improving the accuracy of integration or the selectivity. Figure 6.7 compares nonselective (all peptides are registered) and selective detection (only peptides with aromatic amino acids are registered).

Tungsten–halogen lamps emit in the near-UV and visible ranges (*ca.* 340–850 nm) and make an ideal supplement to deuterium lamps.

The spectral band width in both of these models must not be as narrow as that in recording spectrophotometers, a value of 10 nm being the norm (i.e. the filtered light is not monochromatic but covers a range of about 280–290 nm for a chosen wavelength of 285 nm). This has the advantage of supplying a greater light intensity, which in turn

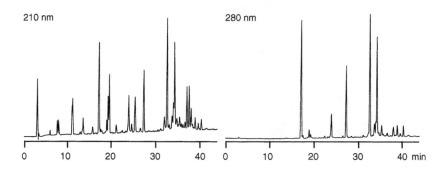

Figure 6.7 Choice of wavelength and selectivity (reproduced by permission of P. von Haller, Department of Chemistry and Biochemistry, University of Bern). Conditions: sample, tryptic digest fragments from lactalbumin; column, 25 cm × 4.6 mm i.d.; stationary phase, Aquapore butyl, 5 μm; mobile phase, 1 ml min⁻¹ water/acetonitrile with 0.1% trifluoroacetic acid, gradient from 0 to 50% acetonitrile in 50 min. 210 nm is nonselective, 280 nm detects only the fragments with aromatic amino acids.

increases the sensitivity. However, the band width should not be too wide, otherwise the linear range of the signal would be restricted.

Fixed-wavelength detectors have disappeared almost completely from the market. They were run with mercury (254 nm), cadmium (229 nm), and zinc lamps (214 nm).

6.3 REFRACTIVE INDEX DETECTORS

Refractive index (RI) detectors are nonselective and often used to supplement UV models. They record all eluting zones which have a refractive index different to that of the pure mobile phase. The signal is the more intense the greater is the difference between the refractive indices of the sample and eluent. In critical cases, the detection limit can be improved by a suitable choice of solvent.

RI detectors are about 1000 times less sensitive than UV detectors (detection limit ca. 5×10^{-7} g of sample per millilitre of eluate under the most favourable conditions, in contrast to 5×10^{-10} g ml^{-1} with UV detectors). As the refractive index of a liquid changes by about 5×10^{-4} units per °C, the cell must be well thermostatically controlled, especially if the flow rate exceeds 1 ml min^{-1}. The cell is incorporated in a metal block which acts as a heat buffer. The eluate passes through a steel capillary before reaching the cell (beware of the extra-column volume!) for its temperature to adjust to that of the detector. A reference cell filled with pure mobile phase is also essential. This makes RI detectors unsuitable for gradient elution unless much time and money are spent on ensuring that the change in composition is the same at any time in each cell. Any fluctuation in the mobile phase flow rate greatly increases the noise level, so pulsation must be meticulously suppressed. Cells in commercially available equipment are not pressure-tight.

Several alternative refractive index detectors have been described, two of which are most used.

Deflection Refractometer

This cell (Figure 6.8) is separated into two parts by an oblique dividing wall. One side is filled with reference fluid and eluate passes through the other. The deflection of the light beam changes when the refractive indices differ in both cells, i.e. some eluted sample is flowing through.

The light passes through a mask, twice through the lens (there and back) and twice through the double cell. When pure mobile phase is in the measuring cell, then the mobile zero-glass can be adjusted to direct the light beam on to the slit in front of the photocell. Any change in refractive index deflects the light beam so that it no longer passes through the slit. The photodiode changes the resistance and a suitable bridge circuit alters the voltage and produces a signal.

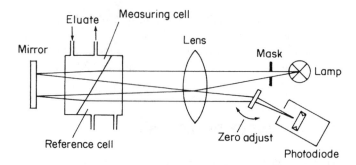

Figure 6.8   Deflection refractometer.

Interferometer

The light passes through a birefractive beam splitter and is divided into two beams of equal intensity, one of which passes through a measuring cell and the other through a reference cell and the two are superimposed in a second beam splitter, i.e. brought to interference. If the refractive indices of the solutions in the measuring and reference cells differ, this means that the two light beams are not moving at the same velocity and will be partly extinguished in the second beam splitter. Hence, the passage of a sample component is recorded as light attenuation in the same way as with other refractive index detectors.

The interferometer (Figure 6.9) is ten times more sensitive than other refract-ometers (with an analytical cell; however, a preparative cell with a shorter path length is also available). The high sensitivity is a drawback in that this detector is prone to disturbances such as changes in flow rate and incomplete column conditioning. This means that considerable care is required to achieve a low detection limit. The most sensitive cell has a volume of 15 µl and the smallest, less sensitive 1.5 µl.

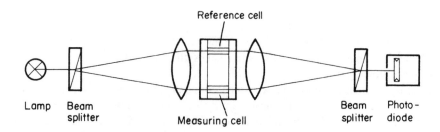

Figure 6.9   Interferometer.

6.4 FLUORESCENCE DETECTORS

Compounds that fluoresce or of which fluorescing derivatives can be obtained are picked up with high sensitivity and specificity by this detector. The sensitivity may be up to 1000 times greater than with UV detection. Light of a suitable wavelength is passed through the cell and the higher wavelength radiation emitted is detected in a right-angled direction. The light intensity and hence the sensitivity are increased by using a relatively large cell (20 µl or greater). Simple units as shown in Figure 6.10 have a fixed excitation wavelength for which band width must not be too narrow and a fixed wavelength range for fluorescent light detection. The excitation wavelength can be selected in the more expensive models and the most advanced equipment has a monochromator for excitation and fluorescent light, providing a highly specific (but less sensitive) level of detection.

Care should be taken to ensure that no compounds such as unsuitable solvent or oxygen in the mobile phase which could quench the fluorescence are present. The linear range depends on the system (sample, solvent, accompanying components) and may be relatively small.

Figure 6.11 illustrates the simplification of the chromatogram if fluorescence can be used instead of UV detection.

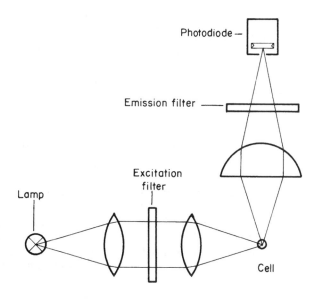

Figure 6.10   Fluorescence detector.

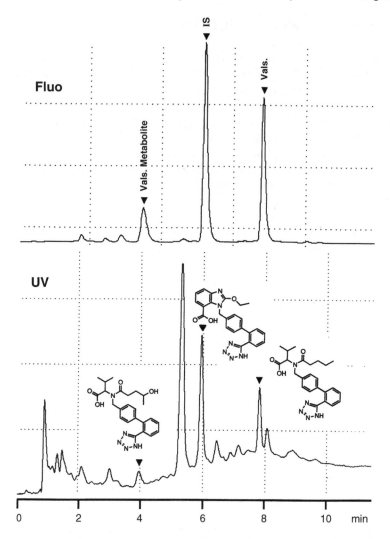

Figure 6.11 Comparison of fluorescence and UV detection [reproduced with permission from G. Iriarte *et al.*, *J. Sep. Sci.*, **29**, 2265 (2006)]. Conditions: sample, 20 μl plasma extract with 13 ng standard; column, 10 cm × 3.9 mm i.d.; stationary phase, Waters Atlantis dC18, 3 μm; mobile phase, 1.3 ml min^{-1} phosphate buffer 5 mM, pH 2.5-acetonitrile, 0.025% TFA, gradient; temperature, 40 °C; detector, UV 234 nm or fluorescence 234/378 nm. Peaks: valsartan metabolite, internal standard, and valsartan (drug for the treatment of hypertension).

6.5 ELECTROCHEMICAL (AMPEROMETRIC) DETECTORS[3]

Electrochemistry provides a useful means of detecting traces of readily oxidizable or reducible organic compounds with great selectivity. The detection limit can be extraordinarily low and the detectors are both simple and inexpensive. The detector cell in which an electrochemical reaction takes place has three electrodes (Figure 6.12). The potential between the working and reference electrodes may be selected. The current arising from electrochemical reaction is drawn by an auxiliary electrode so that it cannot influence the reference electrode potential. The working electrode is made up of glassy carbon, carbon paste or amalgamated gold. Frequently a silver/silver chloride electrode is used as the reference. The steel block represents the auxiliary electrode. It draws the current generated by the electrochemical reaction, thus keeping a constant potential in the cell.

Cyclic voltammetry can be used to show which compounds can be detected by this method and the best potentials to use. Aromatic hydroxy compounds, aromatic amines, indoles, phenothiazines and mercaptans can all be detected oxidatively (with a positive potential). Reductive detection (with a negative potential) is rarely used, as dissolved oxygen and heavy metals (e.g. from steel capillaries) can cause problems. The method can be used for nitrosamines and a large number of pollutants.

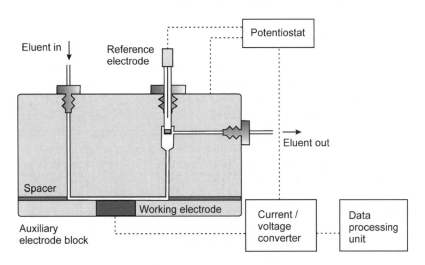

Figure 6.12 Electrochemical detector.

[3] R.J. Flanagan, D. Perrett and R. Whelpton, *Electrochemical Detection in HLPC*, Royal Society of Chemistry, Cambridge (2005); B. E. Erickson, *Anal. Chem.*, **72**, 353A (2000).

The cell medium must favour electrochemical reaction. The mobile phase must be conducting but not necessarily aqueous; nonpolar solvents and hence adsorption chromatography are not compatible with electrochemical detection. The mobile phase must not contain any chlorides or hydroxycarboxylic acids. The electrochemical yield is not higher than between 1 and 10% so that most of the sample leaves the cell unchanged. A conversion level of 100% is known as *coulometric detection*[4], but the equipment is not more sensitive than amperometric systems.

A special technique is *pulsed amperometric detection,*[5] useful for -COH compounds and especially carbohydrates. These analytes can be oxidized on an argent or gold working electrode at high pH but the surface of the electrode becomes immediately poisoned by the reaction products. Therefore its potential is not kept constant but is varied in well defined steps within 0.1 to 1.0 s. One of these steps is used for the detection of the analytes (e.g. during 200 ms), the others are needed to clean the surface.

The working and reference electrodes of electrochemical detectors are prone to contamination and their lifetime is limited. The reference electrode must be replaced after 3 to 12 months. The surface of the working electrode has to be polished from time to time with a slurry of alumina in order to remove any deposits.

6.6   LIGHT-SCATTERING DETECTORS[6]

The evaporative light-scattering detector (ELSD) is an instrument for the nonselective detection of nonvolatile analytes. The column eluate is nebulized in a stream of inert gas. The liquid droplets are then evaporated, thus producing solid particles which are passed through a laser, LED, or polychromatic light beam. The resulting scattered light is registered by a photodiode or photomultiplier (Figure 6.13).

From such a principle it becomes clear that the mobile phase must be volatile, including its buffers and additives. Volatile buffers can be prepared with formic, acetic and trifluoroacetic acid; all these compounds must be of high purity. "Gradient grade" is not necessarily pure enough for ELSD detection. The nebulizer gas is usually nitrogen, helium, or compressed air.

The detector response is a complex function of the injected amount of analyte and not of its chemical composition or of the presence of certain functional groups. The baseline is not influenced by the UV or refractive index properties of the mobile phase or by a changing eluent composition, making the light-scattering detector fully

---

[4] Y. Takata, *Methods Chromatogr.*, **1**, 43 (1996).

[5] W.R. LaCourse, *Pulsed Electrochemical Detection in High Performance Liquid Chromatography*, John Wiley & Sons, Inc., New York (1997); T.R.I. Cataldi, C. Campa and G.E. De Benedetto, *Fresenius J. Anal. Chem.*, **368**, 739 (2000).

[6] C.S. Young and J.W. Dolan, *LC GC Eur.*, **16**, 132 (2003) and **17**, 192 (2004) or *LC GC North Am.*, **21**, 120 (2003) and **22**, 244 (2004). Troubleshooting, see Section 26.7.

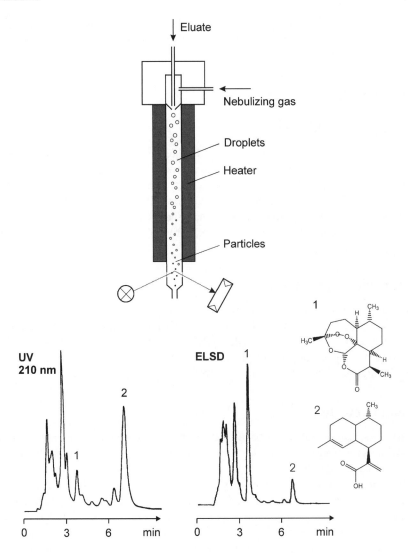

Figure 6.13 Principle of an evaporative light-scattering detector and its application for plant constituents. (Chromatograms after J. L. Veuthey, Analytical Pharmaceutical Chemistry, University of Geneva.) Conditions: sample, extract of *Artemisia annua* (sweet wormwood); column, 12.5 cm × 4 mm i.d.; stationary phase, Nucleosil 100 C_{18}, 5 μm; mobile phase, 1 ml min^{-1} water with trifluoroacetic acid (pH 3)/acetonitrile (39 : 61). Peaks: 1 = artemisinin; 2 = artemisinic acid.

gradient compatible. Its linear range is markedly smaller than the dynamic one. The detection limit is approx. 10- to 100-fold lower than with RI detection, depending on the analyte.

6.7 OTHER DETECTORS

Conductivity Detectors

This is the classical ion chromatography detector and measures the eluate conductivity, which is proportional to ionic sample concentration (provided that the cell is suitably constructed). Its sensitivity decreases as the specific conductivity of the mobile phase increases. The active cell volume of $2\,\mu l$ is very small. Good conductivity detectors have automatic temperature compensation (conductivity is highly temperature-dependent) and electronic background conductivity suppression. The linear range is not large.

Photoconductivity Detectors

These are sensitive, selective detectors for organic halogen and nitrogen compounds. The eluate is split up as it leaves the column. One half passes through the reference cell of a conductivity detector and the other half is irradiated with 214 or 254 nm UV light whereupon suitable sample molecules become dissociated into ionic fragments. The ensuing high level of conductivity is recorded in the measuring cell.

Infrared Detectors

Every organic molecule absorbs infrared light at one wavelength or another. When an IR detector is used, the mobile phase chosen must not be self-absorbent at the required wavelength. Hexane, dichloromethane and acetonitrile are suitable mobile phases for ester detection whereas ethyl acetate is not. The sensitivity is no greater than that of refractive index detectors. The most common wavelengths are given in Table 6.1.

TABLE 6.1 Commonly used wavelengths for IR detectors

Wavelength (μm)	Excited bond	For detection of:
3.38–3.51	C–H	Alkanes
3.24–3.31	C–H	Alkenes
5.98–6.09	C=C	Alkenes
5.76–5.81	C=O	Esters
5.73–6.01	C=O	Ketones

Laser-Induced Fluorescence LIF

Fluorescence can also be induced with laserlight. The signal-to-noise ratio is very favourable, thus the detection limits are low. However, there are only a few wavelengths which can be used, the lowest being 325 nm (rather high). Therefore the analytes need to be derivatized with suitable functional groups (labels).

Radioactivity Detectors[7]

These are used especially for detecting the β-emitters ^{3}H, ^{14}C, ^{32}P, ^{35}S and ^{131}I. The scintillator required for this relatively weak radiation is either added as a liquid between the column and the detector or is contained as a solid in the cell.

Reaction detectors are described in Section 19.8.
Coupling with spectroscopic techniques (HPLC-UV, HPLC-FTIR, HPLC-MS, HPLC-NMR), see Section 6.10.

6.8 MULTIPLE DETECTION

Diode array detectors (see Section 6.10) can measure the eluate at several different wavelengths simultaneously in addition to the ratio between two extinctions. This extracts more information from the chromatogram and provides important data for the qualitative analysis of unknown samples. This is a good solution to use for the problem demonstrated in Figure 6.7. Electrochemical detectors that monitor the eluate simultaneously at two different potentials are also marketed.

Two different detectors may be arranged in series (Figure 6.14), one example of this being the analysis of toxic amines that may occur in food. Not all of these amines show fluorescence; the ratio of UV absorption to fluorescence can be used for positive peak identification.

The eluate should pass through the detector with the smallest cell volume first, except for detectors such as the electrochemical type in which sample substances are altered, in which case this detector must be arranged last of all.

In special circumstances, detectors can be arranged in parallel by means of a splitter. The parallel triple detection of catecholamines is a good example of this,[8] an electrochemical process being accompanied by two different methods of derivatization and subsequent fluorescence detection.

---

[7] A.C. Veltkamp, H.A. Das, R.W. Frei and U.A.T. Brinkman, *Eur. Chromatogr. News*, **1**(2), 16 (1987). For an example see Figure 11.6 with radioactive ^{14}C.
[8] H. Yoshida, S. Kito, M. Akimoto and T. Nakajima, *J. Chromatogr.*, **240**, 493 (1982).

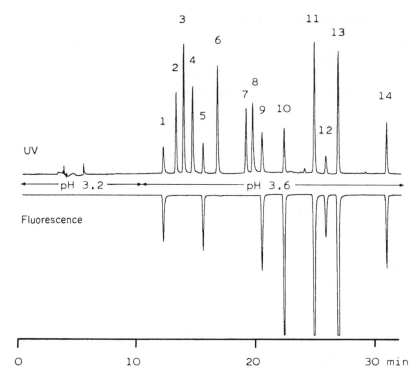

Figure 6.14 Detectors arranged in series [reproduced with permission from G.A. Gross and A. Grüter, *J. Chromatogr.*, **592**, 271 (1992)]. Condition: sample, heterocyclic aromatic amines; column, 25 cm × 4.6 mm i.d.; stationary phase, TSK gel ODS80, 5 µm; mobile phase, 1 ml min^{-1} water with 0.01 M triethylamine, pH 3.2 or 3.6, and acetonitrile, gradient; UV 263 nm and fluorescence detectors. Peaks: 1 and 5 = pyrido-imidazoles; 2 and 4 = imidazo-chinolines; 3, 6, 7, 8 = imidazo-chinoxalines; 9, 10, 11, 13, 14 = pyrido-indoles; 12 = an imidazopyridine.

6.9   INDIRECT DETECTION

Indirect detection is possible in all cases in which the mobile phase itself has a property perceptible by a selective detector. The simplest approach of this kind is indirect UV detection.[9] For this purpose a UV-absorbing mobile phase is used, which can consist of a suitable organic solvent or a solution of any UV-absorbing compound. By using this type of eluent, nonabsorbing sample components will be detected as negative peaks because there will be less light absorbed when they pass through the detector cell. An

---

[9] E.E. Lazareva, G.D. Brykina and O.A. Shpigun, *J. Anal. Chem.*, **53**, 202 (1998).

example of indirect UV detection can be found in Section 13.4 by means of UV-absorbing ion-pair reagents. The mobile phase should have an absorbance of 0.4 at the selected wavelength in order to obtain maximum accuracy of quantitative analysis. If the absorbance is too high (more than approximately 0.8, depending on the instrument used) the detector is no longer working in the linear range and quantitative analysis is impossible. There are guidelines for reagent choice at a given separation problem.[10]

Indirect detection is possible with all selective detection principles, e.g. with fluorescence detection (if the mobile phase itself is fluorescent) or electrochemical detection (if the mobile phase can act as an electrochemical reaction partner). Even indirect detection with atomic absorption spectrometry has been described, the mobile phase containing lithium or copper and the spectrometer being used with a lithium or copper lamp.[11]

System peaks are always observed with indirect detection and need attention (see Section 19.9). In quantitative analysis it is important to note that peak areas are not only a function of sample mass but also of k values (in relation to the k value of the system peak) and of the concentration of the detectable component in the mobile phase. The effects were explained by Schill and Crommen.[12]

6.10   COUPLING WITH SPECTROSCOPY

HPLC can be combined with numerous other analytical techniques but the most important coupling principle is the one with spectroscopy. Chromatography and spectroscopy are orthogonal techniques, i.e. their types of information are very different. Chromatography is a separation method and spectroscopy is a technique which yields a 'fingerprint' of molecules. Coupling with atomic spectrometry is rarely used although it allows the detection of toxic metals in environmental samples or of metalloproteins.[13] Four other techniques, HPLC-UV, HPLC-FTIR, HPLC-MS and HPLC-NMR are more important because excellent spectra are obtained with them, thus allowing structure elucidation.

HPLC-UV: The Diode Array Detector[14]

Compared to conventional UV detectors, the diode array detector is built with inverse optics. Figure 6.15 shows the principle of this instrument. The full light first goes through the detector cell and is subsequently divided spectrally in a polychromator (which is a grating). The spectral light then reaches the diode array, a chip with a large

[10] E. Arvidsson, J. Crommen, G. Schill and D. Westerlund, *J. Chromatogr.*, **461**, 429 (1989).

[11] S. Maketon, E.S. Otterson and J.G. Tartier, *J. Chromatogr.*, **368**, 395 (1986).

[12] G. Schill and J. Crommen, *Trends Anal. Chem.*, **6**, 111 (1987).

[13] P.C. Uden, *J. Chromatogr. A*, **703**, 393 (1995); A. Sanz-Medel, *Anal. Spectrosc. Libr.*, **9**, 407 (1999).

[14] L. Huber and S.A. George, eds., *Diode Array Detection in HPLC*, Dekker, New York (1993).

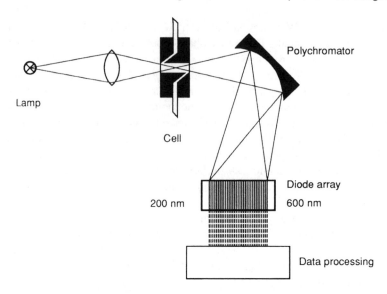

Figure 6.15 Principle of a diode array detector.

number (100–1000) of light-sensitive diodes that are arranged side by side. Each diode only obtains a well defined fraction of the information which is read by the electronics for data processing.

This type of detector allows a wealth of information to be obtained from UV spectra. The most important possibilities are presented in Figure 6.16:

(a) It is possible to obtain and store spectra of individual peaks during a chromato-gram. This process does not take more than half a second, including data processing; therefore it is possible to obtain several spectra, even from narrow peaks. For identification the spectra can be compared on-line with a library. In order to improve correlations the computer calculates the second derivative which shows more maxima and minima than the original spectrum.

(b) The knowledge of spectra of the compounds involved then allows detection at selected wavelengths. Interfering peaks can be eliminated (see also Figure 6.7). Due to the proper choice of detection conditions it is often possible to obtain an accurate quantitation of a compound even if the resolution in principle is poor. The detection wavelength can be altered during a chromatographic run.

(c) Simultaneous detection at two different wavelengths allows calculation of the absorbance ratio.[15] If this ratio is constant over the whole width of a peak it can be assumed that it is pure (with the exception of the exact coincidence of two peaks

---

[15] I.N. Papadoyannis and H.G. Gika, *J. Liquid Chromatogr.*, **27**, 1083 (2004).

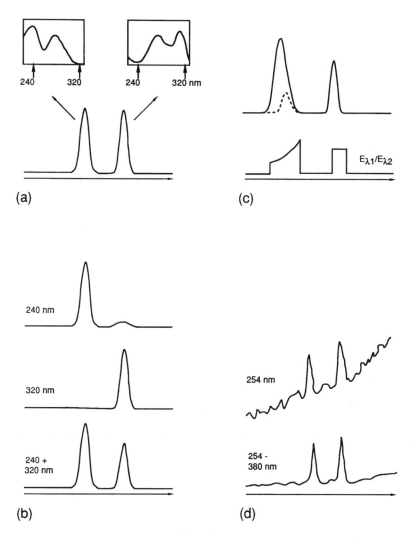

Figure 6.16   Possibilities of the diode array detector. (a) Registration of spectra during the separation; (b) registration of a chromatogram at different wavelengths; (c) peak purity control; (d) chromatogram with and without subtraction of a reference wavelength.

with identical shape or of an impurity with identical UV spectrum). If the ratio is not constant this is proof that the peak is not pure.

(d) Subtraction of two wavelengths allows baseline drift to be reduced during gradient elution as well as noise. The reference wavelength can be chosen arbitrarily in a region where none of the compounds of interest shows absorption.

HPLC-FTIR[16]

The coupling with Fourier-transform infrared spectroscopy allows spectra to be obtained. The detection limit is rather high (i.e. poor) if a flow-through cell is used. The interfaces with solvent elimination are more interesting and flexible but technically more demanding; they can even register spectra of trace analytes. A volatile mobile phase is needed; it may be aqueous.

HPLC-MS[17]

Mass spectrometers for HPLC consist of three different parts: the interface where the eluate enters the MS and the ions are generated, the mass analyzer, and the detector, an electron multiplier which determines the ion beam intensity. Thanks to the different possibilities for ion generation at atmospheric pressure, HPLC-MS is possible with rather low expenditure. The main two ionization techniques are APCI and ESI.

APCI: Atmospheric pressure chemical ionization (Figure 6.17).

(a) Ions are generated by corona discharge (3–6 kV).
(b) Yields molecule ions $(M + H)^+$ (no spectra), negative ionization is also possible.
(c) Suitable for small, medium and nonpolar molecules; however, they need some proton affinity and volatility.
(d) Not suitable for thermolabile analytes.
(e) For aqueous and nonaqueous mobile phases, flow must be at least $1 \, ml \, min^{-1}$.
(f) With nonaqueous eluents no additives are necessary; reactions with the solvent are possible during ionization.
(g) With aqueous eluents an additive may be necessary for efficient ionization.
(h) Mass-sensitive signal.

ESI or APESI: (Atmospheric pressure) electrospray ionization (Figure 6.18).

(a) Ions are generated by 'coulomb explosion' (disintegration) of electrically charged droplets.
(b) Yields ions with single or multiple charge; in the latter case, spectra with many peaks are obtained which must not be mixed up with classical spectra showing molecule fragments.
(c) Suitable for thermolabile analytes and macromolecules, including biopolymers.

[16] G.W. Somsen, C. Gooijer and U.A.T. Brinkman, *J. Chromatogr. A*, **856**, 213 (1999).

[17] *LC GC Eur., Guide to LC-MS*, Advanstar, Dec. 2001; R. Willoughby, E. Sheehan and S. Mitrovich, *A Global View of LC/MS*, Global View Publishing, Pittsburgh, 2nd ed., 2002; R.E. Ardrey, *Liquid Chromatography - Mass Spectrometry: An Introduction*, John Wiley & Sons, Ltd, Chichester (2003); M.C. McMaster, *LC-MS A Practical User's Guide*, Wiley-Interscience, Hoboken, 2005.

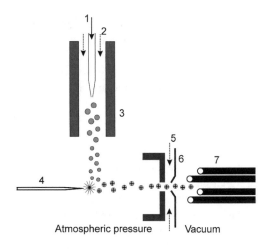

Figure 6.17 HPLC-MS interface for APCI. 1 = from HPLC column; 2 = nebulizing gas; 3 = heater; 4 = corona discharge needle; 5 = drying gas; 6 = skimmer; 7 = quadrupole.

(d) Suitable for aqueous eluents; flow must be small, therefore useful for micro HPLC.

(e) For positive ionization a pH of ca. 5 is suitable, additives are formic and acetic acid, perhaps together with ammonium acetate:

$$\text{Analyte} + \text{HA} \rightarrow \text{AnalyteH}^+ + \text{A}^-$$

(f) For negative ionization a pH of ca. 9 is suitable, additives are ammonia, triethylamine and diethylamine, perhaps together with ammonium acetate:

$$\text{AnalyteH} + \text{B} \rightarrow \text{Analyte}^- + \text{HB}^+$$

(g) Concentration-sensitive signal.

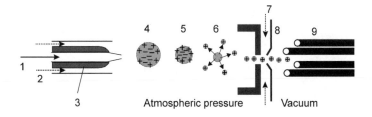

Figure 6.18 HPLC-MS interface for ESI. 1 = from HPLC column; 2 = nebulizing gas; 3 = high voltage; 4 = charged droplet (drawn too large); 5 = evaporation; 6 = coulomb explosion; 7 = drying gas; 8 = skimmer; 9 = quadrupole.

Quadrupoles and ion traps are the most frequently used mass analyzers. Both systems are rather low-priced and robust but their resolution is limited to ca. 1 Da.

The ions can be sent through several mass analysers and fragmentations, setup in series, thus yielding more information about the structure. These techniques are known as MS/MS (two steps) or MS^n (n steps).

HPLC-NMR[18]

Nuclear magnetic resonance can be coupled with HPLC if the solvents are chosen properly (one solvent signal can be suppressed but the others must be avoided by using deuterated solvents) and if the analyte concentration is not too low. Detection limit depends on radiofrequency, therefore a 500–800 MHz instrument must be used. If the analyte concentration is high enough, it is possible to run online HPLC-NMR separations, but usually the individual peaks are stored in loops and spectra are obtained offline with a long enough measuring time.

LC-UV, LC-MS and LC-NMR can be used in combination[19] as shown in Figure 6.19 with the structure elucidation of the components from a plant extract.

[18] K. Albert *et al.*, *J. High Resolut. Chromatogr.*, **22**, 135 (1999); K. Albert, *J. Chromatogr.* A, **856**, 199 (1999); K. Albert, ed., *On-line LC-NMR and Related Techniques*, Wiley-VCH, Weinheim, 2002.
[19] J.L. Wolfender, K. Ndjoko and K. Hostettmann, *J. Chromatogr. A*, **1000**, 437 (2003).

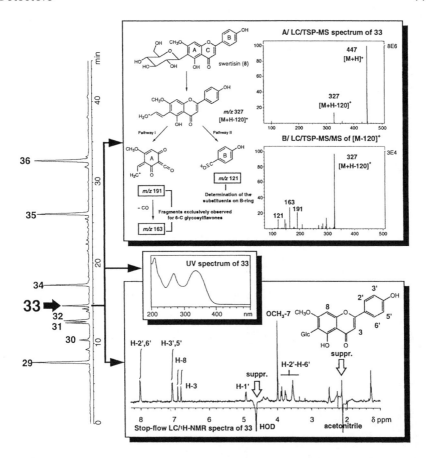

Figure 6.19 Peak identification by LC-MS, LC-MS-MS, LC-UV and LC-NMR [reproduced with permission from J.L. Wolfender, S. Rodriguez and K. Hostettmann, *J. Chromatogr. A*, **794**, 299 (1998)]. Chromatogram (vertical): sample, extract from *Gentiana ottonis*; column, 15 cm × 3.9 mm i.d. and precolumn; stationary phase, Nova-Pak C$_{18}$, 4 μm; mobile phase, 1 ml min^{-1} water/acetonitrile with 0.05% trifluoroacetic acid, gradient from 5 to 65% acetonitrile in 50 min; detector, UV 254 nm. MS: interface, thermospray; instrument, quadrupole. UV: diode array. NMR: D$_2$O instead of H$_2$O; stop-flow; 500 MHz. Peak 33 is the glucosylflavone swertisin.

7 Columns and Stationary Phases

7.1 COLUMNS FOR HPLC

Most HPLC columns are made of 316 grade *stainless steel*, which is austenitic chromium–nickel–molybdenum steel, USA standard AISI, resistant to the usual HPLC pressure and also relatively inert to chemical corrosion (chloride ions and lithium ions at low pH being important exceptions). The inside of the column should have no rough surfaces, grooves or microporous structures, so the steel tubes must be either precision drilled or polished or electropolished after common manufacturing, e.g. by drawing.

Glass tubes are smooth, chemically inert (they do not alter the sample substances at all) and do not corrode. For HPLC they are used as glass-lined steel tubes which are pressure-resistant.

Tantalum is rarely used. It is less prone to corrosion than steel and its surface is smooth, but it does not give, so that fittings have to be glued rather than pressed on. It is very expensive.

Peek, the pressure-resistant plastic mentioned in Section 4.4, cannot only be used for capillaries but also for column tubing.

Flexible *polyethylene tubes* compressed by a hydraulic fluid in a suitable casing represent a special type of column system (Waters). The column wall adapts to suit the packing, thus preventing wall effects and the formation of channels.

Columns, of i.d. 2–5 mm are generally used for analytical purposes. Wider columns of i.d. between 10 mm and 1 in (25.4 mm) may be used for preparative work. Some companies even market preparative columns of i.d. 30 cm and more. Micro and capillary columns are discussed in Section 23.1.

Columns 5, 10, 15 or 25 cm long are common if microparticulate stationary phases of 10 μm or less are used. If higher plate numbers are needed it is usually better to use a

Practical High-Performance Liquid Chromatography, Fifth edition   Veronika R. Meyer
© 2010 John Wiley & Sons, Ltd

packing with smaller particles than to lengthen the column. A longer column increases the retention volume, thus decreasing the concentration of the peak in the eluate (see Section 19.2) and impairing the detection limit. Yet for preparative purposes columns up to 1 m in length are used.

Figure 7.1 depicts various column terminators. The first commercial columns had an end fitting with male thread (design A). This configuration can last for years if treated with care, i.e. by not tightening more than necessary. If too much force is applied, the fitting will leak relatively quickly. Type B is a further development which causes fewer problems, although the column quality itself is not influenced by either model.

Version C is a *cartridge* design. The column is kept in position by means of an outer casing and two retaining nuts. The actual column has no fittings and is therefore reasonably priced, added to which it is readily changed without using tools and can be coupled to other columns (using suitable adapters) with no extra-column volume to worry about.

The column is closed off with steel frits of smaller pore size than the particle diameter of the column packing. The standard pore diameter is 2.0 μm but packings with particles of 3.5 μm or finer need 0.5 μm frits. If a frit is clogged it is best to replace it; if you cannot afford this, try ultrasonic cleaning.

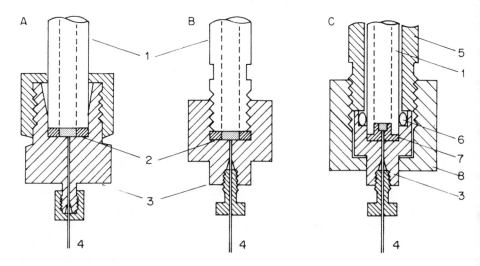

Figure 7.1 Various column terminators. (A) Union with male thread; (B) union with female thread; (C) cartridge system (simplified). 1 = column; 2 = sintered metal and Kel-F frit; 3 = union; 4 = 1/16 in capillary; 5 = casing; 6 = high-pressure seal; 7 = column seal with frit; 8 = retaining nut.

Advantages of Small Diameter Columns

Narrow bore columns offer two main advantages compared to wider ones:

(a) Less solvent consumption (and less waste). The retention volume V_R is proportional to the *square* of the column diameter d_c:

$$V_R \sim d_c^2 \left(\text{because } V_R = (k+1) \frac{\varepsilon \cdot d_c^2 \cdot \pi \cdot L_c}{4} \right).$$

If 10 ml of mobile phase are needed for the elution of a certain peak from a 4.6 mm column this corresponds to:

> 4.8 ml from a 3.2 mm column,
> 1.9 ml from a 2.0 mm column,
> 0.5 ml from a 1.0 mm column (20 × less solvent!).

(b) Better signal height-to-sample mass ratio. The peak maximum concentration c_{max} is proportional to the *inverse square* of the column diameter d_c:

$$c_{max} \sim \frac{1}{d_c^2} \left(\text{because } c_{max} = \frac{c_i \cdot V_i}{V_R} \sqrt{\frac{N}{2\pi}}, \text{ see Section 19.2, and } V_R \sim d_c^2 \right)$$

If, e.g., 1 µg of an analyte gives a signal of 0.1 mV height with a 4.6 mm column, this corresponds to:

> 0.21 mV with a 3.2 mm column,
> 0.53 mV with a 2.0 mm column,
> 2.1 mV with a 1.0 mm column (signal 20× higher!).

7.2   PRECOLUMNS

As Figure 7.2 shows there are two possible uses for precolumns.

Precolumns positioned prior to the sample injector are used for mobile phase conditioning. Their packing will partially dissolve, therefore this type is known as a scavenger column. Any solvent dissolves silica to give silicic acid, the effect being increased by increasing the polarity and ionic strength and by high mobile phase pH values. A scavenger column filled with coarse silica can be used to protect the separation column from dissolution when silica or a silica-based bonded phase is used as the stationary phase. This ensures that the mobile phase is already saturated with silicic acid on entering the column, thus increasing its lifetime.

In contrast, short guard columns are used as a protection between the injector and column. They are packed with the same or a chemically similar stationary phase as the main column and prevent impurities such as strongly retained compounds from

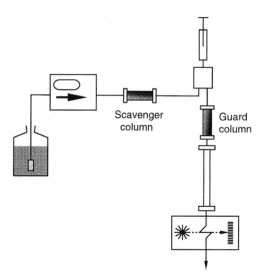

Figure 7.2 Precolumns.

contaminating it. This is particularly important when biological fluids are involved or for direct beverage analysis, as shown in Figure 1.2. If chosen and mounted properly a guard column does not affect the separation performance. Guard columns are replaced at regular time intervals or after a certain number of injections to ensure a maximum lifetime of the separation column.

7.3   GENERAL PROPERTIES OF STATIONARY PHASES[1]

As already discussed in Section 2.2, a large number of theoretical plates can only be achieved by ensuring short diffusion paths in the stationary phase pores; hence HPLC tends to favour microparticles. A sample molecule cannot penetrate more than 2.5 µm into a 5.0 µm particle. The two phenol separation chromatograms in Figure 7.3 show how performance is increased by using small particle diameters: a column with 10 µm material needs a length of 20 cm (below), whereas one with 3 µm material is only 6 cm long (above) to give a total of 7000 theoretical plates each. Analysis is completed in 15 min in the former case and takes only one-tenth of this time in the latter. This enormous saving of analysis time is due to the short column length and to the fact that the small particle packing reaches its van Deemter minimum at a higher flow rate than

[1] M.R. Buchmeiser, *J. Chromatogr. A*, **918**, 233 (2001).

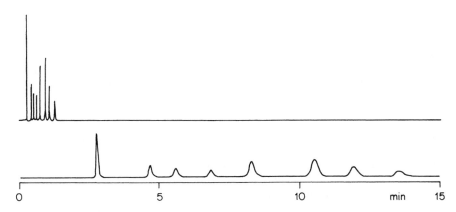

Figure 7.3 Comparison of performance between stationary phases of different particle diameter. Above: 6 cm × 4.6 mm i.d. column filled with 3 μm ODS-Hypersil, flow rate 2 ml min^{-1}. Below: 20 cm × 4.6 mm i.d. column filled with 10 μm ODS-Hypersil, flow rate 0.7 ml min^{-1}. Mobile phase: acetonitrile–water (1 : 1). Sample: 1 μl phenol mixture (Hewlett–Packard).

the larger particle variety. However, despite the short column, the pressure will be approximately ten times higher than in the separation with the coarse material.

The particle size distribution[2] should be as narrow as possible (with a ratio of diameters of the smallest and the largest particles of 1 : 1.5 or 1 : 2, as already mentioned). This is because the smallest particles determine the column permeability and the largest particles fix the plate number. Small particles produce a high flow resistance and large particles are responsible for a high degree of band broadening. For this reason, a particle size analysis accompanies many products.

Example

$d_{10} = 4$ μm, $d_{50} = 5$ μm, $d_{90} = 7$ μm. This means that:

> 10% of the material is less than 4 μm;
> 50% of the material is less than 5 μm;
> 90% of the material is less than 7 μm;

but it says nothing about how large or how small the particles are at the two extremes. This particle size distribution representation may appear unusual, but is based on practical particle size analysis methods such as sieving, air separation, sedimentation, optical methods and the Coulter counter.

---

[2] R.E. Majors, *LC GC Int.*, **7**, 8 (1994) or *LC GC Mag.*, **11**, 848 (1993).

The properties of the packing materials may vary from batch to batch of the same product, so a large amount of one batch should be bought for series analysis over a long period of time.

Various types of stationary phases are in use: Porous particles, nonporous particles of small diameter, porous layer beads, perfusive particles, and monolithic materials.

Porous Particles

This is the usual type of HPLC stationary phase. These materials are 1.8, 3.0, 3.5, 5.0 or 10 μm in size. As a rule of thumb, their performance, i.e. the plate number per unit length, doubles each time from 10 to 5 and 3 μm, whereas the pressure drop increases each time by a factor of four. Their internal structure is fully porous and can best be compared with the appearance of a sponge (however, in contrast to a sponge, the structure is very rigid). Within the pores the mobile phase (and the analytes) does not flow but moves only by diffusion.

Nonporous, Small-diameter Particles

If the stationary phase is nonporous, the mass transfer component of band broadening, the so-called C term of the van Deemter curve (see Sections 2.2 and 8.6) disappears or becomes very small because diffusion within pores does not occur. As a consequence, the curve becomes very flat to the right of its minimum and fast chromatography is possible without a loss in separation performance. An example is presented in Figure 1.1 with the separation of eight compounds in 70 s. In order to maintain a certain sample capacity (i.e. specific surface) it is necessary to use small particles with a diameter of 1–2 μm. The capacity is about 50 times lower than compared to conventional porous phases and pressure drop is high. The retention volume of the peaks is low because the pore volume is missing; the fraction of liquid within the column (the column porosity ε) is only ca. 0.4 compared to 0.8 of a column with porous packing. Therefore extra-column volumes and time constants need to be kept low. Chromatography of biopolymers seems to be more advantageous on these stationary phases with regard to denaturation and recovery.

Porous Layer Beads

These are large particles with a diameter in the 30 μm range which allows to pack them dry. They consist of a nonporous core (e.g. from glass) which is covered with a 1–3 μm layer of a chromatographically active material. Porous layer beads (PLBs) are rarely used nowadays but can be found in guard columns or as repair material for deteriorated columns with collapsed packing.

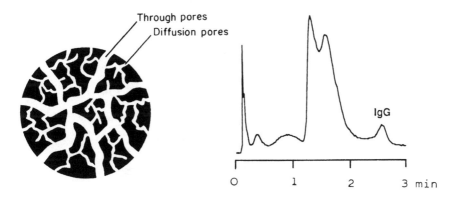

Figure 7.4 Perfusion chromatography. Left: particle of the stationary phase; right: fast separation of immunoglobulin G from cell culture [reproduced with permission from N.B. Afeyan, S.P. Fulton and F.E. Regnier, *J. Chromatogr.*, **544**, 267 (1991)]. Conditions: column, 10 cm × 4.6 mm i.d.; stationary phase, Poros M for hydrophobic interaction, 20 μm; mobile phase, gradient from 2 to 0 M ammonium sulfate in water in 5 min, 10 ml min^{-1}; UV detector, 280 nm.

Perfusive Particles

In analogy to the nonporous phases the perfusive packings have also been developed for the fast separation of biopolymers. They consist of highly cross-linked styrene–divinylbenzene with two types of pores: very large *throughpores* with a diameter of 600–800 nm and narrow *diffusion pores* of 80–150 nm. The active stationary phase (e.g. a reversed phase, ion exchanger or affinity phase) fully covers the external and internal surface of the particles. The throughpores are wide enough to allow the mobile phase to flow through, whereas it stagnates in the diffusion pores. The analytes are rapidly transported in and out by the flowing eluent and diffusion paths in the narrow pores are short, therefore separation is fast and the van Deemter curve is very favourable (Figure 7.4). It is not necessary (and also not possible) to prepare really small particles: the typical diameter of perfusive particles is 20 μm. Their flow resistance is very low.

Monolithic Stationary Phases[3]

It is possible to synthesize stationary phases which consist of one single piece of porous material such as organic polymers or silica (Figure 7.5). With this concept, the chromatographic bed is not a packing of particles but a porous rod which totally fills

---

[3] T. Ikegami and N. Tanaka, *Curr. Opinion Chem. Biol.*, **8**, 527 (2004); F. Svec, *J. Sep. Sci.*, **27**, 747 (2004) (organic polymers); I. Ali, V.D. Gaitonde and H.Y. Aboul-Enein, *J. Chromatogr. Sci.*, **47**, 432 (2009) (silica-based materials); K. Mistry and G. Grinberg, *J. Liq. Chromatogr. Rel. Techn.*, **28**, 1055 (2005); G. Guiochon, *J.Chromatogr. A*, **1168**, 101 (2007).

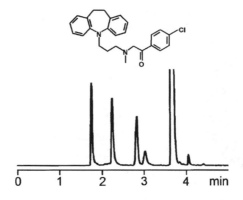

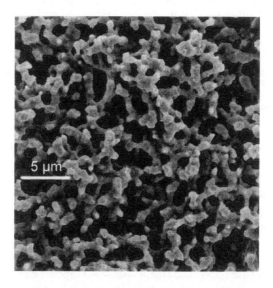

Figure 7.5 Monolithic stationary phase (reproduced by permission of Merck). Top: Separation of Gamonil and byproducts. Conditions: column, 8.3 cm × 7.2 mm i.d.; stationary phase, SilicaROD RP 18 e; mobile phase, water with 20 mM phosphoric acid–acetonitrile, combined solvent and flow gradient, 10–50% acetonitrile and 3–9 ml min^{-1}; UV detector, 256 nm. Bottom: Scanning electron micrograph of the stationary phase.

the cylindrical volume of the column. The diameter of the large pores (where the mobile phase flows through) is e.g. 2 μm, the mean diameter of the skeleton structure is e.g. 1.6 μm, and the diameter of the mesopores is e.g. 12 nm. Such materials have a porosity of more than 0.8, their separation performance is similar to 3 μm porous

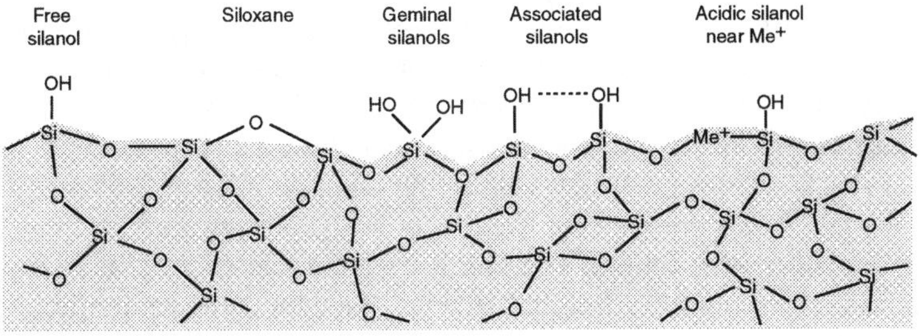

Figure 7.6   Chemical structure of silica.

particles, and their van Deemter curve is very favourable. Monolithic phases do not need frits. Every column is a solitaire, there are no columns "of the same batch". The direct injection of "dirty" samples (serum, food extracts) is often possible without a loss of separation performance or column lifetime.

7.4   SILICA[4]

Silica is an adsorbent with outstanding properties. It may also be used for size-exclusion chromatography and forms the base of numerous chemically bonded stationary phases. It consists of silicon atoms bridged three-dimensionally by oxygen atoms. Figure 7.6 shows that the lattice is saturated at the surface with OH groups, the so-called silanol groups. Because the material is amorphous with a heterogeneous surface it is not easy to synthesize well defined silicas. Although all functional groups at the surface act as adsorptive centres, the various types have different properties:

(a) Free silanols are slightly acidic, therefore basic compounds will preferably adsorb here. This effect can give rise to chemical tailing.
(b) Geminal silanols are not acidic.
(c) Associated (vicinal) silanols are not acidic. Compounds with OH groups tend to adsorb here.
(d) Silanols near metal cations are strongly acidic. They increase the heterogeneity of the surface and can badly affect the separation of basic compounds.
(e) Siloxanes are the product from the condensation of associated silanols. Heat treatment of silica increases the amount of siloxanes and decreases the silanol concentration.

Silica can be produced by a number of different syntheses such as the complete hydrolysis of sodium silicate or the polycondensation of emulsified polyethoxy-

---

[4] K.K. Unger, *Porous Silica,* Elsevier, Amsterdam, 1979; J. Nawrocki, *J. Chromatogr. A,* **779**, 29 (1997).

siloxane, followed by dehydratation. The gels obtained are irregular or spherical in shape, depending on the process; moreover, the properties of the end product depend on the reaction conditions (pH, concentration, additives, even stirring speed and vessel size). One of the most important parameters is the metal content of the starting materials because this will determine the concentration of acidic silanols. 'Conventional' silicas are contaminated with up to 250 ppm sodium and 150 ppm aluminium besides other cations. Their surface reaction can even be basic, depending on the type and concentration of embedded ions. 'Modern', low-metal silicas have not more than 1 ppm of sodium, calcium, magnesium and aluminium, plus slightly higher contents of iron. Only these materials are suited for the separation of basic compounds in both the normal-phase mode (as silicas) or the reversed-phase mode (as chemically bonded phases).

Besides the shape (irregular or spherical), other physical properties can also vary between different brands:

(a) *Pore width*: It should be larger than 5 nm (50 Å). Macromolecules must be separated on wide-pore materials (30 nm = 300 Å or more).

(b) *Specific surface area*: This is inversely proportional to the pore width and is specified in $m^2 g^{-1}$: ca. $100 m^2 g^{-1}$ for 30 nm material, $300 m^2 g^{-1}$ for 10 nm material, and $500 m^2 g^{-1}$ for 6 nm material. The smaller the specific surface area, the lower are the retention factors at constant chromatographic conditions.[5]

(c) *Pore width distribution*: a narrow pore width distribution is essential for symmetrical peaks.

(d) *Density*: the density of silica is ca. $2.2 g cm^{-3}$; the packing density in the column varies according to the product within the range $0.3–0.6 g cm^{-3}$.

The pH stability of silicas is restricted to the range of approximately pH 1–8. Silicas with pore widths up to 400 nm (and a particle size of 10 μm) are produced for use in size-exclusion chromatography. These stationary phases must have a well defined pore width and pore-width distribution and no adsorptive properties.

7.5 CHEMICALLY MODIFIED SILICA[6]

Silica carries OH groups (silanol groups) on its surface and may be chemically modified at these points to give stationary phases with specific properties. Figure 7.7 shows some of the possible reactions.

I. The silanol group can be esterified with an alcohol, ROH, where R may be an alkyl or any other functional group. For steric reasons, reaction is confined to the silica

[5] J.R.K. Huber and F. Eisenbeiss, *J. Chromatogr.*, **149**, 127 (1978).
[6] R.P.W.Scott, *Silica Gel Bonded Phases: Their Production, Properties, and Use in LC*, John Wiley & Sons, Ltd, Chichester, 1993; B. Buszewski *et al.*, *J. High Resolut. Chromatogr.*, **21**, 267 (1998); C. Stella *et al.*, *Chromatographia Suppl.*, **52**, S-113 (2001).

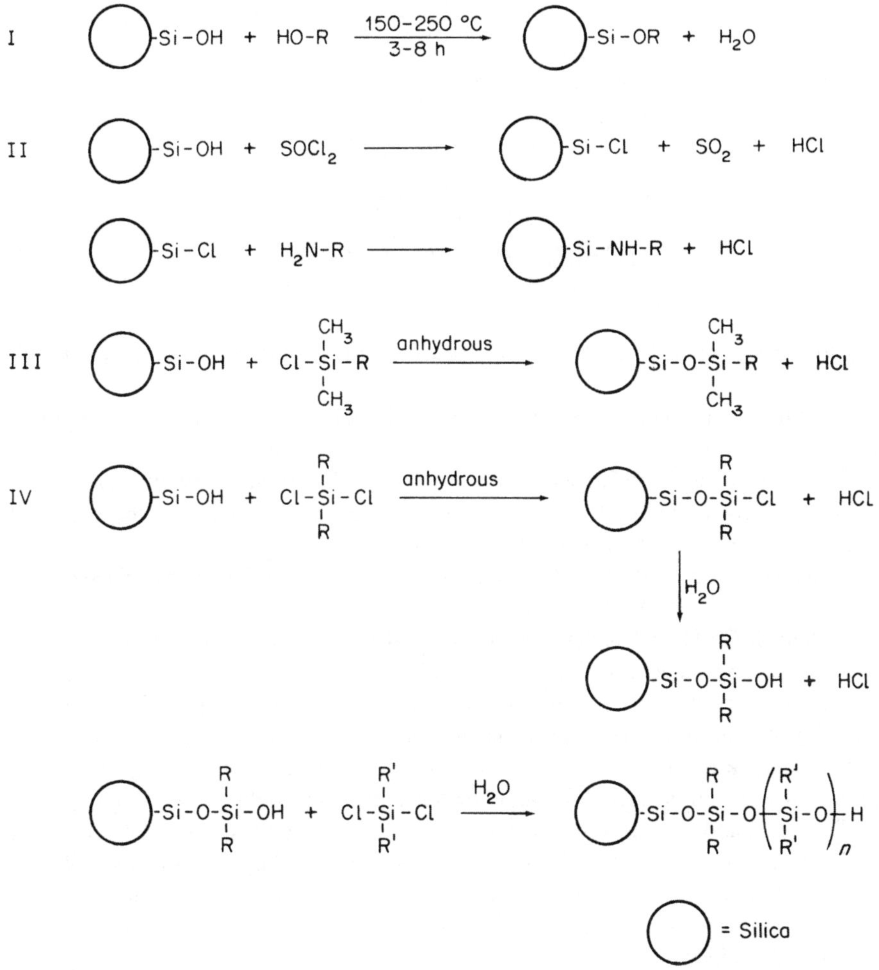

Figure 7.7 Chemical modification of silica.

surface (this obviously includes the 'inner' surface of the pores) and does not involve the solid body interior, so that the bonded residues project from the silica, like tails, or 'brushes' (Figure 7.8).

Esterified silica is prone to hydrolysis and so cannot be used with mobile phases containing water or alcohol, products with Si$-$O$-$Si$-$C bonds generally being preferred.

II. Reaction with thionyl chloride SOCl$_2$ produces chlorides which combine with amines to give an Si$-$N bond. R can be chosen at random. These products have better hydrolytic stability.

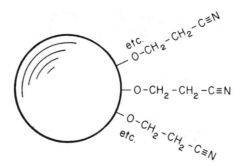

Figure 7.8 Silica with brush surface (propionitrile groups given as an example).

III. Silicas in which the functional group is bonded by an Si$-$O$-$Si$-$C bond (reaction with mono- or dichlorosilane) are the most stable of all.

Octadecylsilane ODS, in which R $= -(CH_2)_{17}CH_3$, is the most widely used of these chemically modified products. It is extremely nonpolar and is the preferred choice for use in reversed-phase chromatography.

Method for the production of chemically bonded phases:[7] 100 ml of dry toluene $+4$ g of dried silica $+2.5$ ml of dry pyridine are mixed with a fourfold excess of silane compound $RSi(CH_3)_2Cl$ (calculated for silica with a surface coverage of 2.4 groups nm^{-2}). The temperature is maintained at 10 °C below that of the boiling point of the most volatile component for between 40 and 150 h (depending on the group R). The mixture is carefully shaken from time to time. Finally, the gel is filtered, washed and dried.

Subsequent treatment with trimethylchlorosilane may reduce the number of silanol groups that remain unreacted for steric reasons, this being known as *end-capping*.

IV. These and similar reactions produce polymer structures (not brushes), which are, in fact, bonded silicone oils known as polysiloxanes. Any cross- linked polymer layer thickness can be chosen. Multifarious methods of synthesis are known.[8] Very advantageous is the shielding of the silica backbone by the polymeric layer, leading to a good pH stability.

Available functional groups are listed in Table 7.1.[9,10,11] Silica can also be functionalized with groups suitable for ion exchange, affinity chromatography or

[7] G.E. Berendsen, K.A. Pikaart and L. de Galan, *J. Liq. Chromatogr.*, **3**, 1437 (1980).

[8] M. Petro and D. Berek, *Chromatographia*, **37**, 549 (1993).

[9] L.C. Sander, K.E. Sharpless and M. Pursch, *J. Chromatogr. A*, **880**, 189 (2000).

[10] Example of a stationary phase with 'polar embedded group', different selectivity for polar analytes than the classical alkyl phases, for eluents with high water content. U.D. Neue *et al.*, *Chromatographia*, **54**, 169 (2001); H. Engelhardt *et al.*, *Chromatographia Suppl.*, **53**, S–154 (2001).

[11] M.R. Euerby, A.P. McKeown and P. Petersson, *J. Sep. Sci.*, **26**, 295 (2003); M. Przybyciel, *LC GC Eur.*, **19**, 19 (2006) or *LC GC North Am.*, **23**, 554 (2005).

TABLE 7.1   Functional groups in chemically modified silicas

Group	Formula	Group	Formula
Triacontyl[9]	$-(CH_2)_{29}CH_3$	Amino	$-NH_2$
Docosyl	$-(CH_2)_{21}CH_3$	Nitro	$-NO_2$
Octadecyl	$-(CH_2)_{17}CH_3$	Nitrile (Cyano)	$-C \equiv N$
Octyl	$-(CH_2)_7CH_3$	Oxypropionitrile	$-OCH_2CH_2C \equiv N$
Hexyl	$-(CH_2)_5CH_3$		$-CH-CH_2$
Trimethyl	$-Si(CH_3)_3$	vic-Hydroxyl (diol)	$\quad\mid\quad\mid$
Alkylcarbamate[10]	$-CO(CO)NH-$		$\quad OH\quad OH$
	$(CH_2)_nCH_3$		
Cyclohexyl	$-C_6H_{11}$	Fluoroalkyl[11]	$-(CF_2)_nCF_3$
Phenyl	$-C_6H_5$	Polycaprolactam	$-[NH(CH_2)_5C=O]_n-$
Diphenyl	$(-C_6H_5)_2$	(polyamide, nylon)	
Dimethylamino	$-N(CH_3)_2$		

the separation of enantiomers. The choice of special phases for the separation of biopolymers is immense (Figure 7.9).

Chemical Stability of Bonded Phases[12]

High pH leads to the dissolution of the silica backbone. Low pH leads to the hydrolysis of the siloxane bond: $SiO-Si-R \rightarrow Si-OH + HO-Si-R$. The end-capped small groups are the preferred site of the attack. Therefore the end-capped phases can alter their properties when used at pH < 3. Long-chain alkyl phases (e.g. octadecyl) are more stable than short-chain ones (e.g. dimethyl).

Chemical stability is higher with materials which were prepared from highly pure silica with low metal content, which have a dense coverage with bonded phase, which are end-capped, and for which sterically protecting derivatization reagents were used (Figure 7.10). Column lifetime is decreased at high temperature, at high buffer concentration and by the use of less suited buffers: inorganic buffers such as phosphate or carbonate are more aggressive than organic ones such as tris or glycine.

7.6   STYRENE-DIVINYLBENZENE[13]

Cross-linked polystyrene is a versatile stationary phase resulting from copolymerization of styrene and divinylbenzene (Figure 7.11). The amount of divinylbenzene

[12] H.A. Claessens and M.A. van Straten, *J. Chromatogr. A*, **1060**, 23 (2004); L. Ma and P.W. Carr, *Anal. Chem.*, **79**, 4681 (2007).
[13] H.W. Stuurman, J. Köhler, S.O. Jansson and A. Litzén, *Chromatographia*, **23**, 341 (1987); L.L. Lloyd, *J. Chromatogr.*, **544**, 201 (1991).

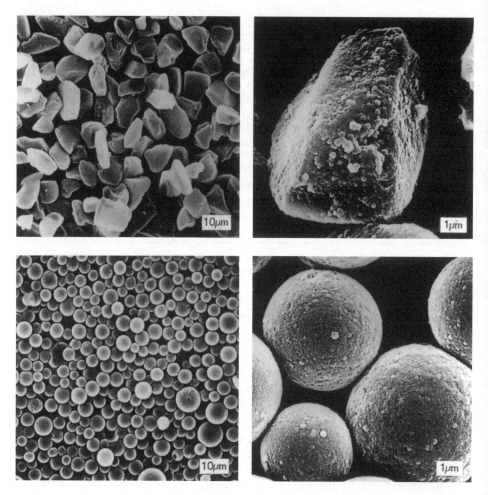

Figure 7.9 Irregular and spherical HPLC material (scanning electron micrographs, taken in the Electron Microscopy Laboratory of the Inorganic Chemistry Institute, University of Berne). Magnification: ×700 (left) and ×7000 (right). Above: irregular silica, mean particle diameter 5 μm (Baker silica). Below: spherical silica gel, mean particle diameter 5 mm (Spherisorb ODS, i.e. chemically modified silica).

added for the reaction determines the degree of cross-linking and hence the pore structure. Resins containing less than 6% of divinylbenzene are not pressure-stable and cannot be regarded as HPLC materials. *Semi-rigid polystyrenes* with 8% divinylbenzene are stable up to approximately 60 bar. They change their bed

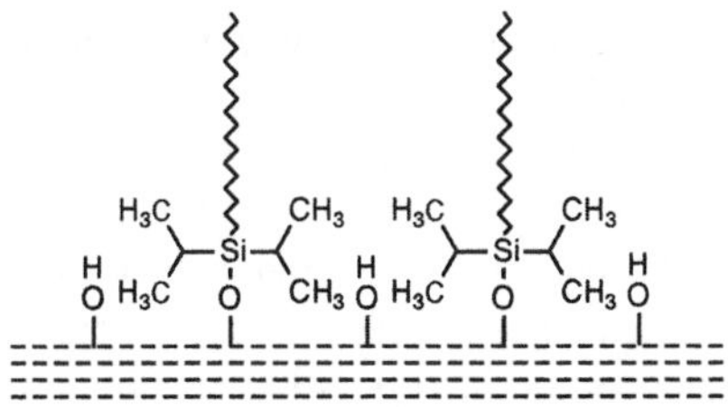

$$H_3C \underset{O}{\overset{CH_3}{\underset{\underset{H_3C}{|}}{\overset{|}{Si}}}} \underset{CH_3}{\overset{|}{CH_3}} \quad H_3C \underset{O}{\overset{CH_3}{\underset{\underset{H_3C}{|}}{\overset{|}{Si}}}} \underset{CH_3}{\overset{|}{CH_3}}$$

Figure 7.10 Bulky side groups protect the bonded phase from attack by chemical agents.

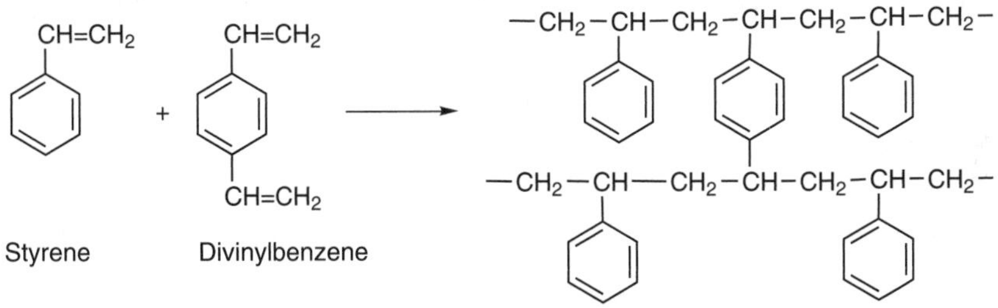

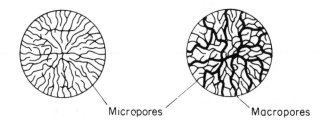

Styrene   Divinylbenzene

Figure 7.11   Styrene–divinylbenzene resin.

volume depending on solvent and ion strength (e.g. they swell at low and shrink at high ionic strength); therefore, once selected, the mobile phase should not be changed. *Rigid polystyrenes* are true high-performance materials. They are highly cross-linked and hence do not swell at all and are stable at pressure up to 350 bar.

Styrene–divinylbenzene phases may be just microporous or contain a mixture of micro- and macropores (Figure 7.12). The presence of macropores with a diameter of several 10 nm facilitates access of larger molecules to the active sites.

Micropores   Macropores

Figure 7.12   Pore structure of styrene–divinylbenzene resin.

In contrast to silica, styrene–divinylbenzene is stable in the pH range 1–13. Styrene–divinylbenzene can be used as a reversed-phase material in combination with aqueous mobile phases; the excellent pH resistance allows a good deal of scope for selecting the best eluent composition, this being a distinct advantage over stationary phases derived from silica (see Figure 7.13). With less polar solvents a chromatographic system of the normal-phase type is obtained.

Ion exchangers are obtained by incorporating suitable groups into the matrix, as shown in Figure 7.14. In a similar manner, it is possible to insert $C_{18}H_{37}$ groups; this type of octadecyl phase does not contain any unreacted OH groups as with reversed phases based on silica.

After cross-linking to give well defined pores, styrene–divinylbenzene represents the most important stationary phase for size-exclusion chromatography.

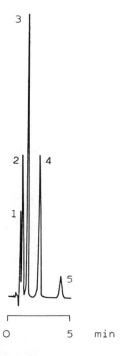

Figure 7.13 Separation of chlorophenols on styrene–divinylbenzene resin (reproduced by permission of Hamilton). Conditions: column, 15 cm × 4.2 mm i.d.; stationary phase, PRP-1, 10 μm; mobile phase, 0.1 N Na_3PO_4 (pH 12)/acetonitrile (2:5), 2 ml min^{-1}; UV detector, 254 nm. Peaks: 1 = 4-chlorophenol; 2 = 2,4-dichloro-phenol; 3 = 2,4,5- and 2,4,6-trichlorophenol; 4 = 2,3,5,6-tetrachloro-phenol; 5 = pentachlorophenol.

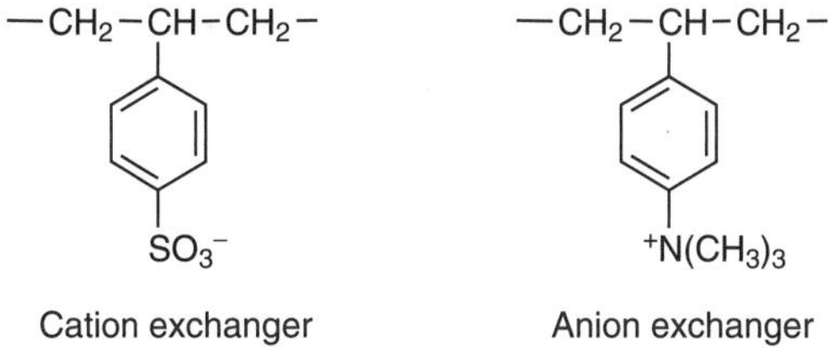

Figure 7.14   Ion-exchange groups in resin.

7.7   SOME OTHER STATIONARY PHASES

Alumina[14]

Alumina is basic but suitable processing can produce neutral and acidic types, neutral alumina being preferred for adsorption chromatography. Basic alumina is a weak cation exchanger, and the acidic form is a weak anion exchanger. Alumina is particularly effective in separating polycyclic aromatic hydrocarbons and structural isomers. In contrast to silica, it is stable over the pH range 2–12. It is more prone to chemisorption problems than is silica, particularly when acidic components are involved and tailing may result. Alumina should not be heated above 150 °C. It has a bulk density of ca. 0.9 g cm^{-3}, a density of ca. 4.0 g cm^{-3} and a packing density of ca. 0.94 g cm^{-3}. Alumina columns have fewer plates than comparable silica columns. Chemical derivatization is possible.[15]

Magnesium Silicate

This must be used with caution as aromatics, amines, esters and many others may be chemisorbed.

Controlled-Pore Glass[16]

This is produced by controlled demixing (or more precisely deglassing) the components of borosilicate glass. Thereby small droplets of B_2O_3 are segregated, which are removed from the SiO_2 matrix with steam. The material is known as controlled-pore glass (CPG) and is pressure-stable, has excellent chemical stability (except towards strong alkalis) and can be sterilized. The surface OH groups can be chemically

[14] C. Laurent, H.A.H. Billiet and L. de Galan, *Chromatographia*, **17**, 253 (1983).
[15] J.J. Pesek and M.T. Matyska, *J. Chromatogr. A*, **952**, 1 (2002).
[16] R. Schnabel and P. Langer, *J. Chromatogr.*, **544**, 137 (1991).

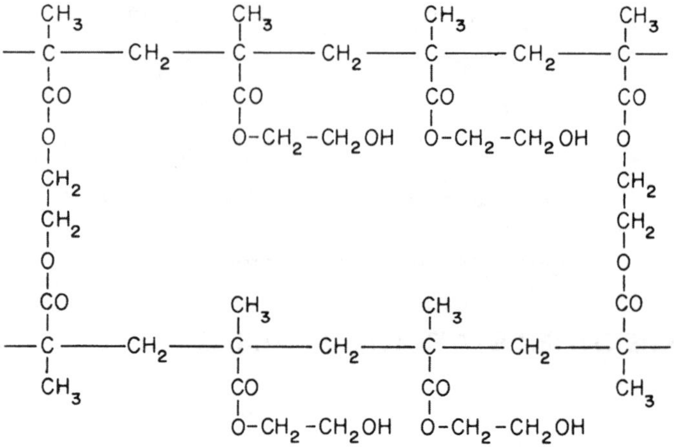

Figure 7.15 Structure of hydroxyalkylmethacrylate gel.

modified to give a material that is as versatile as silica for adsorption, partition, ion exchange, size exclusion and affinity chromatography.

Methacrylate Gels[17]

The large number of OH groups in methacrylate gels (Figure 7.15) make these products very polar and they can be used for gel filtration on their own and for affinity chromatography in derivative form. The hydroxyalkyl residue can be derived from glycol or glycerine.

Hydroxylapatite[18]

This is hexagonally crystallized calcium phosphate, $Ca_{10}(PO_4)_6(OH)_2$. It can be prepared in pressure-stable form (up to 150 bar) and is suitable for the separation of proteins and other biopolymers.

Agarose[19]

Agarose is a cross-linked polysaccharide and is stable over the whole pH range 1–14. It can be derivatized to give, for example, stationary phases for affinity chromatography.

[17] M.J. Beneš, D. Horák and F. Svec, *J. Sep. Sci.*, **28**, 1855 (2005).
[18] T. Kawasaki, *J. Chromatogr.*, **544**, 147 (1991); S. Doonan, *Methods Mol. Biol. (Totowa)*, **59**, 211 (1996).
[19] S. Hjerten, in: *HPLC of Proteins, Peptides and Polynucleotides*, M.T.W. Hearn (ed.), VCH, New York, 1991, pp. 119–148.

Porous Graphitic Carbon[20]

Porous graphitic carbon (PGC) is a fully porous, spherical stationary phase with unique reversed-phase properties but it can also be used in normal-phase mode. PGC has a crystalline graphite surface. It is chemically stable from 10 M acid to 10 M alkali and can withstand high temperatures. The selectivity (i.e. the elution order) is different from silica-based reversed phases. PGC is recommended for the separation of highly polar and ionized compounds as well as of stereoisomers. It is suitable for large, fairly rigid molecules with a multitude of polar groups such as many natural compounds and new drugs. Figure 7.16 shows the separation of two isomers of a prodrug with the sum formula $C_{50}H_{80}N_8O_{20}PK$.

Titania

Titania is crystalline TiO_2 with basic OH groups on its surface (this is in contrast to silica) therefore it is stable at high pH. It can be used in both the normal-phase and the reversed-phase mode.

Zirconia[21]

Zirconia ZrO_2 has similar properties as titania but can also interact with the analytes by ligand exchange because it is a Lewis acid. Chemical derivatization is possible. Coating with polybutadiene or polystyrene gives phases which are stable over a pH range of at least 1–13 and at temperatures higher than 200 °C.

Restricted Surface Access Phases[22]

These special types of stationary phases have been developed for the analysis of drugs and metabolites in serum (and other body fluids). If pH and ionic strength of the mobile phase are chosen appropriately they do not retain proteins, but the selectivity for small, medium-polar molecules is good. Therefore the direct injection of serum on to the column is possible without denaturation of the proteins and without deteriorating the packing by irreversible adsorption. One possibility for the design of this type of stationary phase is shown in Figure 7.17. The pore width is small enough to prevent proteins from penetrating the particles (this is similar to the principle of size exclusion; see Chapter 15); moreover, the outer surface is neither adsorptive nor denaturing. The very stationary phase responsible for the selective separation of small sample molecules is present at the inner surface of the pores.

---

[20] P. Ross and J.H. Knox, *Adv. Chromatogr.*, **37**, 121 (1997).
[21] C.J. Dunlap, C.V. McNeff, D. Stoll and P.W. Carr, *Anal. Chem.*, **73**, 599A (2001).
[22] K.S. Boos and A. Rudolphi, *LC GC Int.*, **11**, 84 and 224 (1998).

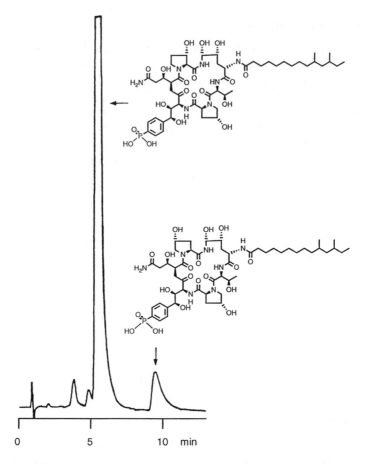

Figure 7.16 Separation of prodrug isomers on porous graphitic carbon
[reproduced with permission from C. Bell, E.W. Tsai, D.P. Ip and D.J. Mathre,
J. Chromatogr. A, **675**, 248 (1994)]. Conditions: column, 10 cm × 4.6 mm i.d.;
stationary phase, Hypercarb, 5 µm; mobile phase, 1.5 ml min^{-1} 0.02 M potassium
phosphate pH 6.8/acetonitrile (55:45); UV detector, 220 nm.

7.8 COLUMN CARE AND REGENERATION

When not in use, HPLC columns should be kept filled with solvent, air-tight and free
from vibration. Water is generally regarded as an unsuitable conservation medium in
view of potential fungal growth. Salt solutions (buffers) may crystallize and cause
clogging.

Do not forget to label the column with the type of mobile phase.

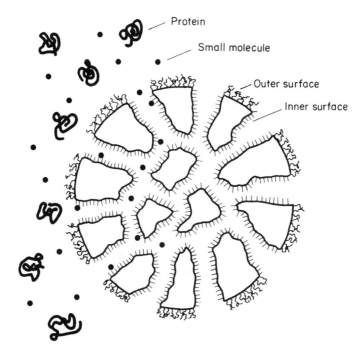

Figure 7.17   Phase with restricted surface access.

After a long period of use or insufficient precautions, the column may become fouled by a build-up of adsorbed materials, especially in the upper packing zones, impairing the overall performance. Before discarding the column, an attempt can be made to regenerate it (albeit with no guarantee of success!). Guard columns must be removed before any such trial begins. It may be a good idea to reverse the column.

Silica Columns

The following solvents are consecutively pumped through at a rate of $1-3\,ml\,min^{-1}$:

75 ml of tetrahydrofuran;
75 ml of methanol;
75 ml of 1–5% aqueous acetic acid if basic impurities are present;
75 ml of 1–5% aqueous pyridine if acidic impurities are present;
75 ml of tetrahydrofuran;
75 ml of *tert*-butylmethyl ether;
75 ml of hexane (if hexane is subsequently used for chromatography, otherwise stop at ether).

If the separation performance should decrease as a result of the silica adsorbing too much water, the latter can be removed by chemical means[23] the reagent either being self-formulated or bought as a ready-made version (Alltech Associates, Deerfield, Illinois, USA).

Octadecyl, Octyl, Phenyl and Nitrile Columns[24]

The following solutions should be consecutively pumped through at a rate of $0.5–2.0\,\text{ml}\,\text{min}^{-1}$:

75 ml of water $+\,4 \times 100\,\mu\text{l}$ of dimethyl sulfoxide injected;
75 ml of methanol;
75 ml of chloroform;
75 ml of methanol.
Water $\rightarrow$ 0.1 M sulfuric acid $\rightarrow$ water is a further possibility.

If a column has lost its chemically bonded phase, then silane compounds may be used in an attempt to bond it to the silica again.[25]

Anion Exchange Columns

The following solvents should be consecutively pumped through at a rate of $0.5–2.0\,\text{ml}\,\text{min}^{-1}$:

75 ml of water;
75 ml of methanol;
75 ml of chloroform;
methanol $\rightarrow$ water.

Cation Exchange Columns

The following solvents should be consecutively pumped through at a rate of $0.5–2.0\,\text{ml}\,\text{min}^{-1}$:

75 ml of water $+\,4 \times 200\,\mu\text{l}$ of dimethyl sulfoxide injected;
75 ml of tetrahydrofuran;
water.

The following sequence works well for anion and cation exchangers:

75 ml of water;
75 ml of a 0.1–0.5 M solution of buffer used previously (to increase the ionic strength);

---

[23] R.A. Bredeweg, L.D. Rothman and C.D. Pfeiffer, *Anal. Chem.*, **51**, 2061 (1979).
[24] R.E. Majors, *LC GC Eur.*, **16**, 404 (2003) or *LC GC North Am.*, **21**, 19 (2003).
[25] C.T. Mant and R.S. Hodges, *J. Chromatogr.*, **409**, 155 (1987).

75 ml of water;
75 ml of 0.1 M sulfuric acid;
75 ml of water;
75 ml of acetone;
75 ml of water;
75 ml of 0.1 M EDTA sodium salt;
75 ml of water.

For styrene–divinylbenzene-based cation exchangers: 0.2 N sodium hydroxide is pumped through at 70 °C overnight (this removes bacteria which could be present at the surface of the particles).

Styrene–Divinylbenzene Columns

The following solvents should be pumped through at a rate of 0.5–2.0 ml min^{-1}:

40 ml of toluene or peroxide-free (<50 ppm) tetrahydrofuran;
200 ml of 1% mercaptoacetic acid in tetrahydrofuran or toluene, if the column is blocked by stryrene-butadiene rubber or natural or synthetic rubber.

Special regeneration procedures are recommended for columns used for the separation of biopolymers.[26]

Filling up the Packing

The column may also be opened to inspect the chromatographic bed. If the packing has collapsed (Figure 7.18a), then the column is repaired in the following way: (b) the excess packing is removed with a fine spatula and the bed smoothed over; (c) dead volumes are filled with glass beads of about 35 µm or suitable porous layer beads and the column is resealed. The frit is replaced by a new one.[27]

Columns with 3 µm packing are often equipped with different frits, e.g. 2 µm on the top and 0.5 µm at the outlet.

Obviously, if enough care is taken (pulse damping, correct preparation of the mobile phase and sample, prevention of precipitation inside the column, etc.), then regeneration and repair work may not be necessary. See the excellent papers by Rabel and by Nugent and Dolan on the subject.[28]

The cavity left by the collapsed packing in a cartridge with an axially compressed chromatographic bed can be removed by simply drawing up the end fittings.

---

[26] C.T. Wehr and R.E. Majors, *LC GC Int.*, **1** (4), 10 (1988) or *LC GC Mag.*, **5**, 942 (1987).
[27] J.W. Dolan, *LC GC Int.*, **6**, 736 (1993).
[28] RM. Rabel, *J. Chromatogr. Sci.*, **18**, 394 (1980); K.D. Nugent and J.W. Dolan, *J. Chromatogr.*, **544**, 3 (1991).

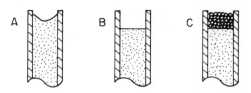

Figure 7.18   Repair of columns with collapsed packing.

Column Emptying

Column contents should not be removed with a spatula, wire, etc., otherwise the internal walls may become scratched and the packing material itself damaged. The best way is to remove the end frit and fitting and use the pump to force the contents out.

Direction of Flow

It is best to use a HPLC column always in the same direction *or* to know why the direction of flow has been changed. Many columns have an arrow which indicates the direction; in this case it is not clear if it is possible to reverse the column. If the frits at both ends are identical in porosity it is no technical problem to run the column in the other direction, however, frits may differ in pore width and the wide-pore frit would be at the entrance. Flow reversal could not be recommended under these circumstances. If a column has no arrow it is good advice to draw one oneself. The column is then always used in the same direction, therefore fines and non-eluted compounds are concentrated at the inlet. The column is only turned for the purpose of regeneration.

8 HPLC Column Tests

8.1 SIMPLE TESTS FOR HPLC COLUMNS

Any newly packed or acquired columns should be automatically put through *at least* the simple tests described in this section. Any suspicion of reduction in plate number, increased tailing or poor permeability must be verified by means of a suitable test.

Number of Theoretical Plates

Calculations for this can be found in Section 2.3.

Compounds that produce no chemical tailing in the particular chromatographic system used are taken as test substances (see Section 8.4). The plate number which can be calculated from the peak of a component that is not or only very slightly retained can be taken as a measure of packing quality and of extra-column volumes. A late-eluted peak will also be affected by mass transfer. If this second plate number is much lower than the first, this is a sign of hindered mass transfer and the chromatographic system is not suitable for separating the mixture in question. If it is much higher, it can be a sign of too large extra-column volumes in the instrument.

The number of theoretical plates is often used to characterize a column but it should not be overvalued as a result. The statement 'my column has 10000 plates' must also be accompanied by the following information:

(a) length and internal diameter of the column;
(b) sample compound and its retention factor;
(c) mobile and stationary phases;
(d) mobile phase flow rate;
(e) sample size;
(f) strictly temperature also.

Practical High-Performance Liquid Chromatography, Fifth edition   Veronika R. Meyer
© 2010 John Wiley & Sons, Ltd

The number of theoretical plates must be determined only under isocratic conditions! Solvent gradients will compress the peaks.

Peak Symmetry

Calculations are given in Section 2.7.

The column can be tested for its characteristics towards acids and bases by including the respective components (phenols, amines) in the test mixture. Pronounced tailing in one case or another shows that the column is unsuitable for the type of analyte.

Permeability

The permeability, K, of a column gives information on the quality of the packing plus any blockages in the system such as clogged frits or tubings. K is defined as:

$$K = \frac{uL_c}{\Delta p}$$

$$u = \frac{L_c}{t_0}$$

$$K = \frac{L_c^2}{\Delta p t_0} \, (\text{mm}^2 \, \text{s}^{-1} \, \text{bar}^{-1})$$

where u is the linear flow velocity of the mobile phase (mm s^{-1}), L_c is the column length (mm) and Δp is the pressure difference between the column inlet and outlet (bar).

If K is too large, then the column packing is poor; if it is too small, this indicates a blockage.

Together with u, L_c and Δp, the specific permeability, $K°$, is also linked with the solvent viscosity, η, and the column packing porosity, ε:

$$K° = \frac{u\eta L_c \varepsilon}{\Delta p} = K\eta\varepsilon$$

$$K° = 10^{-8} K\eta\varepsilon \, (\text{mm}^2)$$

with η in mPa s, ε being about 0.8 for silica phases, $ca.$ 0.65 for chemically bonded phases, and 0.4 for nonporous packings.

Substitution of $u = 4F/d_c^2\pi\varepsilon$ in the equation for K° allows calculation from more accessible variables:

$$K^\circ = \frac{4F\eta L_c}{d_c^2\pi\Delta p}$$

$$K^\circ = 21 \times 10^{-8}\frac{F\eta L_c}{d_c^2\Delta p} \, (\text{mm}^2)$$

where F is the mobile phase flow rate (ml min^{-1}; determined using a measuring flask and a stopwatch) and d_c is the internal diameter of the column (mm).

Problem 18

A 25 cm × 3.2 mm i.d. column is packed with 5 μm silica. A pressure of 70 bar is recorded for an eluent of 0.33 mPa s viscosity (hexane) and a breakthrough time of 60 s. Calculate K and K°.

Solution

$$K = \frac{L_c^2}{\Delta p t_0} = \frac{250^2}{70 \times 60}\,\text{mm}^2\text{s}^{-1}\text{bar}^{-1} = 15\,\text{mm}^2\text{s}^{-1}\text{bar}^{-1}$$

$$\varepsilon = 0.8\,(\text{silica})$$

$$K^\circ = K\eta\varepsilon = 10^{-8} \times 15 \times 0.33 \times 0.8\,\text{mm}^2 = 4.0 \times 10^{-8}\,\text{mm}^2$$

or:

$$K^\circ = 21 \times 10^{-8}\frac{F\eta L_c}{d_c^2\Delta p} = 21 \times 10^{-8}\frac{1.6 \times 0.33 \times 250}{3.2^2 \times 70}\,\text{mm}^2$$

$$= 3.9 \times 10^{-8}\,\text{mm}^2$$

for a flow rate of 1.6 ml min^{-1}.

The calculated plate number and pressure drop for the column may also be roughly compared with Figures 2.29 and 2.30. Something must be wrong if five- or tenfold deviations are observed (the column must be tested not far from its van Deemter minimum).

8.2 DETERMINATION OF PARTICLE SIZE

The particle size, d_p, of a column may not be known at all, or information regarding it may be lost. In this case, the column may be opened up and d_p measured under a microscope. Alternatively, the Kozeny–Carman equation can be used:

$$K° \approx \frac{d_p^2}{1000}$$

Hence the particle diameter can be calculated as:

$$d_p \approx \sqrt{1000 K°}$$

The factor 1000 applies for all types of particle-packed columns if they are packed well and not clogged.

Problem 19

A specific permeability of 4×10^{-8} mm^2 was calculated in Problem 18. To what particle diameter does this $K°$ value correspond?

Solution

$$4 \times 10^{-8} \text{ mm}^2 = 4 \times 10^{-2} \mu\text{m}^2$$
$$d_p = \sqrt{1000 K°} = \sqrt{4 \times 10^{-2} \times 10^3}$$
$$= \sqrt{40} = 6.3 \ \mu\text{m}$$

According to details supplied by the manufacturer, the mean particle size is about 5 μm (spherical silica).

8.3 DETERMINATION OF BREAKTHROUGH TIME

The linear velocity, u, of the mobile phase is needed for calculating permeability. Information on theoretical plates is not strictly correct without involving u, as the plate height is a function of flow rate (see Figure 2.7). Thus u is calculated as follows:

$$u = \frac{L_c}{t_0}$$

Note: $u \neq 4F/d_c^2\pi$ as the stationary phase matrix takes up a specific amount of column volume. The true equation is $u = 4F/d_c^2\pi\varepsilon$ or, with $\varepsilon = 0.65$:

$$u(\text{mm s}^{-1}) = 33 \frac{F(\text{ml min}^{-1})}{d_c^2(\text{mm}^2)}$$

The breakthrough time, t_0, therefore must be known. (t_0 is also needed for the calculation of retention factors k.) It can be determined by measuring the retention time of the first peak on the chromatogram. The question is whether the latter has definitely been transported with the velocity of the mobile phase. Even the slightest retention of this initial peak produces too high a 'breakthrough time' value. Excluded components (and these are not uncommon in reversed-phase chromatography, especially in mobile phases with a high water content) are first picked up by the detector and simulate too low a breakthrough time.[1]

Porosity can be used to verify the breakthrough time. The total porosity, ε, of a column is the volume fraction taken up by the mobile phase:

$$\varepsilon = \frac{V_{\text{column}} - V_{\text{packing material}}}{V_{\text{column}}}$$

ε can be determined simply by:

$$\varepsilon = \frac{4Ft_0}{d_c^2 \pi L_c}$$

$$= 21 \frac{F(\text{ml min}^{-1})t_0(\text{s})}{d_c^2(\text{mm}^2)L_c(\text{mm})}$$

where F is the flow rate of the mobile phase, t_0 is the breakthrough time, d_c the column inner diameter and L_c the column length. ε is about 0.7–0.8 for silica, 0.6–0.7 for bonded phases and about 0.4 for nonporous particles.

Verification of Breakthrough Time

If columns filled with a bonded phase packing (e.g. C_{18} phase) have an $\varepsilon = 0.65$, then the calculated breakthrough time is correct; if $\varepsilon > 1$, then the components in question do not migrate at the same rate as the mobile phase but are effectively retained by the stationary phase and eluted later; if $\varepsilon < 0.5$, then the components are excluded from the pores of the stationary phase (cf. size-exclusion chromatography) and migrate faster than the breakthrough time would indicate.

In the last two cases, the test substance used is unsuitable for the determination of breakthrough time.

A similar diagnosis pertains for silica or for nonporous particles, the critical values of ε then being 0.75 or 0.4, respectively.

Problem 20

Calculate the porosity of a column of 150 mm $\times$ 4.6 mm i.d. which gives a break-through time of 67 s if operated at 1.4 ml min^{-1}.

---

[1] The problems of the determination of breakthrough time were discussed by C.A. Rimmer, C.R. Simmons and J.G. Dorsey, *J. Chromatogr. A*, **965**, 219 (2002).

Solution

$$\varepsilon = \frac{4Ft_0}{\pi d_c^2 L_c} = 21\frac{1.4 \times 67}{4.6^2 \times 250} = 0.62$$

On the other hand, t_0 can be calculated if ε is known. If the porosity is assumed to be 0.65 (bonded phase), the following ready-to-use equation can be derived:

$$t_0(s) = 0.03\frac{d_c^2(\text{mm}^2)L_c(\text{mm})}{F(\text{ml min}^{-1})}$$

Problem 21

Estimate the breakthrough time with the data given in Problem 20 (bonded phase).

Solution

$$t_0(s) = 0.03\frac{d_c^2(\text{mm}^2)L_c(\text{mm})}{F(\text{ml min}^{-1})} = 0.03\frac{4.6^2 \times 150}{1.4} = 68\,\text{s}$$

A calculation of this type can verify if a baseline fluctuation or a peak really marks the breakthrough time. A quick estimation is also possible for columns of 4.6 mm i.d. which are run at a flow of 1 ml min^{-1}: t_0 (min) $\approx 0.1\,L_c$ (cm) or V_0 (ml) $\approx 0.1\,L_c$ (cm).

8.4   THE TEST MIXTURE

A specific test mixture may be prepared for working in a special area, e.g. an aflatoxin mixture for aflatoxin determination. (However, a test mixture for the determination of the plate number should not include any peptides or proteins. These compounds *need* to be eluted by gradient, see Figure 18.7, whereas the plate number is calculated from an isocratic chromatogram.) In all other cases, a test mixture that satisfies the following criteria is needed:

(a) good quality, readily obtainable;
(b) stable;
(c) not too toxic;
(d) no chemical tailing;
(e) good UV absorption or, as a general rule, good detectability.

As the column must never be overloaded, not more than 1 μg of each component per gram of stationary phase should be injected; the amount permitted in analytical columns is so small as to make refractive index detection critical or even impossible.

The test mixture must contain a compound by means of which the breakthrough time can be measured accurately; it must be neither retained nor excluded. These conditions are satisfied when the product has a low molar mass and is as closely related to the eluent as possible, e.g. pentane for elution with hexane. Pentane gives no UV absorption but just a small refractive index peak. The plate number cannot be determined from this signal, so another test compound such as toluene or xylene which is rapidly eluted ($k \approx 0.2$) and which produces a true UV absorption peak is recommended. The theoretical plate number of a little or nonretained substance reflects directly on the packing quality and the extra-column volumes, whereas mass transfer properties are equally important in peaks that are eluted later. In reversed-phase systems, t_0 can be determined with uracil. The test mixture should also contain compounds for which $k \approx 1$ and 3–5.

Test Mixture for Silica Columns

0. Pentane (nonretained, solvent for the following compounds)
1. *p*-Xylene
2. Nitrobenzene
3. Acetophenone
4. 2,6-Dinitrotoluene

Mobile phase: hexane–*tert*-butylmethyl ether, appropriate mixture as to yield suitable k values.

A test chromatogram is shown in Figure 8.1.

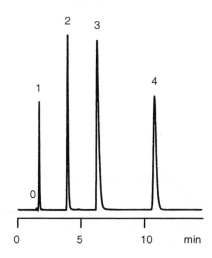

Figure 8.1 Test chromatogram for silica columns. Conditions: column, 25 cm × 3.2 mm i.d.; stationary phase, LiChrosorb SI 60, 5 µm; mobile phase, 1 ml min^{-1} hexane-*tert*. butylmethyl ether (95 : 5); detector, UV 254 nm. Peaks: 0 = pentane, 1 = *p*-xylene, 2 = nitrobenzene, 3 = acetophenone, 4 = 2,6-dinitrotoluene.

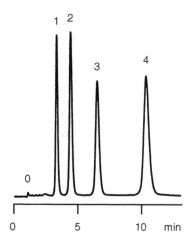

Figure 8.2 Test chromatogram for reversed-phase columns. Conditions: column, 25 cm × 3.2 mm i.d.; stationary phase, Spherisorb ODS, 5 μm; mobile phase, 1 ml min^{-1} water–methanol (50 : 50); detector, UV 254 nm. Peaks: 0 = water–methanol with less than 50% methanol, 1 = methylparabene, 2 = ethylparabene, 3 = propylparabene, 4 = butylparabene.

Test Mixture for Reversed-phase Columns

0. Methanol–water or acetonitrile–water with a different (weaker) composition than the mobile phase (nonretained, solvent for mixture)
1. Methylparabene
2. Ethylparabene
3. Propylparabene
4. Butylparabene

Mobile phase: methanol–water or acetonitrile–water, appropriate mixture as to yield suitable k values.

A test chromatogram is shown in Fig. 8.2.

8.5 DIMENSIONLESS PARAMETERS FOR HPLC COLUMN CHARACTERIZATION[2]

Information on the number of theoretical plates, N, or plate height, H, can be problematic, as outlined in Section 8.1. Dimensionless values (numbers without

---

[2] More details of this and the following section can be found in P.A. Bristow and J.H. Knox, *Chromatographia*, **10**, 279 (1977).

units) are much more suitable for absolute comparison of columns, a whole range of different types of columns (long or short, thick and thin, normal or reversed phase, large or microparticles, with various types of packing) being covered. Dimensionless values are also easy to remember.

First dimensionless parameter: reduced plate height, h, replaces plate height, H. H is compared with the mean particle diameter, d_p:

$$h = \frac{H}{d_p}$$

$$h = \frac{1000 L_c (\text{mm})}{5.54\, d_p (\mu\text{m})} \left[\frac{w_{1/2}(\text{mm or s})}{t_R(\text{mm or s})} \right]^2$$

Second dimensionless parameter: reduced velocity, v, replaces linear flow velocity u. u is set in relation to the diffusion coefficient, D_m, of the analyte in the mobile phase and to the particle diameter d_p.

$$v = \frac{d_p u}{D_m} = \frac{4 d_p F}{d_c^2 \pi \varepsilon D_m}$$

$$v = 1.3 \times 10^{-2} \frac{d_p (\mu\text{m}) F (\text{ml min}^{-1})}{\varepsilon D_m (\text{cm}^2\ \text{min}^{-1}) d_c^2 (\text{mm}^2)}$$

Porosity ε is ca. 0.8 (silica) or 0.65 (bonded phase) for a packing with porous particles. The average diffusion coefficient (see Section 8.7) can be taken as $2.5 \times 10^{-3}\ \text{cm}^2\ \text{min}^{-1}$ for small molecules in low-viscosity eluents (normal phase) or as $6 \times 10^{-4}\ \text{cm}^2\ \text{min}^{-1}$ for small molecules in high-viscosity eluents (reversed phase). Therefore the ready-to-use equations emerge:

$$\text{Normal phase}: \qquad v_{NP} \approx 6.4 \frac{d_p (\mu\text{m}) F (\text{ml min}^{-1})}{d_c^2 (\text{mm}^2)}$$

$$\text{Reversed phase}: \qquad v_{RP} \approx 33 \frac{d_p (\mu\text{m}) F (\text{ml min}^{-1})}{d_c^2 (\text{mm}^2)}$$

Note that both equations are only valid for the conditions given above (small molecules, porous particles).

Third dimensionless parameter: dimensionless flow resistance, Φ, replaces permeability K:

$$\Phi = \frac{\Delta p d_p^2 d_c^2 \pi}{4 L_c \eta F}$$

$$\Phi = 4.7 \frac{\Delta p(\text{bar}) d_p^2 (\mu m^2) d_c^2 (\text{mm}^2)}{L_c(\text{mm}) \eta(\text{mPas}) F(\text{ml min}^{-1})},$$

where η is the viscosity of the mobile phase. Φ is related to the permeability, K, by:

$$\Phi = \frac{d_p^2}{K \eta \varepsilon}$$

(see Section 8.1).

Fourth dimensionless parameter: dimensionless separation impedance, E (also known as 'efficiency'), which embraces retention time, pressure drop, plate number, eluent viscosity and retention factor, is taken as a measure of quality:

$$E = \frac{\Delta p t_0}{N^2 \eta}$$

$$E = \frac{10^8}{5.54^2} \frac{\Delta p(\text{bar}) t_0(\text{s})}{\eta(\text{mPa s})} \left[\frac{w_{1/2}(\text{mm or s})}{t_R(\text{mm or s})} \right]^4$$

$$E = h^2 \Phi \varepsilon$$

The greater the value of E, the worse the performance of the column.

The values of h, v, Φ and E derived from these parameters fully characterize a column. No serious HPLC operator should be satisfied with the parameters described in Section 8.1, but should also incorporate these four dimensionless values into the overall calculations.

A good column has $h = 2$–5 (2 is the lower limit and means that complete chromatographic equilibrium is obtained over only two layers of stationary phase packing; $h = 2$ produces 30 000 theoretical plates in a 30 cm long column with 5 mm particles).

For optimum efficiency v must be between 3 and 20.

Φ is 1000 for slurry-packed columns (i.e. the Kozeny-Carman factor). A much greater value of Φ (e.g. 2000) indicates a blocking somewhere (frits or tubing, too much packing abrasion). A lower value indicates a poor packing with voids or channels.

A good separation impedance is lower than 10 000; e.g. $h = 3$, $\Phi = 1000$ and $\varepsilon = 0.65$ for a chemically bonded phase, giving $E = 5850$. If E is greater than 10 000, then HPLC can no longer be regarded as a 'high-performance' process.

Problem 22

How good is the quality of my column of $15\,\text{cm} \times 3\,\text{mm}$ i.d. which has 4700 theoretical plates? The stationary phase is C_{18}, $5\,\mu\text{m}$.

Solution

$$h = \frac{H}{d_p} = \frac{L_c}{Nd_p}$$

$$h = \frac{150}{4700 \times 0.005} = 6.4$$

Such a large reduced plate height is inadequate for a reversed-phase column.

Problem 23

Which flow rate is recommendable for the column of Problem 22? The mobile phase is water–acetonitrile (1 : 1).

Solution

This is a reversed-phase system.

$$v_{RP} = 33\frac{d_p(\mu\text{m})F(\text{ml min}^{-1})}{d_c^2(\text{mm}^2)}, \text{ should be 3 (or slightly higher)}$$

$$F = \frac{v d_c^2}{33 d_p} = \frac{3 \times 3^2}{33 \times 5} = 0.2\,\text{ml min}^{-1}$$

Problem 24

If the column of Problem 22 is used with $0.5\,\text{ml min}^{-1}$ of methanol, a pressure of 115 bar is obtained. Is this what can be expected?

Solution

The viscosity of pure methanol is 0.6 mPa s (Section 5.1). With $\varepsilon = 0.65$:

$$\Phi = 4.7 \frac{\Delta p(\text{bar})d_p^2(\mu\text{m}^2)d_c^2(\text{mm}^2)}{L_c(\text{mm})\eta(\text{mPa s})F(\text{ml min}^{-1})} = 4.7 \frac{115 \times 5^2 \times 3^2}{150 \times 0.6 \times 0.5} = 2700$$

This reduced flow resistance is much too high. The column, frit, or a part of the instrument is clogged.

Note: The separation impedance of this column is poor: $E = h^2 \Phi \varepsilon = 6.4^2 \times 2700 \times 0.65 = 72\,000$!

8.6 THE VAN DEEMTER EQUATION FROM REDUCED PARAMETERS AND ITS USE IN COLUMN DIAGNOSIS

The empirical van Deemter equation (see Section 2.2) with dimensionless values is as follows:

$$h = Av^{0.33} + \frac{B}{v} + Cv$$

where A is the eddy diffusion and flow distribution component, B is the longitudinal diffusion component and C is the mass transfer component. This modified van Deemter equation is known as Knox's equation.

A, B and C are constants; $A = 1$, $B = 2$ and $C = 0.1$ for good (not excellent) columns,[3] van Deemter's equation then being written as (see Figure 8.3):

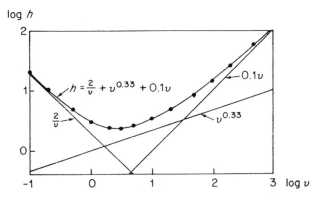

Figure 8.3   Van Deemter curve with reduced values (reproduced by permission of Vieweg Publishing, from P.A. Bristow and J.H. Knox, *Chromatographia*, **10**, 282 (1977). Its shape and position are independent of the particle diameter of the stationary phase.

---

[3] In cases of excellent mass transfer C can be 0.05 or less.

$$h = v^{0.33} + \frac{2}{v} + 0.1v$$

A is a function of packing quality (and also of k) and is greater than 2 (up to 5) in poorly packed columns. B is a function of the sample diffusion coefficient and is not altered by anything else. C is a function of the packing material, i.e. of its mass-transfer properties. Hence the reason for poor performance in columns with a small number of theoretical plates is more likely to be found if A and C are known. However, A, B and C cannot be calculated at all unless many $h(v)$ pairs have been determined with sufficient accuracy, so a qualitative assessment of the $h(v)$ curve is much more convenient:

(a) If h_{min} is between $v = 3–5$ and is less than 3 then the column is well packed. However, if the minimum position remains the same but h_{min} exceeds 10, then the column is poorly packed.
(b) If h is less than 10 at $v = 100$, then the stationary phase has good mass transfer properties for the injected material (C is small); the column is also likely to be well packed.
(c) If the curve rises sharply on the right-hand side, then C is large and the stationary phase has poor mass-transfer properties (the steep slope may also be caused instrumentally).
(d) If the minimum is high and is not very pronounced, then the column is poorly packed and A is large.

No such details can be given for the $H(u)$ curve as its position is a function of particle diameter and diffusion coefficient alone.

If assessment of the whole $h(v)$ curve is too time-consuming, even one pair of values can provide a certain amount of information:

When $v = 3$, h should not be greater than 3–4.
When $v = 100$, h should not be greater than 10–20.
When $v \approx 3$ and h is higher, then the packing is likely to be poor.
When $v = 20–100$ and h is higher, then either poor mass transfer or poor packing is suspected and a few neighbouring points should be determined to find out which, a weak rise indicating poor packing and a steep rise poor mass transfer.

The curve minimum is found at a reduced flow velocity of about 3. As mentioned in Sections 2.2 and 8.5, it is not worth operating at much lower flow rates as longitudinal diffusion then becomes a detrimental factor.

8.7 VAN DEEMTER CURVES AND OTHER COHERENCES

Fig. 8.4 clarifies some coherences.

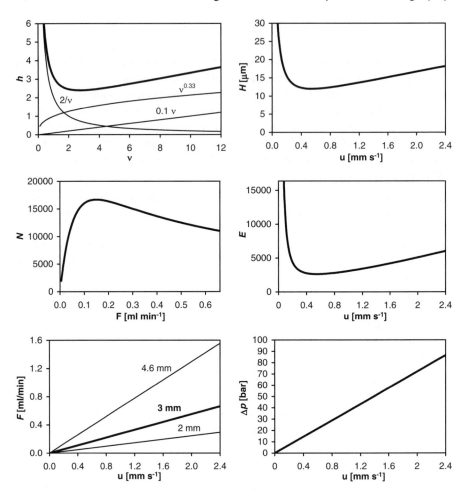

Figure 8.4 Various van Deemter curves and coherences. For details see Section 8.7. The h/v curve in the upper left is valid for many types of HPLC packings. The other graphs represent only the properties of well-defined phase systems and column dimensions.

The h/v curve (upper left) depends on the equation behind it. One of broad validity for HPLC is the one already mentioned in Section 8.6:

$$h = v^{0.33} + \frac{2}{v} + 0.1v$$

It is independent of column dimensions and particle diameter but depends on the packing quality, the diffusion coefficient and the mass transfer properties of the system analyte – mobile phase – stationary phase. Fig. 8.3 shows the same curve but with logarithmic axes.

The more common H/u curve (the classical van Deemter relationship, upper right) is derived from the h/v curve by taking into account the particle diameter and the diffusion coefficient: $H = hd_p$ and $u = vD_m/d_p$. The curve shown here is valid for a particle diameter of 5 μm and a diffusion coefficient of $1 \times 10^{-9}\,\mathrm{m^2\,s^{-1}}$, typical for reversed-phase separations. For packings of smaller particles the curve is lower. For normal-phase separations its minimum is more to the right because the diffusion coefficients are higher (see Section 8.8). The ranges of both axes correspond to the ones of the h/v curve.

Despite van Deemter's approach, chromatographers most often think of numbers of theoretical plates N and flow rates F (middle left) instead of plate heights H and linear flow rates u. The curve shown here is valid for columns of 20 cm length ($N = L_c/H$), 3 mm inner diameter and a porosity of 0.65, typical for bonded phases ($F = ud_c^2\pi\varepsilon/4$ plus the conversion from seconds to minutes). Under these circumstances, 2.4 mm s^{-1} corresponds to 0.66 ml min^{-1}. The N/F curve has a maximum.

The dimensionless separation impedance E depends on u (or F) and has a minimum like a van Deemter curve (middle right). With the Kozeny-Carman factor of $\Phi = 1000$ we come to the relationship:

$$E = h^2 \times 1000 \times \varepsilon$$

The lowermost graphs show linear relationships. The flow rate is proportional to the linear flow velocity (lower left). It depends from the porosity and the column inner diameter. Again, $\varepsilon = 0.65$ for bonded phases. With $\varepsilon = 0.8$, typical for bare silica, the lines would be steeper. The graph shows the data for three different column diameters.

Finally (and rather trivial), the pressure is proportional to the linear flow velocity (lower right) or the flow rate. The graph is valid for $\phi = 1000$, $\eta = 1$ mPa s (water-acetonitrile), $L_c = 20$ cm, $\varepsilon = 0.65$ and $d_p = 5$ μm.

8.8   DIFFUSION COEFFICIENTS[4]

Diffusion coefficients can be calculated by Wilke and Chang's equation:[5]

$$D_m(\mathrm{m^2\,s^{-1}}) = \frac{7.4 \times 10^{-12}\sqrt{\Psi MT}}{\eta(\mathrm{mPa\ s})[V_s(\mathrm{cm^3 mol^{-1}})]^{0.6}}$$

---

[4] Experimental determination of diffusion coefficients using HPLC: E. Grushka and S. Levin, in *Quantitative Analysis Using Chromatographic Techniques*, E. Katz (ed.), John Wiley & Sons, Inc., New York, 1987, pp. 360–374.

[5] C.R. Wilke and P. Chang, *Am. Inst. Chem. Eng. J.*, **1**, 264 (1955).

where Ψ is the solvent constant (2.6 for water, 1.9 for methanol, 1.5 for ethanol, 1.0 for other (nonassociated) solvents), M is the molar mass (molecular weight) of the solvent, T is the absolute temperature, η is the solvent viscosity and V_s is the molar volume of the test compound [calculated from the molar mass $(g \, mol^{-1})$ divided by the density $(g \, cm^{-3})$]. The accuracy is $\pm 20\%$.

Some diffusion coefficients of small molecules at $20 \, °C$ are:

Nitrobenzene in hexane	$3.8 \times 10^{-9} \, m^2 \, s^{-1} = 2.3 \times 10^{-3} \, cm^2 \, min^{-1}$
Phenol in methanol	$1.9 \times 10^{-9} \, m^2 \, s^{-1} = 1.1 \times 10^{-3} \, cm^2 \, min^{-1}$
Phenol in water	$9.7 \times 10^{-10} \, m^2 \, s^{-1} = 5.8 \times 10^{-4} \, cm^2 \, min^{-1}$
Phenol in water–methanol $(1:1)$	$6.1 \times 10^{-10} \, m^2 \, s^{-1} = 3.7 \times 10^{-4} \, cm^2 \, min^{-1}$

As a rule of thumb, the diffusion coefficients of small molecules in normal-phase solvents can be assumed to be $2.5 \times 10^{-3} \, cm^2 \, min^{-1}$, in reversed-phase solvents to be $6 \times 10^{-4} \, cm^2 \, min^{-1}$.[6]

Problem 25

Calculate the diffusion coefficient of phenetol in methanol–water $(1:1)$ at $20 \, °C$. The viscosity of this eluent is $1.76 \, mPas$.

Solution

Calculation of Ψ and M of the solvent mixture is based on its molar fraction:

1 part by volume of water, density $1 \, g \, cm^{-3} = 1$ part by mass;
1 part by volume of methanol, density $0.79 \, g \, cm^{-3} = 0.79$ parts by mass;
1 part by mass of water, $18 \, g \, mol^{-1} = 0.56$ molar parts;
0.79 parts by mass of methanol, $32 \, g \, mol^{-1} = 0.025$ molar parts;
Total $= 0.081$ molar parts.

$$\text{Water molar fraction} = 0.056/0.081 = 0.69$$
$$\text{Methanol molar fraction} = 0.025/0.081 = 0.31$$
$$\Psi = 0.69(2.6) + 0.31(1.9) = 2.38$$
$$M = 0.69(18) + 0.31(32) = 22.3$$
$$T = 293 \, K$$

Phenetol: $C_6H_5OCH_2CH_3 = C_8H_{10}O$, molar mass $= 122 \, g \, mol^{-1}$, density $= 0.967 \, g \, cm^{-3}$

---

[6] $1 \times 10^{-9} \, m^2 \, s^{-1} = 1 \times 10^{-5} \, cm^2 \, s^{-1} = 6 \times 10^{-4} \, cm^2 \, min^{-1}$.

$$V_{\text{phenetol}} = \frac{122 \text{ g cm}^3}{0.967 \text{ g mol}} \ 126 \text{ cm}^3 \text{ mol}^{-1}$$

$$D_m = \frac{7.4 \times 10^{-12}\sqrt{\Psi M T}}{\eta V^{0.6}}$$

$$D_m = \frac{7.4 \times 10^{-12}\sqrt{2.38 \times 22.3 \times 293}}{1.76 \times 126^{0.6}} \text{ m}^2 \text{ s}^{-1}$$

$$= 4.9 \times 10^{-10} \text{ m}^2 \text{ s}^{-1} = 2.9 \times 10^{-4} \text{cm}^2 \text{ min}^{-1}$$

Macromolecules

Macromolecules have a large molar volume (their molar mass is high but does not affect the density by it, to a first approximation). For example, a molecule with the same density as phenetol but with 100 times its molar mass has a molar volume of $1.26 \times 10^4 \text{ cm}^3 \text{ mol}^{-1}$, giving a low diffusion coefficient of $3.1 \times 10^{-11} \text{ m}^2 \text{ s}^{-1}$. Indeed, Wilke and Chang's equation does not apply for macromolecules and the following correction factors must be incorporated:

Molar mass	Multiply D_m by:
10^3	1.3
10^4	2.0
10^5	3.2

Hence, the diffusion coefficient of the hypothetical macromolecule is $3.1 \times 10^{-11} \text{ m}^2 \text{ s}^{-1} \times 2 = 6.2 \times 10^{-11} \text{ m}^2 \text{ s}^{-1} = 3.7 \times 10^{-5} \text{ cm}^2 \text{ min}^{-1}$.

The fact that macromolecules have small diffusion coefficients is of vital consequence to chromatographic practice.

Problem 26

How high should the value of the mobile phase linear flow velocity, u, be for chromatographing phenetol ($D_m = 0.49 \times 10^{-9} \text{ m}^2 \text{ s}^{-1}$) and a hypothetical macromolecule ($D_m = 6.2 \times 10^{-11} \text{ m}^2 \text{ s}^{-1}$) with an optimum reduced flow rate of $v_{\text{opt}} = 3$? The particle diameter d_p of the stationary phase is $5 \, \mu\text{m}$.

Solution

$$v = \frac{u d_p}{D_m}$$

$$u = \frac{v D_m}{d_p}$$

Phenetol:

$$u = \frac{3 \times 0.49 \times 10^{-9} \, m^2}{5 \times 10^{-6} \, m \, s}$$
$$= 0.29 \times 10^{-3} \, m \, s^{-1}$$
$$= 0.29 \, mm \, s^{-1}$$

Macromolecule:

$$u = \frac{3 \times 6.2 \times 10^{-11} \, m^2}{5 \times 10^{-6} \, m \, s}$$
$$= 3.7 \times 10^{-5} \, m \, s^{-1}$$
$$= 0.037 \, mm \, s^{-1}$$

The breakthrough time is 344 s (less than 6 min) in the first case and 2700 s (45 min) in the second for a 10 cm column.

Hence, macromolecules must be chromatographed slowly!

9 Adsorption Chromatography: Normal-Phase Chromatography

This method is explained with silica, the most important chromatographic adsorbent.

9.1 WHAT IS ADSORPTION?[1]

As already mentioned in Section 7.4, the silica structure is saturated with silanol groups at the ends. These OH groups are statistically distributed over the whole of the surface (Figure 9.1).

The silanol groups represent the active sites in the stationary phase (in alumina, the Al^{3+} centres are the most prominent, followed by the linking O^{2-} moieties). They form a weak type of 'bond' with any molecule in the vicinity when any of the following interactions are present:

dipole-induced dipole;[2]
dipole-dipole;
hydrogen bonding;
π-complex bonding.

These situations arise when the molecule has one or several atoms with lone electron pairs or a double bond (for π-complex interaction), in other words, when it is

---

[1] Standard textbook: L.R. Snyder, *Principles of Adsorption Chromatography*, Dekker, New York, 1968.
[2] A molecule or a functional group acts as a dipole when the electric charge distribution is not homogeneous throughout the whole molecule or group. The dipole is said to be 'induced' when a charge is moved solely under the influence of an external field.

Practical High-Performance Liquid Chromatography, Fifth edition   Veronika R. Meyer
© 2010 John Wiley & Sons, Ltd

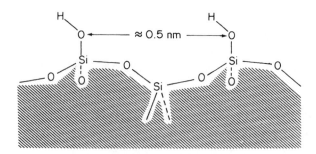

Figure 9.1   Silica surface.

unsaturated or carries a so-called functional group. On the other hand, alkanes cannot interact, as they are saturated and consist solely of C and H atoms which carry no lone pairs.

The adsorption strengths and hence k values (elution series) increase in the following order:

> saturated hydrocarbons < olefins < aromatics ≈ organic halogen compounds < sulphides < ethers < nitro compounds < esters ≈ aldehydes ≈ ketones < alcohols ≈ amines < sulphones < sulphoxides < amides < carboxylic acids.

If a molecule has several functional groups, then the most polar one (the last one in the above series) determines the retention properties. This weak 'bond formation' is called adsorption and its dissolution which is just as rapid is known as desorption.

Three conclusions can be drawn from this concept:

(a) The silica gel in the chromatographic bed is surrounded on all sides by mobile phase. The solvent occupies all active sites more or less strongly. A sample molecule can only be adsorbed if it interacts more strongly than the solvent with the adsorbent.

(b) All sample molecules (and also those of the solvent) are arranged on the silica surface so that their functional group or double bond is close to the silanol groups (Figure 9.2). Any possible hydrocarbon 'tails' are diverted from the silica. Hence the adsorbent cannot distinguish between molecules that are identical apart from their aliphatic moiety. A mixture of hexanol, heptanol and octanol cannot be separated very satisfactorily by adsorption chromatography.

(c) The strength of interaction depends not only on the functional groups in the sample molecule but also on steric factors. Molecules with a different steric structure, i.e. isomers, are eminently suitable for separation by adsorption chromatography.

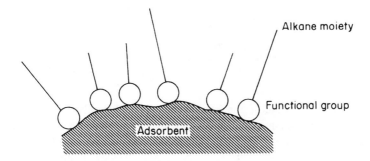

Figure 9.2   During adsorption, functional groups are directed towards the silica surface.

The separation of azo dyes with the following structures:

1.

2.

3.   *p*-Isomer

4.   *m*-Isomer

5.   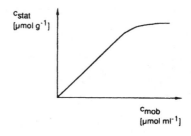 o-Isomer

represents a good example of this (see Figure 2.11 and Problem 5).

This qualitative concept can also be used in a more quantitative manner. If the equilibrium concentrations of the sample in the mobile and stationary phases are known the *distribution coefficient, K*, can be calculated (see Section 2.1). K in general is not a constant but depends on the sample size and temperature. The sample concentration in the stationary phase as a function of sample concentration in the mobile phase at constant temperature is described by the *adsorption isotherm*, which is specific for a given chromatographic system (sample–mobile phase–stationary phase).[3]

9.2 THE ELUOTROPIC SERIES

It is clear from the above that not every solvent elutes sample molecules at the same speed. If an aliphatic hydrocarbon constitutes the mobile phase, it becomes very difficult to displace the sample molecules from the active adsorbent sites and the solvent is classified as *weak*. Tetrahydrofuran, on the other hand, competes strongly for the active sites, leaving less time for sample molecules to be adsorbed and hence they are rapidly eluted. This type of solvent is comparatively *strong*.

The influence of mobile phase strength on the separation is shown in Figure 9.3. Hexane is too weak a solvent for the analysis of this test mixture and the separation needs an unnecessarily long time. *tert*-Butylmethyl ether is much too strong. A suitable eluent is the mixture hexane-*tert*-butylmethyl ether (9 : 1) (volume parts).

The elution force or strength of the various solvents is determined empirically and recorded numerically, being represented by the symbol ε^0. The series of weak,

---

[3] The most common adsorption isotherm on silica is of the Langmuir type, the ideal shape of which is as follows:

c_{stat}
[µmol g^{-1}]

c_{mob}
[µmol ml^{-1}]

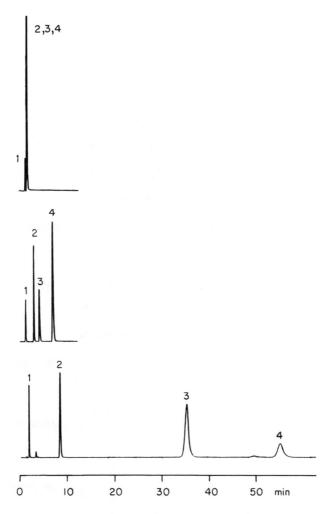

Figure 9.3 Separation of a test mixture on silica with eluents of different strength. Top: *tert*-butylmethyl ether, $\varepsilon^0 = 0.48$; middle: hexane-*tert*-butylmethyl ether (9 : 1), $\varepsilon^0 = 0.29$; bottom: hexane, $\varepsilon^0 = 0$. The attenuation of the hexane chromatogram is only half as much as of the other two. Conditions: column, 25 cm × 3.2 mm i.d.; stationary phase, LiChrosorb SI 60 5 μm; flow rate, 1 ml min^{-1}; UV detector, 254 nm. Peaks: 1 = *p*-xylene; 2 = nitrobenzene; 3 = acetophenone; 4 = 2,6-dinitrotoluene.

medium strength and strong solvents thus derived is referred to as the *eluotropic series*. The eluotropic series of various solvents is given in Section 5.1.

It is possible to obtain a desired elution strength by mixing two solvents. A choice of binary mixtures is shown in Figure 9.4. If, for example, a solvent strength of $\varepsilon^0 = 0.4$ is

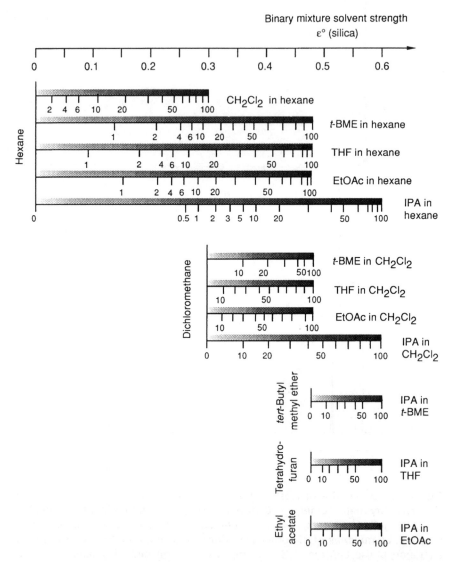

Figure 9.4 Elution strength of binary mixtures as used for adsorption chromatography on silica [reproduced with permission from M.D. Palamareva and V.R. Meyer, *J. Chromatogr.*, **641**, 391 (1993)]. The graph covers the 12 possible mixtures of hexane, dichloromethane, *tert*-butylmethyl ether, tetrahydrofuran, ethyl acetate and isopropanol.

needed, this can be obtained by the following binary eluents:

approximately 60% of *tert*-butylmethyl ether in hexane,
          45% of tetrahydrofuran in hexane,
          50% of ethyl acetate in hexane,
          15% of isopropanol in hexane,
          20% of *tert*-butylmethyl ether in dichloromethane,
          45% of tetrahydrofuran in dichloromethane,
          50% of ethyl acetate in dichloromethane.
or   20% of isopropanol in dichloromethane.

Unfortunately the calculation of the strength of a solvent mixture is tedious and cannot be made without computer aid.[4]

9.3 SELECTIVITY PROPERTIES OF THE MOBILE PHASE

Eluent selectivity is the ability of different mobile phases to change the separation factor α of two or more compounds present in the sample. It has nothing to do with eluent strength ε^0 but is another means that allows a separation to be influenced. Adsorption chromatographic selectivity has two different aspects, localization and basicity.

Localization is a measure of the ability of the solvent molecules to interact with the adsorbent which is used as stationary phase.[5] As already mentioned in Section 9.1, the adsorptive sites of silica are its silanol groups. Molecules that can interact with these sites by means of their polar functional group, such as ethers, esters, alcohols, nitriles and amines, will prefer a specific position with respect to a nearby silanol group. Silica surrounded by such a localizing solvent or sample is covered with a well defined layer of molecules. In contrast to this, a nonlocalizing solvent or sample such as dichloromethane or benzene will interact with silica to a much weaker extent and the coverage is at random. Figure 9.5 clarifies these effects.

Of the solvents listed in Section 5.1, the upper half of the table from fluoroalkane to dichloroethane represents nonlocalizing solvents with the exception of diethyl ether. This solvent as well as the lower half of the table from triethylamine to water are the localizing eluents. Therefore the two solvents dichloromethane (nonlocalizing) and diethyl ether (localizing) differ in this important selectivity property although they have almost identical strength ($\varepsilon^0 = 0.30$ and 0.29, respectively).

---

[4] M.D. Palamareva and H.E. Palamarev, *J. Chromatogr.*, **477**, 235 (1989).
[5] L.R. Snyder, J.L. Glajch and J.J. Kirkland, *J. Chromatogr.*, **218**, 299 (1981).

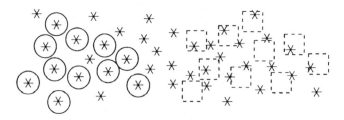

Figure 9.5 Localizing ○ and nonlocalizing ⬚ molecules and their interaction with an adsorbent. The silica surface is drawn in an overhead view and its adsorptive centres (silanol groups) are symbolized by asterisks.

Basicity is one of the axes of the solvent selectivity triangle of Figure 5.1. The most basic of the common HPLC solvents are the ethers.

9.4 CHOICE AND OPTIMIZATION OF THE MOBILE PHASE

The best mobile phase strength for a specific separation problem can be determined by thin-layer chromatography (TLC). It is represented by that solvent or solvent mixture which gives an R_F value of about 0.3 in the thin-layer chromatogram. TLC results are readily transferable to HPLC and k values can be predicted if the stationary phase in both methods is the same product, albeit with a difference in particle size.[6] A developing distance of 5 cm is sufficient for this test, so it can be carried out on small TLC plates in a very short time. Small sealable jam-jars make good developing tanks.

Substances zones can be detected by:

(a) UV light (for UV-absorbing compounds by use of layers with fluorescence indicators);
(b) iodine vapour;
(c) suitable spray reagents should the first two methods fail.

By this TLC test the correct strength (ε^0) can be found but the selectivity may need optimization as individual peaks may remain only partly resolved.

In order to obtain selectivity changes it is necessary to choose solvents which differ in their localization and basicity. In many cases the mobile phase consists of two solvents, A and B. The usual A solvent is hexane which has no strength, localization or basicity. However, heptane (also with $\varepsilon^0 = 0$) is to be preferred because hexane is neurotoxic. For B it is best to use either a nonlocalizing, a nonbasic localizing, or a

---

[6] W. Jost, H.E. Hauck and F. Eisenbeiss, *Kontakte Merck*, 3/84, 45 (1984); F. Geiss, *Fundamentals of Thin-Layer Chromatography*, Hüthig, Heidelberg, 1986, chapter X: 'Transfer of TLC Separations to Columns'; see also P. Renold, E. Madero and T. Maetzke, *J. Chromatogr. A*, **908**, 143 (2001).

basic localizing solvent.[7] For systematic selectivity tests and maximum changes in elution pattern, the separation should be tried with all these types of B solvents.

A typical *nonlocalizing B solvent* is dichloromethane.

Typical *nonbasic localizing B solvents* are acetonitrile and ethyl acetate. Acetonitrile is only slightly miscible with hexane; ethyl acetate has a high UV cutoff of 260 nm.

A typical *basic localizing B solvent* is *tert.* butylmethyl ether.

As the selectivity properties are not always the only ones to be considered in practice, other solvents may also be used. For ecological reasons, halogenated ones should be only used if really necessary. For the preparation of the various solvent mixtures, which should be of similar strength, Figure 9.4 is a help. If the sample includes acidic analytes, it may be necessary to add a small amount of acetic, trifluoroacetic or formic acid to the eluent; for basic analytes a possible additive is triethylamine.

Deactivators

Adsorption sites on silica (or alumina) do not all have the same level of activity. The result is that the column is very soon overloaded and hence the retention factor is influenced by the sample mass, even if low amounts are injected (see Section 2.7). Tailing also occurs.

The adsorbent sample capacity can only be fully utilized if the most highly adsorptive centres are consciously occupied, i.e. deactivated. This can be achieved by adding a small amount of a localizing substance to the mobile phase. The deactivator tends to occupy the centres with the greatest 'attraction', so that the sample molecules are confronted with a homogeneous adsorbent surface and approximately the same level of activity at each centre.

Water is the most important deactivator (moderator). It is highly polar and present in all solvents, albeit sometimes only in trace amounts. Its influence on chromatography is less pronounced with more polar mobile phases and on used silica columns (i.e. new ones are more ticklish). The alcohols methanol, ethanol or isopropanol are also possible deactivators. Although this is usually superfluous, activation control may be necessary if pure nonpolar A solvent is used or if the separation is very demanding.[8]

Gradient elution is possible in adsorption chromatography[9] (see Figure 9.7) if the polarity difference of the two solvents is not too large. It can be more difficult, thus less recommended, than in reversed-phase chromatography:

(a) the solvents could demix within the column, i.e. the more polar one could be adsorbed;

---

[7] J.J. Kirkland, J.L. Glajch and L.R. Snyder, *J. Chromatogr.*, **238**, 269 (1982).

[8] D.L. Saunders, *J. Chromatogr.*, **125**, 163 (1976).

[9] V.R. Meyer, *J. Chromatogr. A*, **768**, 315 (1997); P. Jandera, *J. Chromatogr. A*, **965**, 239 (2002).

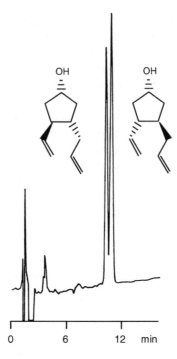

Figure 9.6 Separation of diastereomeric alcohols. Conditions: column, 25 cm × 3.2 mm i.d.; stationary phase, LiChrosorb SI 60, 5 µm; mobile phase, 1 ml min⁻¹ hexane–isopropanol (99 : 1); refractive index detector.

(b) their modifier content may differ;
(c) re-equilibration times may be long;
(d) it is advantageous not to start the gradient at 0% B but with a certain (perhaps very low) amount of the stronger solvent.

9.5 APPLICATIONS

Silica is an excellent stationary phase for the separation of isomers. The cover of this book shows the separation of petasol and isopetasol, two compounds found in *Petasites hybridus* (butterbur) which differ only in the position of one double bond. The mobile phase was pure diethyl ether. Figure 9.6 shows a separation of diastereomers.

As already mentioned, gradient elution cannot generally be recommended for the separation of complex mixtures on silica because the re-equilibration to the starting

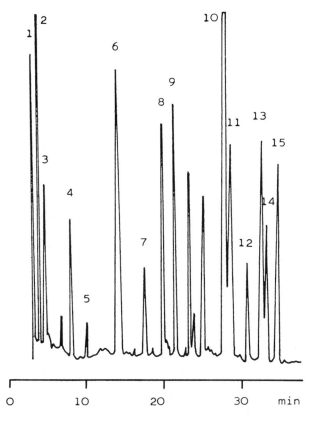

Figure 9.7 Separation of carotenoids from red pepper with gradient [reproduced with permission from L. Almela, J. M. López-Roca, M. E. Candela and M. D. Alcázar, *J. Chromatogr.*, **502**, 95 (1990)]. Conditions: sample, saponificated extract from red pepper fruit; column, 25 cm × 4.6 mm i.d.; stationary phase, Spherisorb 5 μm; mobile phase, 1 ml min⁻¹ petroleum ether–acetone, linear gradient from 5 to 25% acetone in 30 min; visible-range detector 460 nm. Peaks: 1 = β-carotene; 2 = cryptocapsin; 3 = cryptoflavin; 4 = β-cryptoxanthin; 5 = antheraxanthin; 6 = capsolutein; 7 = luteoxanthin; 8 = zeaxanthin; 9 = mutatoxanthin; 10 = capsanthin; 11 = capsanthin-5,6-epoxide; 12 = violaxanthin; 13 = capsorubin; 14 = capsorubin isomer; 15 = neoxanthin.

conditions may take a longer time than with bonded phases. Nevertheless, it is possible to use this technique very successfully as this is documented with Figure 9.7.

Figure 9.8 shows the separation of very polar compounds on silica. The carotenoid crocetin, which is the most nonpolar compound and is eluted first, and its mono- and diglycosyl esters are some of the products isolated from saffron extract. Strictly

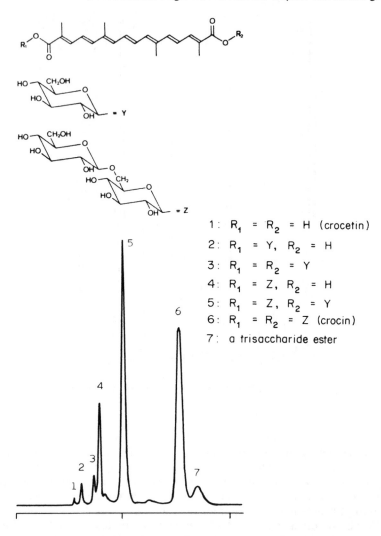

1: R_1 = R_2 = H (crocetin)
2: R_1 = Y, R_2 = H
3: R_1 = R_2 = Y
4: R_1 = Z, R_2 = H
5: R_1 = Z, R_2 = Y
6: R_1 = R_2 = Z (crocin)
7: a trisaccharide ester

Figure 9.8 Separation of purified saffron extract [reproduced with permission from H. Pfander and M. Rychener, *J. Chromatogr.*, **234**, 443 (1982)]. Conditions: column, 25 cm × 4.6 mm i.d.; stationary phase, LiChrosorb SI 60, 7 μm; mobile phase, 0.6 ml min^{-1} ethyl acetate–isopropanol–water (56:34:10); visible-range detector 440 nm.

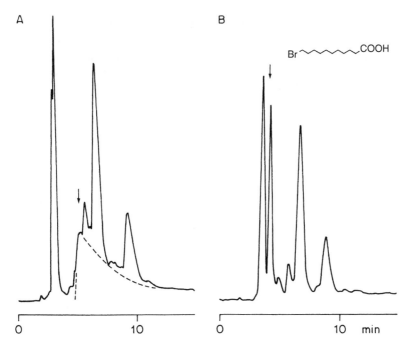

Figure 9.9 Separation of ω-bromoundecanoic acid [reproduced by permission of Marcel Dekker, Inc., from R. Schwarzenbach, *J. Liq. Chromatogr.*, **2**, 205 (1979)]. Conditions: sample, raw material from esterification; column, 25 cm × 3 mm i.d.; stationary phase, LiChrospher SI 100, 5 µm; mobile phase, 1 ml min^{-1} hexane-diethyl ether (2:3); UV detector. (A) Untreated silica; bromoundecanoic acid is eluted with extreme tailing and cannot be quantitatively determined. (B) Silica treated with citrate buffer (pH 2.8): bromoundecanoic acid appears as a distinct peak and the other components remain unaffected.

speaking, the method used here is no longer adsorption chromatography, but with a mobile phase of such extreme polarity a liquid–liquid partition system is created.[10]

The separation of acidic or basic products may cause difficulties in adsorption chromatography. Reversed-phase or ion-exchange chromatography would be more suitable but solubility problems may arise. One answer is to use buffered silica, as shown in Figure 9.9. The silica is treated either as bulk material or in the column with a buffer solution of a suitable pH (acidic buffer for acidic samples, basic buffer for

---

[10] See also: R.W. Schmid and C. Wolf, *Chromatographia*, **24**, 713 (1987).

bases). Compounds that were previously problematical can be eluted without tailing and without affecting the neutral components by chromatography using a nonpolar mobile phase (buffer salts may be flushed out if the solvent is too polar).[11]

Further examples of adsorption chromatography are shown in Figures 2.11, 2.23, 2.28, 8.1, 21.1, 21.2 and 23.4.

[11] R. Schwarzenbach, *J. Chromatogr.*, **334**, 35 (1985).

10   Reversed-Phase Chromatography

10.1   PRINCIPLE

Reversed-phase chromatography is the term used to describe the state in which the stationary phase is less polar than the mobile phase. Chemically bonded octadecyl-silane (ODS), an n-alkane with 18 carbon atoms, is the most frequently used stationary phase. C_8 and shorter alkyl chains and also cyclohexyl and phenyl groups provide other alternatives. Phenyl groups are more polar than alkyl groups.

Water is often described as the strongest elution medium for chromatography, but in fact this is only true for adsorption processes. Water may interact strongly with the active centres in silica and alumina, so that adsorption of sample molecules becomes highly restricted and they are rapidly eluted as a result. Exactly the opposite applies in reversed-phase systems: water cannot wet the nonpolar (hydrophobic = water repellent) alkyl groups and does not interact with them in any way. Hence it is the weakest mobile phase of all and gives the slowest sample elution rate. The greater the amount of water in the eluent, the longer is the retention time.

The chromatogram depicted in Figure 10.1 demonstrates this point. Benzene (2), chlorobenzene (3), o-dichlorobenzene (4) and iodobenzene (5) were eluted on a reversed-phase column by various methanol–water mixtures (1 being the refractive index peak of the solvent used for sample dissolution).

Sample compounds are better retained by the reversed-phase surface the less water soluble (i.e. the more nonpolar) they are. The retention decreases in the following order: aliphatics > induced dipoles (e.g. CCl_4) > permanent dipoles (e.g. $CHCl_3$) > weak Lewis bases[1] (ethers, aldehydes, ketones) > strong Lewis

---

[1] Lewis base = electron donor; Lewis acid = electron acceptor.

Practical High-Performance Liquid Chromatography, Fifth edition   Veronika R. Meyer
© 2010 John Wiley & Sons, Ltd

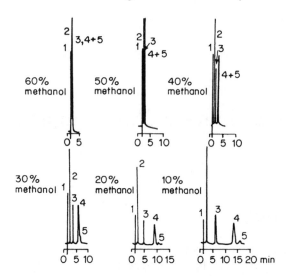

Figure 10.1 Effect of eluent composition in reversed-phase chromatography (reproduced with permission from N.A. Parris, *Instrumental Liquid Chromatography*, Elsevier, 1st ed., 1976, p. 157).

bases (amines) > weak Lewis acids (alcohols, phenols) > strong Lewis acids (carboxylic acids).

Also, the retention time increases as the number of carbon atoms increases, as shown in Figure 10.2 for the separation of dec-1-ene (1), undec-1-ene (2), dodec-1-ene (3), tridec-1-ene (4) and tetradec-1-ene (5) on an ODS column. As a general rule the retention increases with increasing contact area between sample molecule and stationary phase, i.e. with increasing number of water molecules which are released during the 'adsorption' of a compound.

Branched-chain compounds are eluted more rapidly than their corresponding normal isomers.

Yet the retention mechanisms on reversed-phase are complex and not easy to understand.[2]

10.2 MOBILE PHASES IN REVERSED-PHASE CHROMATOGRAPHY

The mobile phase generally consist of mixtures of water or aqueous buffer solutions with various water-miscible solvents, e.g.:

---

[2] A. Vailaya and C. Horváth, *J. Chromatogr. A*, **829**, 1 (1998).

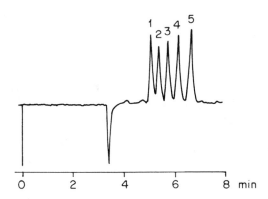

Figure 10.2 Reversed-phase separation of alkene homologues (reproduced with permission of Du Pont). Conditions: stationary phase, Zorbax ODS; mobile phase, 0.75 ml min^{-1} tetrahydrofuran–acetonitrile (10:90) (this is an example of nonaqueous reversed-phase chromatography!); IR detector, 3.4 μm.

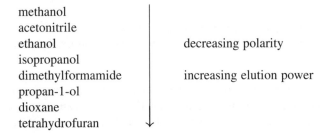

methanol
acetonitrile
ethanol   decreasing polarity
isopropanol
dimethylformamide   increasing elution power
propan-1-ol
dioxane
tetrahydrofuran

However, nonaqueous eluents are needed for the reversed-phase chromatography of highly nonpolar analytes.

As gradients are often involved in reversed-phase chromatography, the solvents used must be extremely pure[3] (see Figure 18.6). A chromatographic test for the purity of both water and organic eluents has been described,[4] consisting of a well defined series of rising and falling gradients which can be used to establish the purity levels of mobile phase components.

The mixtures of water with organic solvents often have a markedly higher *viscosity* than the pure compounds. The situation with the frequently used compositions of methanol–water, tetrahydrofuran–water and acetonitrile–water is presented in Figure 10.3. The viscosity maximum for methanol is obtained with 40% B in water and reaches 1.62 mPa s (25 °C), which is almost the threefold value of methanol or the

[3] S. Williams, *J. Chromatogr. A*, **1052**, 1 (2004).
[4] D.W. Bristol, *J. Chromatogr.*, **188**, 193 (1980).

Viscosity [mPa s]

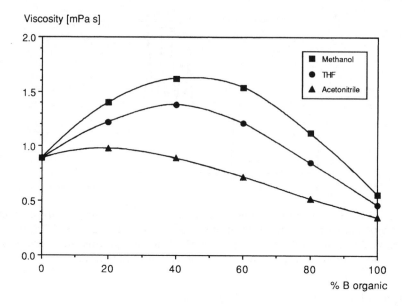

% B organic

Figure 10.3   Viscosity of mixtures of water and organic solvents at 25 °C (numeric values after J.W. Dolan and L.R. Snyder, *Troubleshooting LC Systems*, Humana, Clifton, 1989, p. 85).

double value of water. The pressure drop over the column is proportional to the viscosity; therefore it is not constant during a gradient run. High-viscosity maxima are also found with 80% acetic acid in water (2.7 mPa s at 20 °C) and 40% ethanol in water (2.8 mPa s at 20 °C).

Water[5] for HPLC is, unfortunately, not free as it must be specially purified. Ion exchange water is generally not of sufficient purity and double distillation stills may even increase the content of organics. It is best either to purchase HPLC grade water or to use water which is produced by a multiple-stage purification system. Sterile water is obtained by filtration through a 0.2 μm filter.

Methanol has the disadvantage of producing a relatively highly viscous mixture with water (see above), giving rise to much higher pressures than other mobile phases. If methanol–water mixtures are prepared manually then both components must be weighed or each one volumetrically determined separately. The considerable volume contraction that occurs produces a solution with a methanol content in excess of 50% by volume from an initial 500 ml of water topped up to 1000 ml with methanol. For the same reason the retention times are not identical if a water–methanol elutent is premixed manually or prepared by a high-pressure gradient system; in the latter case the

---

[5] S. Mabic, C. Regnault and J. Krol, *LC GC Eur.*, **18**, 410 (2005) or *LC GC North Am.*, **23**, 74 (2005).

total flow rate will be lower than requested. Methanol is favourable if buffer salts or ion-pair reagents need to be added to the eluent because such additives have better solubility in methanol than in acetonitrile or tetrahydrofuran.

Acetonitrile is very expensive, the highly pure 'far UV' quality being especially so. Viscosity properties cause no problems. Note the azeotrope formed with water as a result of regeneration by distillation: it boils at 76.7 °C and contains 84% acetonitrile.

Tetrahydrofuran can be a very interesting solvent with regard to separation selectivity. Its UV cutoff of 220 nm is relatively high. Column re-equilibration after a gradient run with THF is slower compared to gradients with methanol or acetonitrile. Once a bottle of HPLC grade THF is opened, peroxides are rapidly formed; they can react with analytes and are a safety risk.

In many cases it cannot be recommended to use mobile phases with less than 10% organic solvent in water.[6] Under such conditions many brush-type stationary phases with e.g. C_{18} alkyl chains are in an ill defined conformation and equilibration takes a long time. (Even for gradients it is usually not necessary to start them at 0% B, 10% B is a weak enough eluent.) In an environment with more than 10% organic solvent the chains are more or less expanded straightly whereas they collapse when too much water is present. The conformation is, however, again well defined in a totally aqueous eluent without organic modifier. Under such conditions the chains are totally folded but it takes time to bring them into the straight conformation when organic solvent is added.[7] Specially designed reversed phases which are suitable for eluents with high water content are commercially available from many manufacturers.

10.3   SOLVENT SELECTIVITY AND STRENGTH[8]

The selectivity triangle with numerous solvents is discussed in Section 5.2. Eluent selectivity in reversed-phase separations is directly related to this triangle because localization effects, important in adsorption chromatography, play no role. A look at Figure 5.1 shows that the three solvents methanol, acetonitrile and tetrahydrofuran are a good choice for the optimization of selectivity. Figure 10.4 represents a triangle with these three solvents only.

The usual A solvent in reversed-phase chromatography is water or aqueous buffer. The strength of binary mixtures is not a well defined function of %B but depends on analyte and stationary phase properties.[9] Nevertheless, it is possible to give numerical

---

[6] J.W. Dolan, *LC GC Int.*, **8**, 134 (1995) or *LC GC Mag.*, **13**, 96 (1995); M. Przybyciel and R.E. Majors, *LC GC Eur.*, **15**, 652 (2002) or *LC GC North Am.*, **20**, 516 (2002).

[7] An alternative explanation for the inferior performance of C_{18} phases in eluents with high water content is "pore dewetting".

[8] A. Klimek-Turek, T.H. Dzido and H. Engelhardt, *LC GC Eur.*, **21**, 33 (2008).

[9] This problem is discussed by S. Ahuja, *Selectivity and Detectability Optimizations in HPLC*, Wiley–Interscience, New York, 1989, Sections 6.3 and 6.4.

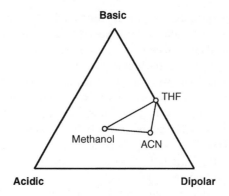

Figure 10.4   Solvent triangle for reversed-phase chromatography.

data which allow to obtain a good approximation of mobile phase composition when it is necessary to try more than one solvent in order to find the best selectivity. The following equation can be used:

$$\Phi_{B1} P'_{B1} = \Phi_{B2} P'_{B2}$$

where Φ is the volume fraction of a certain solvent and P' is its reversed-phase polarity. Different numbers have been proposed as P' values (with the exception of water whose polarity is always 0); the following data set is by Snyder *et al.*:[10]

P'_{water}	0
$P'_{methanol}$	3.0
$P'_{acetonitrile}$	3.1
$P'_{tetrahydrofuran}$	4.4

Alternatively, a nomogram can be used. It is based on numerous experimental data determined with small organic molecules.[11] As a matter of fact, each of these analytes would give a different nomogram; Figure 10.5 shows a mean graphical representation (which may be less valid for large molecules).

Problem 27

If a separation with 70% of methanol gives adequate retention but poor selectivity, which other solvent mixtures could be tried?

---

[10] L.R. Snyder, J.W. Dolan and J.R. Gant, *J. Chromatogr.*, **165**, 3 (1979).
[11] P.J. Schoenmakers, H.A.H. Billiet and L. de Galan, *J. Chromatogr.*, **185**, 179 (1979).

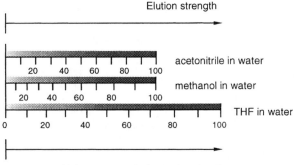

Figure 10.5 Solvent strength of binary mixtures for reversed-phase chromatography.

Solution

By equation:

$$\Phi_{ACN} = \frac{\Phi_{MeOH}P'_{MeOH}}{P'_{ACN}} = \frac{0.7 \times 3.0}{3.1} = 0.68 = 68\%$$

or:

$$\Phi_{THF} = \frac{0.7 \times 3.0}{4.4} = 0.48 = 48\%$$

By nomogram:

$$\Phi_{ACN} = 60\% \quad \text{or} \quad \Phi_{THF} = 45\%$$

It is obvious that both methods can only give a rough estimation. Figure 10.6 gives an example of how a separation pattern can change when different B solvents are used. The sample is a total extract of butterbur *(Petasites hybridus)*. The separation is clearly best with acetonitrile. Tetrahydrofuran gives broad and poorly resolved peaks. With methanol the peaks are tailed and the selectivity in the later part of the chromatogram is worse than with acetonitrile.

Small amounts of other organic solvents may be added as a third component for the fine-tuning of selectivity although such an approach complicates method development and cannot be generally recommended. Possible additives are dichloromethane (for chlorinated analytes) or *N,N*-dimethylformamide (for aromatic amines and *N*-heterocyclics). For ionic samples the adjustment of buffer pH can

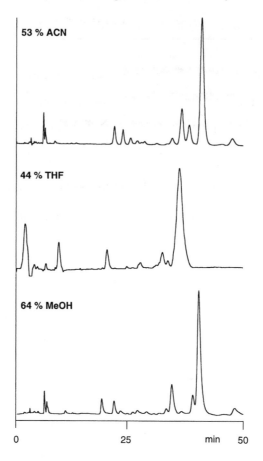

Figure 10.6 Separation selectivity with different B solvents (after S. Jordi, Department of Chemistry and Biochemistry, University of Bern). Conditions: sample, extract of *Petasites hybridus;* column, 25 cm × 4.6 mm i.d.; stationary phase, YMC Carotenoid C30, 3 µm; mobile phase, 0.9 ml min^{-1}, B solvent in water as indicated; UV detector, 230 nm.

be of the utmost importance. If basic compounds need to be separated it can be necessary to add a 'competing base' such as a trace of triethylamine which does interact strongly with silanol groups still accessible to the analytes. In this case it is, however, much better to use a stationary phase which is suited for basic samples (Section 10.4). Similarly, it has been recommended to use ammonium acetate as a 'universal additive' but this should not be necessary with modern types of stationary phase.

10.4 STATIONARY PHASES

Reversed-phase stationary phases are more or less hydrophobic, and the degree of this property is characterized by their hydrophobicity H. As a general rule, retention times are longer the more C atoms the bonded stationary phase contains. (The reason is that the volume taken up by the bonded nonpolar groups, i.e. that required by the actual stationary phase, is greater with long chains than it is with shorter chains; retention is directly proportional to the volume ratio between the stationary and mobile phases; see Section 2.3.) Figure 10.7 demonstrates this effect.

This means that retention is stronger the longer is the alkyl moiety (C_{18} is more retentive than C_8), the higher is the bonding density of the alkyl chains (in groups per nm^2 of the surface), the higher is the degree of end-capping, the thicker is the organic stationary phase (polymeric layers are more retentive than monomeric ones), or, as a summary, the higher is the carbon content of the material, determined by elemental analysis.

Strong retention means that the analysis takes a longer time or that a higher percentage of B solvent is necessary. A highly retentive stationary phase can be advantageous for polar analytes which are not retained enough on a low-carbon phase even with a high percentage of water. For nonpolar analytes it can be advantageous to use a stationary phase with low carbon content because retention times will be shorter and less organic solvent will be necessary for elution.

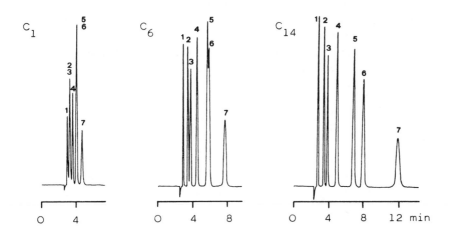

Figure 10.7 Effect of chain length on retention [reproduced with permission from G.E. Berendsen and L. de Galan, *J. Chromatogr.*, **196**, 21 (1980)]. Conditions: mobile phase, methanol–water (60 : 40). Peaks: 1 = acetone; 2 = *p*-methoxyphenol; 3 = phenol; 4 = *m*-cresol; 5 = 3,5-xylenol; 6 = anisole; 7 = *p*-phenylphenol.

Another reversed-phase feature is their *silanol activity*. As mentioned in Section 7.5, it is not possible to derivatize all the silanol groups of the silica surface for sterical reasons. The remaining silanols can be end-capped by a second derivatization reaction or shielded sterically. The various reversed phases which are commercially available differ markedly in their silanol activity because not all of them are end-capped. Although for most separations active silanols are unwanted (especially for the separation of basic compounds) a stationary phase with a distinct silanol activity can be advantageous for the separation of strongly hydrophilic (polar) analytes. Silanol activity has two features: Nonionized silanols (in acidic mobile phase) can act as hydrogen bond donors with a certain degree of acidity A. In contrast, ionized silanols (in neutral or basic mobile phase) possess cation exchange activity C.

Other properties of a reversed phase are its ability to act as a hydrogen bond acceptor, expressed as its basicity B (maybe due to sorbed water) and its steric resistance S towards bulky analytes.

Every reversed phase available on the marked has its own measurable set of characteristic parameters H, A, B, C, and S. They can be quantified[12], listed and compared.[13] These parameters help to identify similar or very different columns compared to a given one. The 'distance' F_S between two phases is calculated by the column matching factor:

$$F_S = \sqrt{\begin{array}{c} [12.5(H_2-H_1)]^2 + [100(S_2-S_1)]^2 + [30(A_2-A_1)]^2 \\ + [143(B_2-B_1)]^2 + [83(C_2-C_1)]^2 \end{array}}$$

Two phases 1 and 2 are similar (and can replace each other) if F_S is low, i.e. ≤ 3. They are different (and column 2 can be tried if column 1 does not give the needed separation) if F_S is high.

The C term is tabulated differently when the column is used with an acidic mobile phase (pH 2.8) or under neutral conditions (pH 7.0). If a sample does not contain acids and/or ionized bases, the B and C terms are irrelevant and are not included in the calculation of F_S.

Problem 28

Calculate the F_S values of Prontosil C18 AQ and of Platinum EPS C18 compared to the octadecyl phase Genesis AQ when used in an acidic mobile phase. Their selectivity parameters are as follows (tabulated by Snyder, Dolan and Carr, *J. Chromatogr. A*[13]):

---

[12] L.R. Snyder *et al.*, *J. Chromatogr. A*, **1057**, 49 (2004).

[13] L.R. Snyder, J.W. Dolan and P.W. Carr, *J. Chromatogr. A.*, **1060**, 77 (2004); the same, *Anal. Chem.*, **79**, 3254 (2007); L.R. Snyder and R.E. Majors, *LC GC Eur.*, **18**, 196 (2005) or *LC GC North Am.*, **22**, 1146 (2004).

	Genesis AQ	Prontosil C18 AQ	Platinum EPS C18
H	0.961	0.973	0.614
S	-0.037	0.011	-0.162
A	-0.155	-0.057	0.330
B	0.008	0.006	0.018
C (2.8)	0.061	0.125	0.720
C (7.0)	0.234	0.288	1.730

Solution

The eluent is acidic, therefore it is necessary to take the term $C(2.8)$ into consideration. Prontosil C18 AQ (column 1) compared to Genesis AQ (column 2):

$$F_S = \sqrt{\begin{array}{l}[12.5(0.973-0.961)]^2 + [100(0.011+0.037)]^2 + [30(-0.057+0.155)]^2 \\ + [143(0.006-0.008)]^2 + [83(0.125-0.061)]^2\end{array}}$$

$$= 7.7$$

Analogously, the F_S value of Platinum EPS C18 compared to Genesis AQ is 58.

The Prontosil column is very similar to, although not identical with, Genesis. It could be used as an alternative if the latter one is not available. On the other hand, the Platinum is very different and cannot be used as a substitute. However, it could be interesting to use it in order to check if all peaks of a mixture are separated in the course of selectivity tests during validation.

The results of these calculations are confirmed by the chromatograms shown in Figure 10.8.

F_S is a measure of the distance of two different stationary phases in the five-dimensional space spanned by H, S, A, B and C. Such a space is unimaginable by humans. Our conceptions are limited to two or three dimensions. Figure 10.9 shows the distribution of some C_8 and C_{18} phases within the two dimensions 'hydrophobicity' and 'silanol activity' (the latter property is not separated into the above-mentioned parameters A and C). The C_8 materials are less hydrophobic than the C_{18} ones, therefore they are located in the left half of the graph whereas the octadecyl phases are mainly located in the right half. 'Modern' phases with excellent bonding chemistry on low-metal silicas are in the lower part of the graph and 'old' phases in the upper part. These 'old' phases are less suited for basic analytes but they can offer interesting selectivities for special separation problems. Such two-dimensional representations may be helpful but the concept of the column matching factor F_S is more comprehensive.

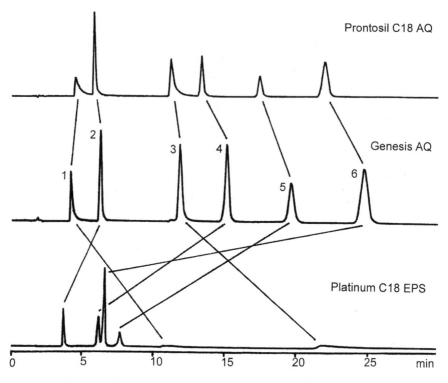

Figure 10.8 Separation of hydrophilic compounds on three different octadecyl columns [redrawn with permission from M.R. Euerby and P. Petersson, *J. Chromatogr. A*, **994**, 13 (2003)]. Stationary phases: Prontosil C18 AQ (similar, $F_S = 7.7$), Genesis AQ (reference) and Platinum EPS C18 (different, $F_S = 58$). Conditions: columns, 15 cm × 4.6 mm i.d.; mobile phase, 1 ml min^{-1} potassium dihydrogen phosphate pH 2.7 in water-methanol (96.7 : 3.3); temperature, 60 °C; UV detector, 210 nm. Peaks: 1 = nicotine; 2 = benzylamine; 3 = procainamide; 4 = terbutaline; 5 = salbutamol; 6 = phenol.

For the separation of basic analytes it is recommendable to use a stationary phase which is specially designed for such samples.[14] Otherwise severe tailing may occur (which perhaps can be suppressed by additives to the mobile phase, but this approach is less elegant). An example can be found in Figure 10.10.

The equilibrium between mobile and stationary phase is rapidly established in reversed-phase chromatography, therefore gradients can easily be performed. Re-equilibration times need to be determined empirically but in many cases five column volumes of the new eluent are sufficient to obtain reproducible results.

---
[14] D.V. McCalley, *J. Sep. Sci.*, **26**, 187 (2003).

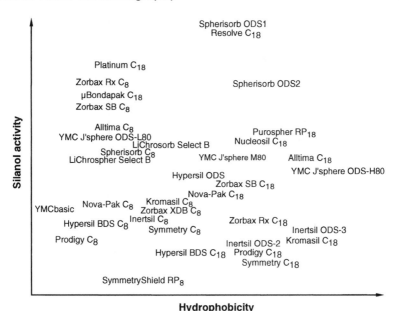

Figure 10.9 Selectivity properties of C_8 and C_{18} phases (redrawn with permission of Waters).

10.5 METHOD DEVELOPMENT IN REVERSED-PHASE CHROMATOGRAPHY[15]

In reversed-phase separations, the first step in method development is usually a solvent gradient from 10 to 100% solvent B. This approach is explained in Section 18.2. What follows here are proposals with a recommended sequence of changes of the various parameters. It is, e.g., more convenient to try another B solvent before changing the stationary phase.

Proposal for nonionic samples[16]

Use a C_8 or C_{18} stationary phase with unbuffered water-acetonitrile at ca. 40 °C if temperature adjustment is possible, otherwise at ambient temperature.

---

[15] U.D. Neue *et al.*, Method Development in RP Chromatography, in: I.D. Wilson, ed., *Bioanalytical Separations*, Elsevier, Amsterdam, 2003, pp. 185–214.

[16] L.R. Snyder, J.J. Kirkland and J.L. Glajch, *Practical HPLC Method Development*, Wiley-Interscience, 2nd ed., 1997, p. 253.

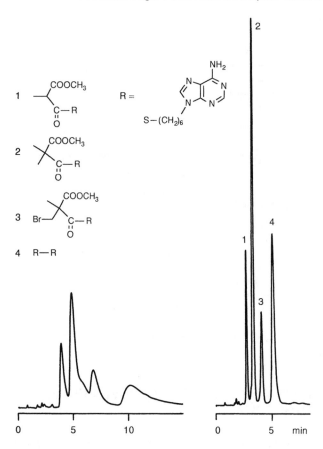

Figure 10.10 Stationary phase for basic compounds. Conditions: sample, adenine derivatives; columns, 25 cm × 4 mm i.d.; stationary phase left, LiChroshper 100 RP-18, 5 μm (with high separation performance and good peak symmetry for neutral compounds); stationary phase right, LiChrospher 60 RP-Select B, 5 μm; mobile phase, 1.5 ml min^{-1} water–methanol (25:75); UV detector, 260 nm.

1. Adjust %B or gradient range for retention factors between 1 and 10 (or 1 and 20 for difficult separations). If the separation is inadequate, adjust selectivity in the following order:
2. Change the organic B solvent.
3. Use a mixture of organic B solvents.
4. Change the stationary phase (preferably to a type which has markedly different properties, see Section 10.4). It is probably necessary to start at step 1 again.
5. Change the temperature.
6. Optimize the physical parameters such as column dimensions, particle size or flow rate.

Proposal for Ionic Samples[17]

Use C_8 or C_{18} stationary phase which is suitable for basic analytes with buffer pH 2.5-methanol at 40 °C if possible.

1. Adjust %B or gradient range. If the separation is inadequate:
2. (a) change pH or
   (b) use ion-pair chromatography.
3. Adjust %B.
4. Change the organic B solvent.
5. (a) change pH or
   (b) change pH and ion-pair reagent.
6. Change temperature.
7. Change the stationary phase to phenyl or cyano.
8. Optimize the physical parameters.

If the separation is still inadequate after all these proposals have been tried it is necessary to use another method (ion exchange, adsorption, size-exclusion chromatography).

Basic analytes may cause problems, as already mentioned.[18] Figure 10.11 explains the reason. If the base is charged as well as the still present and accessible silanol groups (neutral conditions), a mixed retention mechanism is the result with both hydrophobic and ionic interactions, thus leading to tailing and unstable separation

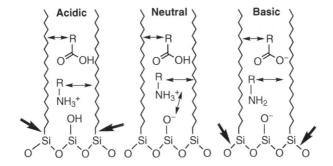

Figure 10.11 The possible interactions between analyte and reversed stationary phase. Horizontal double-headed arrows indicate the hydrophilic interactions whereas the inclined double-headed arrow shows the ionic interaction. The thick arrows mark the sites of attack: under strongly acidic conditions the bonded phase becomes hydrolyzed whereas this happens to the silica backbone at high pH.

---

[17] As ref. 16, but p. 315. Another proposal was published by J.J. Kirkland, *LC GC Mag.*, **14**, 486 (1996).

[18] D.V. McCalley, *LC GC Eur.*, **12**, 638 (1999) or *LC GC Mag.*, **17**, 440 (1999).

conditions. The mixed mechanism can be suppressed in some cases by using a buffered mobile phase. If this does not help it is recommended to work at acidic conditions because many silica-type stationary phases are not stable at high pH. Most promising are the specially designed stationary phases for basic analytes (Figure 10.10); they have a low content of silanol groups or these groups are shielded well, they are based on highly pure silica with a very low concentration of heavy metal cations, or they contain a polar embedded group. Acidic analytes are not critical because ionic interactions are not possible.

As a general rule one must not forget that ionic analytes are hydrophilic, therefore they are eluted early in the chromatogram. At low pH these are the bases, at high pH the acids.

10.6 APPLICATIONS

Aqueous solutions such as those of biological origin, pharmaceutical formulations, drinks, etc., are common samples for analysis. As water is the weakest eluent, aqueous

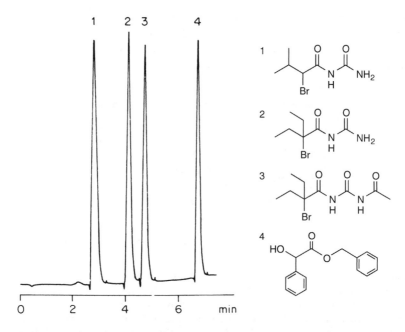

Figure 10.12 Chromatogram of tranquillizers (reproduced with permission of Hewlett-Packard). Conditions: column, 25 cm × 4 mm i.d.; stationary phase, LiChrosorb RP-8, 10 μm; mobile phase, gradient elution, 30% acetonitrile in water up to 90% acetonitrile in water, 16 min; UV detector, 254 nm. Peaks: 1 = bromural; 2 = carbromal; 3 = acetocarbromal; 4 = benzyl mandelate.

solutions may be injected directly without any special preliminary treatment (although filtration or centrifugation is strongly recommended). An example of the injection of 10 µl of drinks for caffeine determination was depicted in Figure 1.2. Figure 10.12 provides a further example. Tranquillizer tablets were dissolved in water, filtered and then injected. All other methods for the analysis of this type of drug take far longer.

Figure 10.13 shows the excellent performance of reversed-phase HPLC in biotechnological research. It represents the separation of the tryptic hydrolysate of the normal form and of a mutant of tissue-type plasminogen activator. This protein is built up of 527 amino acids and has a mass of approximately 67 000 Da. The mutant differs in a single amino acid, which leads to a deviating retention time of this specific fragment.

However, reversed-phase chromatography is not just suitable for polar compounds alone. For example, the 17 polycyclic aromatic hydrocarbons which

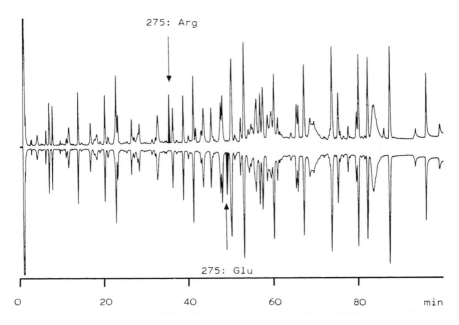

Figure 10.13 Separation of the tryptic hydrolysate of tissue-type plasminogen activator [reproduced with permission from R.L. Garnick, N.J. Solli and P.A. Papa, *Anal. Chem.*, **60**, 2546 (1988)]. Conditions: stationary phase, Nova Pak C18, 5 µm; mobile phase: 1 ml min^{-1} 50 mM sodium phosphate pH 2.8–acetonitrile, step gradient; UV detector 210 nm. Top: the normal protein with arginine at position 275; below: the mutant with glutamic acid at position 275.

are referred to by environmental analysts as 'priority pollutants' can be completely separated within 15 min, as shown in Figure 10.14. Figure 10.15 shows the separation of the complex mixture of aromatic gasoline components with a highly resolved elution pattern usually only expected from capillary gas chromatography.

Further examples of reversed-phase chromatography are given in Figures 1.1, 1.2, 2.12, 2.20, 2.21, 2.24, 2.27, 4.8, 6.5, 6.7, 6.11, 6.13, 6.14, 6.19, 7.5, 8.2, 15.12b, 18.12, 18.13, 18.15, 19.9, 21.5, 22.5, 23.1, 23.3, 23.5 and 23.10.

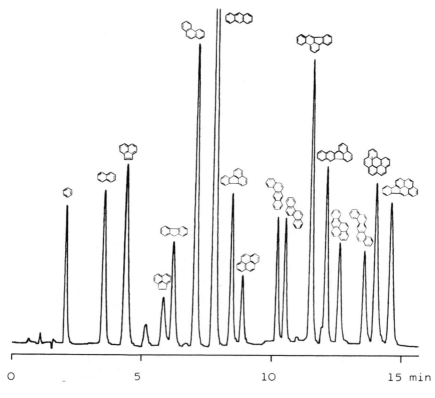

Figure 10.14 Separation of polycyclic aromatic hydrocarbons (reproduced with permission of Separations Group). Conditons: column, 15 cm × 4.6 mm, i.d.; stationary phase, Vydac TP C_{18}, 5 μm; mobile phase, 1.5 ml min^{-1} water–acetonitrile, linear gradient from 50 to 100% acetonitrile between 3 and 10 min; UV detector, 254 nm.

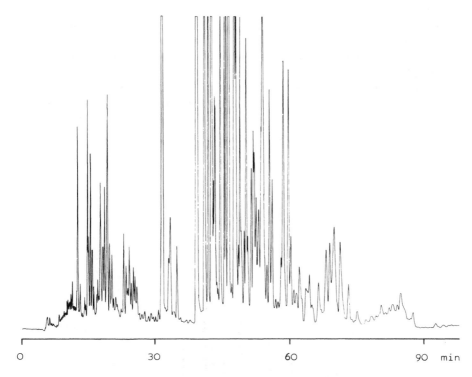

Figure 10.15 Separation of gasoline (Beckman). Conditons: sample, 5 µl of gasoline diluted with 10 ml of acetonitrile; columns, four 25 cm × 4.6 mm i.d.; stationary phase, Ultrasphere-ODS, 5 µm; mobile phase, 1 ml min^{-1} water–acetonitrile, 60% acetonitrile for 30 min then 80% acetonitrile; temperature, 42 °C; UV detector, 254 nm.

10.7 HYDROPHOBIC INTERACTION CHROMATOGRAPHY[19]

The interaction between the sample and the stationary phase is so strong in reversed-phase chromatography that an aqueous eluent is too weak without the addition of an organic solvent. However, organic mobile phases are not allowed in certain cases of protein separation because of the risk of denaturation and subsequent loss of biological activity.

---

[19] K.O. Eriksson, in *Protein Purification*, J.C. Janson and L. Ryden, eds., Wiley-Liss, New York, 2nd ed., 1998, pp. 283–309.

Pure aqueous mobile phases are only suitable for separations on weakly hydrophobic stationary phases; hence materials containing one tenth to one hundredth of the carbon load of classical reversed phases have been developed. This is achieved by low coverages of short-chain groups such as butyl or phenyl. Proteins can then be retained when the eluent has a relatively high salt content (e.g. 1 M or more) and eluted when the salt content drops. This mild method of protein separation, which is a variant of reversed-phase chromatography, is known as hydrophobic interaction chromatography (HIC; Figure 10.16).

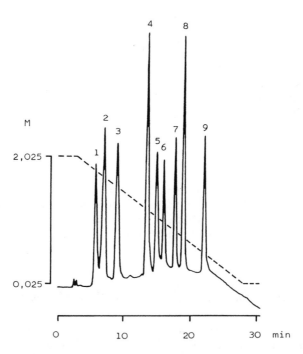

Figure 10.16 Separation of a peptide mixture by hydrophobic interaction chromatography (reproduced with permission from A.J. Alpert, *J. Chromatogr.*, **444**, 269 (1988)]. Conditions: column, 20 cm × 4.6 mm i.d.; stationary phase, polypropyl aspartamide 5 μm; mobile phase, 1 ml min^{-1} of 2 M ammonium sulfate with 0.025 M sodium phosphate, pH 6.5–0.025 M sodium phosphate pH 6.5, linear gradient as indicated; detector, UV 220 nm. 1 = substance P(1–9); 2 = [Arg^8]-vasopressin; 3 = oxytocin; 4 = substance P, free acid; 5 = [Try^8]-substance P; 6 = substance P; 7 = [Tyr^{11}]-somatostatin; 8 = somatostatin; 9 = [Tyr^1]-somatostatin.

TABLE 10.1   Optimization of hydrophobic interaction chromatographic separations[20]

Observation	Recommendation
Protein poorly retained	Increase salt concentration.
	Change salt to one that increases the surface tension.
	Change pH towards the isoelectric point of the protein.
	Change stationary phase to one with smaller hydrocarbonaceous moieties and/or lower ligate density, i.e. reduce phase ratio.
Insufficient selectivity	Change salt.
	Use additives that selectively affect protein, e.g. inhibitors or allosteric effectors.
Protein is not eluted	Decrease salt concentration if it is 'high'; increase if it is 'low'.
	Change salt to one that decreases the surface tension.
	Add amine or another silanophile to the mobile phase.
	Change pH away from the isoelectric point of the protein.
	Change stationary phase to one with longer ligate and/or greater chain density, i.e. increase phase ratio.

The composition of the mobile phase markedly influences the retention behaviour, as can be seen from Table 10.1.

A further example is shown in Figure 7.4.

---

[20] Reproduced with permission from W.R. Melander, D. Corradini and C. Horváth, *J. Chromatogr.*, **317**, 67 (1985).

11 Chromatography with Chemically Bonded Phases

11.1 INTRODUCTION

Section 7.5 described the chemical derivatization of silica. The stationary phases obtained may have any polarity. Because of their importance the nonpolar products have already been described in Chapter 10, so this chapter is devoted to some high and medium polarity phases, the discussion being limited to the most important types.

The separations obtained with these stationary phases are often similar to those on silica, although the selectivity may be different (depending on type), a fact the analyst should take advantage of. These phases have the following advantage over silica:

(a) They can be used for both normal- and reversed-phase chromatography.
(b) The mobile phase needs no deactivator.
(c) The solvent can be changed without regard to the eluotropic series and the subsequent equilibrium time is short.
(d) Gradients are possible.

11.2 PROPERTIES OF SOME STATIONARY PHASES

Diol

The OH groups in this phase make the product comparable to silica. It is of interest for compounds with which it can form hydrogen bonds and is particularly suitable for tetracyclines, steroids, organic acids and biopolymers (proteins), amongst other

Practical High-Performance Liquid Chromatography, Fifth edition   Veronika R. Meyer
© 2010 John Wiley & Sons, Ltd

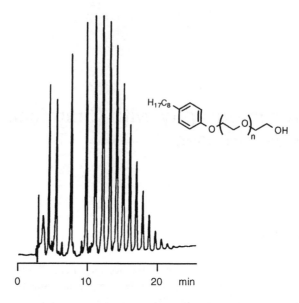

Figure 11.1 Separation of a nonionic surfactant on a diol phase (reproduced with permission of Supelco). Conditions: sample, Triton X-100 with average $n = 10$; column, 25 cm × 4.6 mm i.d.; stationary phase, Supelcosil LC-Diol, 5 μm; mobile phase: 1 ml min^{-1}, nonlinear gradient from 15.5% dichloromethane + 3% methanol in hexane to 40% dichloromethane + 10% methanol in hexane in 35 min; temperature, 35 °C; UV detector, 280 nm.

analytes. Figure 11.1 presents the separation of a nonionic, polar surfactant which consists of a mixture of polyethoxylated alkylphenols. The oligomers differ in the number of ethoxy units; Triton X-100 has an average chain length of ca. $n = 10$.

Nitrile (Cyano)

As it has a lower polarity, this phase often gives separations in a similar way to silica, but the k values are smaller for the same mobile phase.[1] It can be used as a reversed phase as well; in this mode it is less retentive than C_8 or C_{18} phases.[2] It is particularly selective towards components with double bonds and towards tricyclic antidepressants. An application from this latter field is presented in Figure 11.2.

[1] An example of this was given by H. Pfander, H. Schurtenberger and V. R. Meyer, *Chimia*, **34**, 179 (1980).
[2] D.H. Marchand *et al.*, *J.Chromatogr. A*, **1062**, 57 (2005).

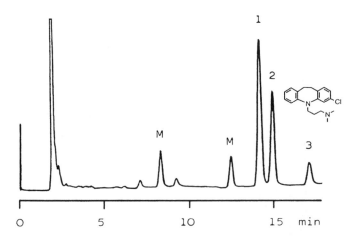

Figure 11.2 Separation of tricyclic antidepressants on a nitrile phase [reproduced with permission from G.L. Lensmeyer, D.A. Wiebe and B.A. Darcey, *J. Chromatogr. Sci.*, **29**, 444 (1991)]. Conditions: sample, serum extract from a patient who received clomipramine; column, 25 cm × 4.6 mm i.d.; stationary phase, Zorbax Cyanopropyl, 5–6 μm; mobile phase, 1.2 ml min⁻¹ water–acetonitrile–acetic acid-*n*-butylamine (600 : 400 : 2.5 : 1.5); temperature, 45 °C; UV detector, 254 nm. Peaks: M = metabolites of clomipramine; 1 = trimipramine (internal standard); 2 = des-methylclomipramine; 3 = clomipramine.

Amino

Sugar and glycoside analysis (Figure 11.3) is the classical use for amino phases. The amino function can act as both a proton acceptor and a proton donor in hydrogen bonding, being both Brønsted base and acid. Organic and inorganic anions (acetate, acrylate, glycolate, formate, nitrite, bromide, nitrate, iodate, dichloroacetate) can be separated on an amino phase with a phosphate buffer of ca. pH 3.[3] Figure 11.4 shows the analysis of cabbage-lettuce containing about 1000 ppm of nitrate and 300 ppm of bromide (from soil treatment with methyl bromide).

---

[3] U. Leuenberger, R. Gauch, K. Rieder and E. Baumgartner, *J. Chromatogr.*, **202**, 461 (1980); H.J. Cortes, *J. Chromatogr.*, **234**, 517 (1982).

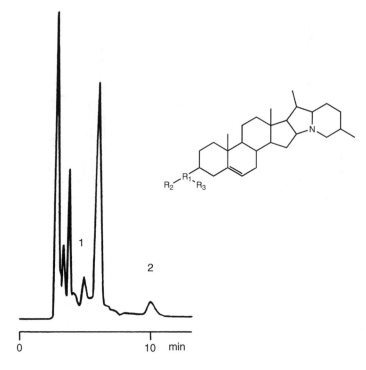

Figure 11.3 Separation of potato tissue glycoalkaloids on an amino phase [reproduced with permission from K. Kobayashi, A.D. Powell, M. Toyoda and Y. Saito, *J. Chromatogr.*, **462**, 357 (1989)]. Conditions: sample, extract from young potato plantlets after solid-phase extraction pretreatment; column, 30 cm × 3.9 mm i.d.; stationary phase, µBondapak NH_2; mobile phase, ethanol-acetonitrile-potassium dihydrogenphosphate (3:2:1); UV detector, 205 nm. Peaks: 1 = α-chaconine; with $R_1 = \beta$-D-glucose, $R_2 = R_3 = \alpha$-L-rhamnose; 2 = α-solanine with $R_1 = \beta$-D-galactose, $R_2 = \beta$-D-glucose, $R_3 = \alpha$-L-rhamnose.

NH_2 is readily oxidized, so peroxides (in diethyl ether, dioxane or tetrahydrofuran) should be strictly avoided. Ketones and aldehydes react to give Schiff bases.[4] Amino silicas are less stable towards hydrolysis than other bonded phases and it is recommended to use a pre-column (scavenger column) to saturate the mobile phase.[5]

---

[4]

$$-NH_2 + O{=}C\diagup^{R}_{\diagdown R' \text{ or } H} \quad \rightarrow \quad -N{=}C\diagup^{R}_{\diagdown R',H} \quad + \quad H_2O$$

For reactivation of an amino column that had become exhausted as a result of reaction with actone, see D. Karlesky, D.C. Shelly and I. Warner, *Anal. Chem.*, **53**, 2146 (1981).

[5] B. Porsch and J. Krátká, *J. Chromatogr.*, **543** 1 (1991).

Figure 11.4 Separation of nitrate and bromide in cabbage-lettuce extract on an amino phase. Conditons: sample, 20 μl of extract after homogenization, protein precipitation and filtration; column, 25 cm × 3.2 mm i.d.; stationary phase, LiChrosorb NH$_2$, 5 μm; mobile phase, 1 ml min^{-1} 1% KH$_2$PO$_4$ in water adjusted to pH 3 with H$_3$PO$_4$; UV detector, 210 nm. Peaks: 1 = nitrile (from nitrate converted during sample treatment); 2 = bromide; 3 = nitrate.

The amino group acts as a weak anion exchanger in aqueous acidic solutions (taking the form of primary ammonium ion, RNH$_3^+$); hence retention in an aqueous mobile phase is a function of pH. 0.1 M ammonia solution can be used to regenerate the RNH$_2$ form.

Nitro

This phase is selective for aromatics and its effectiveness is demonstrated in Figure 11.5. Normal coal-tar pitch (residue from commercial high-temperature coal tar distillation) was analysed and around 90 components could be distinguished, including polycyclic aromatic hydrocarbons and related heterocyclic and oligomer systems. Twenty-four of these were clearly identified by their characteristic UV spectra and comparison with known reference spectra.

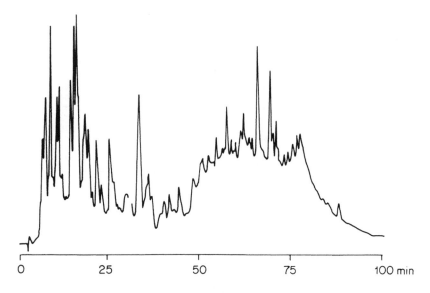

Figure 11.5 Separation of normal coal-tar pitch on a nitro phase [reproduced with permission from G.P. Blümer, R. Thoms and M. Zander, *Erdöl Kohle, Erdgas, Petrochem.*, **31**, 197 (1978)]. Conditions: columns, two 20 cm × 4 mm i.d. plus a guard column; stationary phase, Nucleosil NO₂, 5 μm; mobile phase, hexane-chloroform, gradient elution; UV detector 300 nm; sensitivity, 1 a.u.f.s. and, after 30 min, 0.4 a.u.f.s.

11.3 HYDROPHILIC INTERACTION CHROMATOGRAPHY[6]

Hydrophilic interaction chromatography (HILIC) offers a possilbility to separate very polar analytes in a normal-phase mode. The stationary phase proper is water which is adsorbed on a suitable packing if the eluent is aqueous. Nonpolar analytes are eluted before the polar ones as in classical normal-phase chromatography although the mobile phase contains water. The strongly eluting component of the eluent is water, buffer solution or an alcohol. The weak component can be acetonitrile whose content decreases during the gradient separation (Figure 11.6).

There is a great selection of stationary phases for HILIC:[7] silica, derivatized silica with diol, nitrile, amino or amide groups, silica with a hydrophilic polymeric coating, and various polymers. They all were especially developed for HILIC whereas

---

[6] P. Hemström and K. Irgum, *J. Sep. Sci.*, **29**, 1784 (2006).
[7] P. Jandera, *J. Sep. Sci.*, **31**, 1421 (2008).

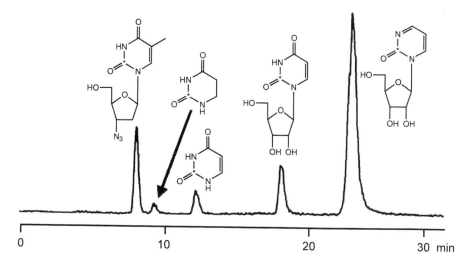

Figure 11.6 Separation of zebularine (last peak) and metabolites with HILIC [reproduced with permission from J.H. Beumer et al., *J. Chromatogr. B*, **831**, 147 (2006)]. Conditions: sample, mouse plasma after sample preparation; column, 25 cm × 4.6 mm i.d.; stationary phase, Zorbax NH_2, 5 µm; mobile phase, 0.6 ml min^{-1} isopropanol (gradient 12–49%)–1 M ammonium formiate (0.24–1.0%)–acetonitrile (88–50%); radioactivity detector. The analytes are labelled with radioactive ^{14}C at the positions with asterisks. Elution order: zidovudine (internal standard), dihydrouracil, uracil, uridine (these three compounds are metabolites), and zebularine (drug).

'conventional' silicas or diol phases usually give poor results when HILIC separations are tried on them.

HILIC is suited for very polar analytes[8] such as glycosylated molecules which are not retained on reversed phases and which cannot be separated on classical adsorbents because they are not eluted by the nonpolar mobile phase.

---

[8] Y. Hsieh, *J. Sep. Sci.*, **31**, 1481 (2008).

12 Ion-Exchange Chromatography

12.1 INTRODUCTION

Ion-exchange chromatography has been used to separate amino acids since 1956.[1] These early separations are comparable to classical column chromatography on silica or alumina as regards both expenditure and output. The resins or gels used were relatively coarse, pressures were low and the chromatography of complex biological mixtures could take several days. Hence, the biochemistry sector felt an urgent need for instrumental analysis accompanied by automation and optimization techniques, the eventual outcome being modern HPLC.

A large number of stationary phases are now pressure-resistant and capable of high performance in ion-exchange chromatography also, so the scope of HPLC can be extended into this field: complex mixtures can be separated over a short period of time.

12.2 PRINCIPLE

The principle of ion-exchange chromatography is not unlike that of adsorption chromatography. In the latter case, the adsorbent bears 'active sites' which interact with molecules in their vicinity to a more or less specifically defined extent. Sample and solvent molecules compete with each other for adsorption.

A stationary phase capable of ion exchange, on the other hand, has electric charges on its surface. Ionic groups such as SO_3^{2-}, COO^-, NH_3^+ or NR_3^+ are incorporated in the resin or gel. Charges are neutralized by mobile counter ions. The mobile phase contains ions, and ionic sample molecules compete with these for a place on the surface of the stationary phase.

---

[1] S. Moore, D.H. Sparkman and W.H. Stein, *Anal. Chem.*, **30**, 1185 (1958).

Practical High-Performance Liquid Chromatography, Fifth edition   Veronika R. Meyer
© 2010 John Wiley & Sons, Ltd

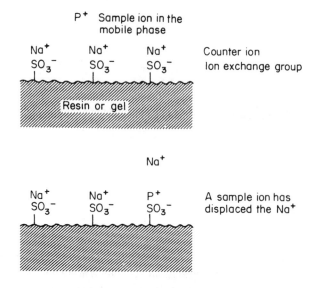

Figure 12.1   Cation exchanger.

Figure 12.1 represents a cation exchanger, as it forms a bond with cations. A resin with SO_3^- groups is a strong cation exchanger and a COO^- resin is a weak cation exchanger. An anion exchanger contains NR_3^+ (strong) or NR_2H^+ or NH_3^+ (weak) groups. It forms a bond with negatively charged anions.

Under these circumstances, how can competition between ions of the mobile phase and those of the sample, which is the essence of chromatographic separation, be stimulated? The analyst must ensure optimum conditions by carefully selecting:

(a) the type of ion exchanger,
(b) the pH of the mobile phase,
(c) the ionic strength (concentration) of the mobile phase and
(d) the type of counter ions in the mobile phase,

and varying any or all of them if necessary. This optimization process is often empirical, although certain rules have to be obeyed, of course. The elution order is often difficult to predict as so many factors are involved.

12.3   PROPERTIES OF ION EXCHANGERS

With ion exchangers, a large number of abbreviations are in use; some more common ones are listed in Table 12.1. Note that the first four abbreviations denote the type of exchanger where the other ones are related to its functional group.

TABLE 12.1 Some abbreviations used with ion exchangers

Abbreviation	Meaning		Type
SAX	Strong anion exchanger		
WAX	Weak anion exchanger		
SCX	Strong cation exchanger		
WCX	Weak cation exchanger		
AE	Aminoethyl	$-CH_2-CH_2-NH_3{}^+$	WAX
CM	Carboxymethyl	$-CH_2-COO^-$	WCX
DEA	Diethylamine	$-NH(CH_2-CH_3)_2{}^+$	WAX
DEAE	Diethylaminoethyl	$-CH_2-CH_2-NH(CH_2-CH_3)_2{}^+$	WAX
DMAE	Dimethylaminoethanol	$-O-CH_2-CH_2-NH(CH_3)_2{}^+$	WAX
PEI	Polyethyleneimine	$-(NH-CH_2-CH_2-)_n-NH_3{}^+$	WAX
QA	Quaternary amine	$-NR_3{}^+$ (R e.g.CH_3)	SAX
QAE	Quaternary aminoethyl	$-CH_2-CH_2-N(CH_2-CH_3)_3{}^+$	SAX
SA	Sulfonic acid	$-SO_3{}^-$	SCX

All these functional groups can be bound on organic resins such as styrene–divinylbenzene or on silica (however, note the limited chemical stability of silica-based bonded phases as described in Section 7.5). The ligand density, which is identical with the maximum ion exchange capacity, is ca. 3 meq (milliequivalents) per gram on S-DVB and ca. 1 meq per gram on silica. However, silica has a higher density, therefore the capacities per volume unit of the HPLC column are similar for both types.

Strong anion and cation exchangers are charged over the full pH range which is used in HPLC, therefore their capacity cannot be altered by changing pH. They are used for the separation of weak acids or bases and the driving force of the separation is the charge of the analytes which are not fully ionized, depending on pH. However, if the sample molecules are strong acids or bases, then their retention behaviour is scarcely affected by changes in the pH of the mobile phase as they are fully charged over a very wide range. Therefore it makes no sense to try a separation of strong acids or bases on strong ion exchangers. At extreme pH values, strong cation exchangers in the H^+ form (pH < 2) and strong anion exchangers in the OH^- form (pH > 10) may catalyse numerous reactions which may alter the sample as a result (ester and peptide hydrolysis, disproportionation of aldehydes).

The capacity, i.e. the retention behaviour, of weak exchangers is influenced by the pH of the mobile phase. A weak cation exchanger is not dissociated if the pH is markedly lower (ca. 2 pH units) than its pK_a value (Figure 12.2a). At such a pH the exchanger is in its hydrogen form with the protons bonded too strongly to change places with sample cations; its capacity is near zero. However, this cation exchanger is fully dissociated if the pH rises markedly over its pK_a value and its ionic groups are

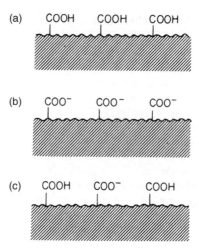

Figure 12.2 (a) Undissociated cation exchanger; (b) dissociated cation exchanger; (c) partly dissociated cation exchanger.

able to interact with the sample molecules at maximum capacity (Figure 12.2b). At a pH close to the pK_a value, the weak cation exchanger is only partly dissociated (Figure 12.2c). Analogous principles are valid for weak anion exchangers but their capacity is high at low pH and low at high pH. The ideal behaviour of weak exchangers as a function of pH is shown in Figure 12.3. As a typical example, pK_a values of 4.2 and 9.0 were chosen.

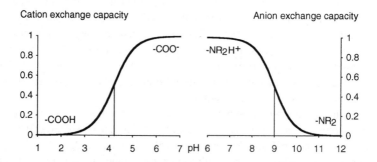

Figure 12.3 Exchange capacity of a weak cation exchanger with pK_a 4.2 (left) and of a weak anion exchanger with pK_a 9.0 (right).

12.4   INFLUENCE OF THE MOBILE PHASE

Several equilibria intermesh in ion-exchange chromatography; these are controlled by equilibrium constants.[2]

As Figure 12.4 shows, cation exchange is a function of:

(a) competition between P^+ sample ions and Na^+ counter (buffer) ions, expressed by K_1;
(b) ionic strength (concentration) of counter ions;
(c) ionic strength of sample ions;
(d) base strength of sample, expressed by K_2 (pK_a value);
(e) acid strength of cation exchanger, expressed by K_3;
(f) mobile phase pH.

The conclusions drawn from this are as follows:

(a) *An increase in counter ion concentration reduces the retention time* (equilibrium constant K_1 is such that the counter ions are preferentially bonded and the P^+ ions must dissolve in the solution).
(b) *An increase in pH reduces the retention time in cation exchange* (the equilibrium constant K_2 shown in Fig. 12.4 is shifted to the right and the undissociated P^+OH^- is not able to interact with the exchanger group).

Exception: weak cation exchangers are more dissociated at high pH, resulting in improved sample interaction and a longer retention time; one of these two effects will be dominant.

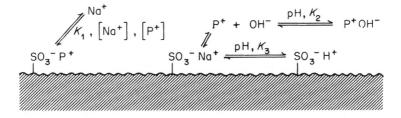

Figure 12.4   Equilibria during cation exchange.

---

[2] General definition of an equilibrium constant: $K = [C][D]/[A][B]$ for reaction $A + B \rightarrow C + D$ (square brackets represent concentrations).

Conversely:

(c) *A decrease in pH reduces the retention time in anion exchange.*

Exception: weak anion exchangers are more dissociated at low pH; hence elution is slower.

Ion exchange may also be influenced by a suitable choice of counter ion. The ion exchanger prefers:

- the ion with higher charge;
- the ion with smaller diameter;
- the ion with greater polarizability (greater potential for moving electric charge, i.e. better dipole induction capacity). Ions that can be readily polarized are called *soft,* those that cannot are referred to as *hard.*

(d) *The retention time in cation exchange increases if a counter ion is exchanged with another in the following sequence:* $Ba^{2+}-Pb^{2+}-Sr^{2+}-Ca^{2+}-Ni^{2+}-Cd^{2+}-Cu^{2+}-Co^{2+}-Zn^{2+}-Mg^{2+}-Mn^{2+}-UO_2^{2+}-Te^+-Ag^+-Cs^+-Rb^+-K^+-NH_4^+-Na^+-H^+-Li^+$. For example, K^+ solutions elute more rapidly than Li^+ solutions (although the exact sequence is dependent on the cation exchanger used). K^+, NH_4^+, Na^+ and H^+ tend to be the most common.

(e) *The retention time in anion exchange increases if a counter ion is exchanged with another in the following sequence:* citrate—sulfate—oxalate—tartrate—iodide—borate—nitrate—phosphate—chromate—bromide—rhodanide—cyanide—nitrite—chloride—formate—acetate—fluoride—hydroxide—perchlorate. For example, nitrate solutions elute more rapidly than chloride solutions (the exact sequence being dependent on the anion exchanger used).

In practice, corrosive, reducing and highly UV-absorbing anions are avoided, the most common buffers being phosphate, borate, nitrate and perchlorate, with sulfate, acetate and citrate occasionally being considered.

12.5 SPECIAL POSSIBILITIES OF ION EXCHANGE

Ion exchange separations can be influenced in many ways other than pure ion exchange. In such cases the mixed retention mechanism is known as *ion-moderated partition chromatography*. It includes one or more of the following mechanisms which may operate at once on the analytes (organic acids and bases, carbohydrates, alcohols, metabolites): ion exchange, ion exclusion, normal- and reversed-phase partition, ligand exchange and size-exclusion chromatography.

Heavy metal cations can be separated as chemical complexes with anion exchangers; e.g. Fe^{3+} forms a stable complex with chloride:

$$Fe^{3+} + 4Cl^- \rightleftharpoons FeCl_4^-$$

Ligand-exchange reactions also form part of ion-exchange chromatography, although the participant ligands do not have to be ionic.[3] This involves coating the resin or gel with copper or nickel by chemical reaction. Amino acids may work as selective ligands, the mobile phase containing ammonium ions (also a good ligand for Cu and Ni) in this case.

Carbohydrates (and sugar alcohols) can form a complex with calcium and other metal ions if three of their OH groups form a series axial–equatorial–axial:

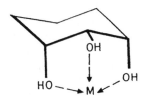

Therefore calcium-loaded cation exchangers are important stationary phases for the analysis of carbohydrates (see Figure 12.5).

Ion exclusion is another separation mechanism that may take place on ion exchangers.[4] The ionic functional groups of the ion exchanger repel ions of the same charge sign by electrostatic force, thus preventing them from penetrating the stationary phase pore system. Separation is based on the size-exclusion principle, details of which can be found in Chapter 15. (Note, however, that the reason for the exclusion effect is different in the two cases.) Ions are eluted within the breakthrough volume of the column under total ion-exclusion conditions. Figure 12.6 shows an example of this method, which can be highly effective, the charge ratios being as follows:

$$-SO_3^- \qquad \leftrightarrow \qquad {}^-OOCR$$
stationary phase   sample (carboxylic acid)

---

[3] A ligand is a molecule or an ion that can form a relatively stable bond with a metal cation. For example, ethylenediaminetetraacetic acid (EDTA) is a good ligand for many cations. The complex formation arises as a result of gaps in the metal ion electron system, which are then filled by the ligand.
[4] J.S. Fritz, *J. Chromatogr.*, **546**, 111 (1991).

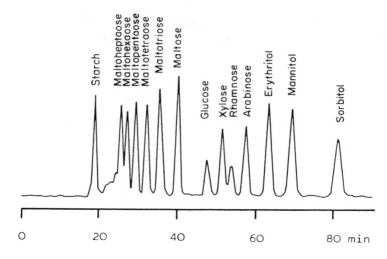

Figure 12.5 Ligand-exchange chromatography of sugars and sugar alcohols [reproduced with permission from J. Schmidt, M. John and C. Wandrey, *J. Chromatogr.*, **213**, 151 (1981)]. Conditions: column, 90 cm × 7.8 mm i.d.; stationary phase: calcium-loaded cation exchanger; mobile phase: 1 ml min^{-1} water; temperature: 85 °C; refractive index detector.

Ion chromatography is an important special case of ion exchange and is described in detail in Chapter 14.

12.6 PRACTICAL HINTS

The *sample size* should be selected so that the maximum column exchange capacity is not utilized by more than 5%.

Problem 29

The cation exchanger LiChrosorb CXS has an exchange capacity of 850 µeq g^{-1}. The column is filled with 15 g of gel. What is the maximum amount of radioactive ^{24}NaCl sample permitted?

Solution

Total exchange capacity $= 0.85$ µeq g^{-1} $\times$ 15 g $= 12.75$ µeq
5% of this $= 0.64$ µeq
Molar mass of ^{24}NaCl $= 59.5$ g mol^{-1}
0.64 mmol of ^{24}NaCl $= 38$ mg

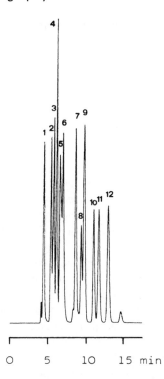

Figure 12.6 Separation of carboxylic acids by ion exclusion on a cation exchanger [reproduced with permission from E. Rajakylä, *J. Chromatogr.*, **218**, 695 (1981)]. Conditions: column, 30 cm × 7.8 mm i.d.; stationary phase, Aminex HPX-87, 9 μm; mobile phase, 0.8 ml min^{-1} of 0.006 N H_2SO_4; temperature, 65 °C; UV detector, 210 nm. Peaks: 1 = oxalic acid; 2 = maleic acid; 3 = citric acid; 4 = tartaric acid; 5 = gluconic acid; 6 = malic acid; 7 = succinic acid; 8 = lactic acid; 9 = glutaric acid; 10 = acetic acid; 11 = levulinic acid; 12 = propionic acid.

If a separation is to be developed, a good starting buffer strength is 50 mM. It should not be lower than 20 mM because otherwise an effective buffering is not guaranteed. The salts used should be of high purity.

The *mobile phase pH* is a function of the sample's acid or base strength, i.e. its pK_a value.[5] A pH of 1.5 units higher than the pK_a value of the base to be separated is advantageous in cation-exchange chromatography. This means that less than 10% of

---

[5] pK_a value of an acid HA: HA dissociates into H^+ and A^-. Equilibrium constant $K_a = [H^+][A^-]/[HA]$. pK_a is the negative logarithm of K_a. If the pH is the same as the pK_a value, then the acid HA is 50% dissociated.

TABLE 12.2 pK_a values and eluent pH values for different types of sample

Sample	pK_a	Approximate eluent pH
Aliphatic amines	Mostly 9.5–11	11–12.5
Aromatic amines	4.5–7	6–8.5
Carboxylic acids	~5	3.5
Phenols	~10	8.5

the sample is in dissociated form, so that small modifications in pH give rise to significant changes in retention times. Conversely, the pH should be about 1.5 units below the pK_a value for the separation of acidic components using anion-exchange chromatography. The best pH value or gradient should be determined empirically. Typical values are given in Table 12.2.

Note that effective buffering is only possible if the deviation from the pK_a value of the salt used is not larger than one pH unit. As an example, phosphate has the pK_a steps 2.1, 7.2 and 12.3. Therefore it is possible to prepare buffers for HPLC in the ranges of pH 1.1–3.1 and 6.2–8.2 (pH 12.3 is much too basic). Citrate has the pK_a steps 3.1, 4.7 and 5.4, acetate has pK_a 4.8. (See also Section 5.4.)

Mobile phase additives: tailing can often be reduced by adding a water-miscible solvent. A small amount of fungicide such as caproic acid, phenylmercury(II) salts, sodium azide, trichlorobutanol, carbon tetrachloride or phenol is often also required in order to prevent fungal growth in ion exchangers based on styrene and other organic resins.

Higher temperatures are often used in ion-exchange chromatography (60–80 °C) in order to obtain a lower eluent viscosity, an increase in plate number and shorter retention times.

A wash with the new solution easily changes the *ion exchanger form* if the new counter ions are chosen for preference over the original ones, e.g.:

$$R^-K^+ + AgNO_3 \rightarrow R^-Ag^+ + KNO_3$$

and:

$$R^+Cl^- + NH_4NO_3 \rightarrow R^+NO_3^- + NH_4Cl$$

If the new counter ion is not preferred, then the H^+ or OH^- form must be the starting point. This is achieved by a column wash with excess acid or base and a subsequent rinse with water to neutrality of the effluent:

$$R^-K^+ + HNO_3 \rightarrow R^-H^+ + KNO_3$$

and:

$$R^+Cl^- + NaOH \rightarrow R^+OH^- + NaCl$$

A solution of the new counter ion is then pumped through the column, the pH being as high as possible for the cationic form and as low as possible for the anionic form of the exchanger. This produces water, which moves the equilibrium in the required direction:

$$R^-H^+ + Li^+ + OH^- \rightarrow R^-Li^+ + H_2O$$

and:

$$R^+OH^- + ClO_4^- + H^+ \rightarrow R^+ClO_4^- + H_2O$$

Caution: care must be taken to prevent precipitation of insoluble salts in the column.

12.7 APPLICATIONS

All lanthanides are separated (as cations) within 30 min under the conditions shown in Figure 12.7. Elution invariably follows the order of decreasing

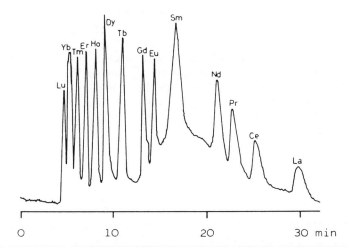

Figure 12.7 Separation of lanthanides by cation exchange [reproduced with permission of Vieweg Publishing from A. Mazzucotelli, A. Dadone, R. Frache and F. Baffi, *Chromatographia*, **15**, 697 (1982)]. Conditions: column, 25 cm × 4 mm i.d.; stationary phase, Partisil SCX, 10 μm; mobile phase, 1.2 ml min^{-1} 2-hydroxyiso-butyric acid in water, gradient from 0.03 to 0.07 M; visible-range detector, 520 nm, after derivatization with 4-(pyridylazo)-resorcinol.

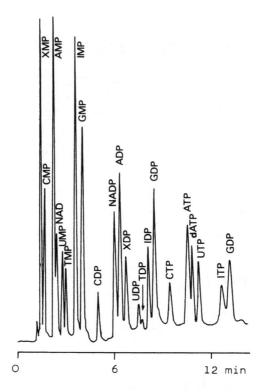

Figure 12.8 Separation of nucleotides and related compounds by anion exchange [reproduced with permission of Vieweg Publishing from D. Perrett, *Chromatographia*, **16**, 211 (1982)]. Conditions: stationary phase, APS-Hypersil, 5 μm; mobile phase, A = 0.04 M KH_2PO_4 (pH 2.9), B = 0.5 M KH_2PO_4 + 0.8 M KCl (pH 2.9), linear gradient from A to B in 13 min; UV detector, 254 nm.

atomic mass. Promethium, which does not occur naturally, is missing and would be eluted in the significant gap between samarium and neodymium.

    Figure 12.8 shows the high efficiency of high-performance ion-exchange chromatography by the rapid separation of 22 nucleotides and related compounds. Figure 12.9 is an example of the separation of RNA, DNA and a plasmid in a cell lysate on a wide-pore (400 nm) amino phase.

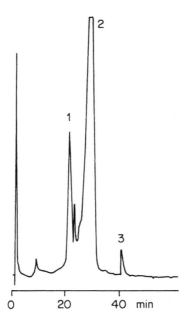

Figure 12.9   Separation of cell lysate from *E. coli* containing the plasmid pBR 322 [reproduced with permission from M. Colpan and D. Riesner, *J. Chromatogr.*, **296**, 399 (1984)]. Conditions: column, 7.5 cm × 6.2 mm i.d.; stationary phase, Nucleogen Dimethylamino 4000, 10 µm; mobile phase, 2 ml min^{-1} of 20 mM potassium phosphate (pH 6.9) with a gradient from 0 to 1.5 M potassium chloride in 50 min; UV detector, 260 nm. Peaks: 1 = cell RNA; 2 = cell DNA; 3 = plasmid.

Further examples are shown in Figures 18.1, 18.3 and 18.17.

13 Ion-Pair Chromatography

13.1 INTRODUCTION

Ion-pair chromatography represents an alternative to ion-exchange chromatography. Many problems can be solved by either method but ion-exchange is not as good for separating mixtures of acids, bases and neutral products under certain circumstances, and this is where ion-pair chromatography then comes into its own. The fact that the reversed phases described in Chapter 10 can be used as stationary phases has added to its popularity.

Ionic samples may be separated by reversed-phase chromatography, provided that they contain only weak acids or only weak bases (in addition to neutral compounds) present in undissociated form, as determined by the chosen pH; this is known as 'ion suppression'.[1] Ion-pair chromatography is an extension of this principle. An organic ionic substance is added to the mobile phase and forms an ion pair with a sample component of opposite charge. This is, in fact, a salt but its chromatographic behaviour is that of a nonionic organic molecule:

$$\text{analyte}^+ + \text{counter ion}^- \rightarrow [\text{analyte}^+ \text{ counter ion}^-]\text{ pair}$$
$$\text{analyte}^- + \text{counter ion}^+ \rightarrow [\text{analyte}^- \text{ counter ion}^+]\text{ pair}$$

Reversed-phase chromatography can be used in this instance. For example, an alkylsulfonate is added to cationic analytes and tetrabutylammonium phosphate to anionic analytes. A sample containing both anionic and cationic components has one type 'masked' by a counter ion and the other suppressed by a suitable pH level.

---

[1] The process cannot be used for strong acids or bases because these are dissociated over a wide pH range and extreme pH levels would be required for ion suppression. Ion chromatography as described in Chapter 14 was developed specifically for cases such as this.

Practical High-Performance Liquid Chromatography, Fifth edition   Veronika R. Meyer
© 2010 John Wiley & Sons, Ltd

The advantages of ion-pair chromatography for separating ionic analytes may be summarized as follows:

(a) Reversed-phase systems can be used for separation.
(b) Mixtures of acids, bases and neutral products and also amphoteric molecules (which have one cationic and one anionic group) can be separated.
(c) Ion-pair chromatography can also be a good choice if the pK_a values of the analytes are very similar.
(d) Selectivity can be influenced by the choice of the counter ion.

13.2 ION-PAIR CHROMATOGRAPHY IN PRACTICE

Ion-pair chromatography is suitable for all ionic compounds but not all counter ions are equally good in every case.[2] Table 13.1 lists some counter ions suitable for reversed-phase chromatography with methanol–water or acetonitrile–water.[3]

There are a number of suitable reagent mixtures in buffer solutions on offer, although many workers prefer to mix their own mobile phase 'cocktails' to suit their own particular problem. A few examples are as follows (Waters):

(a) Tetrabutylammonium phosphate, $^+N(C_4H_9)_4$, buffered to pH 7.5, forms ion pairs with strong and weak acids and buffering suppresses weak base ions.
(b) Alkylsulfonic acids, $^-SO_3(CH_2)_nCH_3$ with $n = 4$–7, buffered to pH 3.5, form ion pairs with strong and weak bases and buffering suppresses weak acid ions. The longer the alkyl chain, the greater is the retention time.

TABLE 13.1 Applicability of different counter ions

Counter ion	Suitable for
Quaternary amines, e.g. tetramethyl-, tetrabutyl-, palmityltrimethylammonium	Strong and weak acids, sulfonated dyes, carboxylic acids
Tertiary amines, e.g. trioctylamine	Sulfonates
Alkyl- and arylsulfonates, e.g. methane- and heptanesulfonate, camphorsulfonic acid	Strong and weak bases, benzalkonium salts, catecholamines
Perchloric acid	Very strong ion pairs with many basic samples
Alkylsulfates, e.g. laurylsulfate	Similar to sulfonic acids, different selectivity

[2] B.A. Bidlingmeyer, *J. Chromatogr. Sci.*, **18**, 525 (1980); J.W. Dolan, *LC GC Eur.*, **21**, 258 (2008) or *LC GC North Am.*, **26**, 170 (2008).
[3] R. Gloor and E.L. Johnson, *J. Chromatogr. Sci.*, **15**, 413 (1977).

Amphoteric molecules such as p-aminobenzoic acid ($H_2NC_6H_4COOH$, where NH_2 is a basic function and COOH is acidic) may be chromatographed with both of the above reagents. The acid function forms an ion pair with the former and the amino group remains undissociated as a result of buffering. The reverse applies with the latter reagent.

The only problems are caused by mixtures of strong and weak acids with strong and weak bases. Either the strong bases or the strong acids are in ionic form. However, this type of mixture is rare.

The k values are proportional to the counter-ion concentration, so that the retention times are not only affected by the type but also by the concentration of the ion-pair reagent. Under certain circumstances, two reagents instead of one may be used for optimizing a separation problem, e.g. a mixture of pentane- and octanesulfonic acid. Of course, retention times can also be adjusted by changing the amount of organic solvent in the mobile phase and all optimization procedures described in Chapter 10 can be used, including the addition of competing compounds.

Quaternary ammonium salts in an alkaline medium are extremely damaging to silica. A scavenger column placed in front of the sample injector, in which the mobile phase is saturated with silica, is highly recommended in such instances.

Gloor and Johnson[3] provided some guidelines for practical ion-pair chromatography:

(a) It should be used when other methods such as reversed-phase and ion- suppression modes have failed or when the sample contains both nonionic and ionic components.
(b) A methanol–water mixture should be used as the mobile phase, thus minimizing counter-ion solubility problems. Acetonitrile is preferred if the selectivity is unsatisfactory, but the solubility properties may be different in this case.
(c) The choice of the correct counter ion is important, the short-chain variety being preferred for samples with very little difference in molecular structure and the longer-chain, hydrophobic types if greater retention is required.
(d) Counter-ion solubility should be checked for all mobile phase mixing ratios that arise during analysis (gradient elution!).
(e) The mobile phase pH should be chosen to provide maximum sample ionization. However, the pH compatibility of silica as the bonded stationary phase base (pH 2.0–7.4) must also be taken into consideration.
(f) A monomeric C_{18} or C_8 reversed phase should be chosen as the stationary phase for the initial test. Shorter alkyl chains seem to lack stability.
(g) The mobile phase should be degassed *prior* to counter-ion addition, to prevent foaming (especially relevant when long-chain detergent-like compounds are used).

(h) Concentration:[4] Ion-pair reagents should be used in the 1–10 mM range (preferably a lower concentration if a long-chain counter ion is used and vice versa). If a buffer is needed to maintain the required pH level, then equal amounts are added to all the solvents involved in the case of gradient elution. Buffer components should have poor ion-pair but good solubility properties.

(i) Optimization of separation by solvent gradient, change in counter-ion or buffer concentrations, etc. If the sample contains both ionic and nonionic components, then the nonionic ones should be optimized first and a counter ion that elutes the ions at favourable retention times is then sought.

(j) The pump should not be turned off as long as counter-ion reagent is still in the tubing and column. At night, the rate may be reduced to 3–5 ml h^{-1} to prevent salt precipitation.

(k) The UV transparency of the counter ion needs to be checked.

Five further hints:

(l) Larger samples possibly are better separated if the sample is in ion-pair form before addition to the column. An approximately stoichiometric amount of counter ion is added to the sample solution for this purpose, care being taken to ensure maximum ionization by suitable pH adjustment.

(m) As a general rule, ion-pair reagents tend to be harmful for silica.

(n) Columns used for ion-pair chromatography should be reserved strictly for this purpose.

(o) Equilibration of ion-pair systems is slow. If the mobile phase composition needs to be changed it is necessary to pump at least 20 column volumes through the column.

(p) Ion-pair equilibria are temperature dependent, therefore temperature control is necessary.

13.3 APPLICATIONS

Examples of the separation of cations (with an anionic reagent) and anions (with a cationic reagent) are given in Figures 13.1 and 13.2. A further example of ion-pair chromatography is shown in Figure 22.2.

---

[4] L.R. Snyder, J.J. Kirkland and J.L. Glajch, *Practical HPLC Method Development*, Wiley–Interscience, New York, 2nd ed., 1997, Chapter 7.4, pp. 317–341.

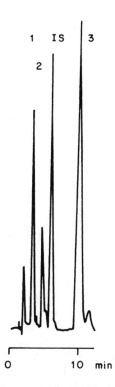

Figure 13.1 Analysis of catecholamines in urine (reproduced with permission of Bioanalytical Systems). Conditions: stationary phase, octylsilane; mobile phase, citrate-phosphate buffer (pH 4) containing 7% methanol and 80 mg l^{-1} sodium octyl sulfate; electrochemical detector, +700 mV; for sample preparation see R.M. Riggin *et al.*, *Anal. Chem.*, **49**, 2109 (1977). Peaks (with concentrations in urine): 1 = norepinephrine (160 ng ml^{-1}); 2 = epinenphrine (31 ng ml^{-1}); 3 = dopamine (202 ng ml^{-1}); IS = 3,4-dihydroxybenzylamine (internal standard).

13.4 APPENDIX: UV DETECTION USING ION-PAIR REAGENTS[5]

Significant effects are achieved if the aqueous mobile phase contains both hydrophobic, UV-absorbing ions and hydrophilic ions, e.g. salts. The distribution equilibrium of the UV-absorbing ions between the mobile and stationary phases is specifically altered following sample injection. The sample components in the UV detector then produce a signal, despite being non-UV-absorbing themselves. Hence,

---

[5] M. Denkert, L. Hackzell, G. Schill and E. Sjögren, *J. Chromatogr.*, **218**, 31 (1981); L. Hackzell and G. Schill, *Chromatographia*, **15**, 437 (1982).

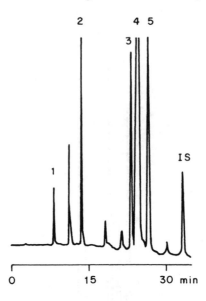

Figure 13.2 Analysis of α-keto acids in plasma [after T. Hayashi, H. Tsuchiya and H. Naruse, *J. Chromatogr.*, **273**, 245 (1983)]. Conditions: sample, extract from 50 µl of human plasma, derivatized with *o*-phenylenediamine; column, 25 cm × 4 mm i.d.; stationary phase, LiChrosorb RP-8, 5 µm; mobile phase, 1 ml min^{-1} of 1 mM tetrapropylammonium bromide in 50 mM phosphate buffer-acetonitrile, gradient from 5 to 60% acetonitrile; temperature, 50 °C; fluorescence detector, 350/410 nm. Peaks: 1 = α-ketoglutaric acid; 2 = pyruvic acid (α-ketopropionic acid); 3 = α-ketoisovaleric acid; 4 = α-ketoisocaproic acid; 5 = α-keto-β-methylvaleric acid; IS = α-ketocaprylic acid (internal standard).

this method can be regarded as an alternative to refractive index detection and is a type of indirect detection, as mentioned in Section 6.9.

Chromatograms for such systems show positive and negative peaks depending on the charge and k values of the sample components, and also one or more 'system peaks' (those that cannot be ascribed to a particular component) (Figure 13.3). These phenomena have been described in detail and substantiated in the original papers cited.[5]

To obtain good results the following recommendations for the design of a mobile phase have been given:[6]

(a) Optimum sensitivity is obtained if the k values of the analyte of interest and of the ion-pair reagent system peak are similar. For the same reason the molar absorptivity of the reagent should be high.

---

[6] E. Arvidsson, J. Crommen, G. Schill and D. Westerlund, *J. Chromatogr.*, **461**, 429 (1989).

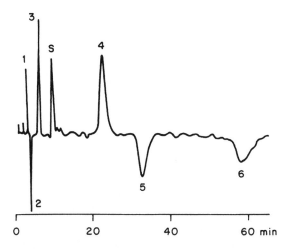

Figure 13.3 Separation of UV-inactive anionic and cationic compounds [reproduced with permission of Elsevier Science Publishers BV from M. Denkert *et al.*, *J. Chromatogr.*, **218**, 31 (1981)]. Anions detected before the system peak produce positive peaks and cations produce negative peaks, whereas the reverse applies for compounds following the system peak. Conditions: column, 10 cm × 3.2 mm i.d.; stationary phase, μBondapak Phenyl, 10 μm; mobile phase, 0.5 ml min^{-1} of 4×10^{-4} M naphthalene-2-sulfonate in 0.05 M phosphoric acid; UV detector, 254 nm. Peaks: 1 = butylsulfate; 2 = pentylamine; 3 = hexanesulfonate; 4 = heptylamine; 5 = octanesulfonate; 6 = octylsulfate; S = system peak.

(b) To avoid problems in sensitivity and peak shape it is important to use a mobile phase with as simple a composition as possible, preferably only prepared with ion-pair reagent and a hydrophilic buffer.

(c) The ion-pair reagent should be aprotic, i.e. it should not form hydrogen bridges.

(d) The buffer capacity and concentration must not be low in order to keep the distribution coefficient of the reagent constant.

14   Ion Chromatography[1]

14.1   PRINCIPLE

Ion chromatography is a special technique which was developed for the separation of inorganic ions and organic acids. The common detection principle is the monitoring of eluate conductivity[2] although some analytes can also be detected with UV or, perhaps after derivatization, with visible light. Typical applications are the analysis of:

(a) anions in drinking and process water (chloride, nitrate, sulfate and hydrogen carbonate are the most important ones);
(b) nitrate in vegetables;
(c) fluoride in toothpaste;
(d) bromide, sulfate and thiosulfate in fixing baths;
(e) organic acids in beverages;
(f) ammonium, potassium, nitrate and phosphate in soil and fertilizers;
(g) sodium and potassium in clinical samples such as body fluids and infusions.

Stationary phases for ion chromatography have a lower exchange capacity than the ones used for classical ion-exchange separations as described in Chapter 12. Therefore, the ionic strength of the eluent can be low and 1 mM solutions are not uncommon. Diluted mobile phases have low conductivity which facilitates detection. However, even with mobile phases of low electrolyte concentration, the background conductivity is too high to allow detection without special techniques. Two principles

---

[1] J.S. Fritz and D.T. Gjerde, *Ion Chromatography*, Wiley-VCH, Weinheim, 4th ed., 2009; C. Eith, M. Kolb, A. Seubert and K.H. Viehweger, *Practical Ion Chromatography — An Introduction*, Metrohm, Herisau, 2001.
[2] R.D. Rocklin, *J. Chromatogr.*, **546**, 175 (1991); W.W. Buchberger, *Trends Anal. Chem.*, **20**, 296 (2001).

Practical High-Performance Liquid Chromatography, Fifth edition   Veronika R. Meyer
© 2010 John Wiley & Sons, Ltd

for the elimination of background conductivity are used, namely electronic and chemical suppression.

14.2 SUPPRESSION TECHNIQUES

Electronic suppression is possible if the eluent is properly chosen for extra-low conductivity which is e.g. the case with phthalate buffers. The detector must be able to compensate for the background conductivity by its electronics. Such a set-up gives good calibration linearity over a wide range but its detection limit is rather poor. For historical reasons it is also termed *single-column ion chromatography*.

Chemical suppression works with chemical elimination of the buffer ions between column and detector. The eluate flows through a small, packed column with ion-exchanger resin or through an ion-permeable membrane tube or hollow fibre. Buffer cations are replaced by H^+, anions by OH^-, and water is formed which has a very low conductivity. Figure 14.1 explains the reactions on the suppressor column for the separation of cations with diluted acids as eluents (left) and for the separation of anions with diluted bases as eluents (right). The suppressor column can be regenerated with an appropriate buffer solution or by electrochemically generated OH^- or H^+ ions, respectively. If two of these columns are used in a parallel design, one of them can be used for eluate suppression while the other one is regenerated. For the next chromatographic run the first column is regenerated and the other one acts as the suppressor. If a membrane or fibre suppressor is used it is continuously flushed with a basic (for cation analysis) or an acidic (for anion analysis) solution on its outer surface to replace the penetrating buffer ions by water. It is obvious that a system with chemical suppression has also been termed *dual-column ion chromatography*. Its linear calibration range is limited but the detection limits are lower than with electronic suppression.

Conductivity depends strongly on temperature, therefore the detector needs to be thermostatted carefully.

14.3 PHASE SYSTEMS[3]

Mobile and stationary phases need to harmonize with each other; a certain eluent will not necessarily interact well with a given ion exchanger. Typical eluents for separations with electronic suppression are diluted phthalic or benzoic acid, perhaps with a low amount of acetone or methanol (to influence the selectivity) for anions, and nitric, oxalic, tartaric, citric or dipicolinic acid (the latter one for complex formation)

---

[3] C. Sarzanini, *J. Chromatogr. A*, **956**, 3 (2002).

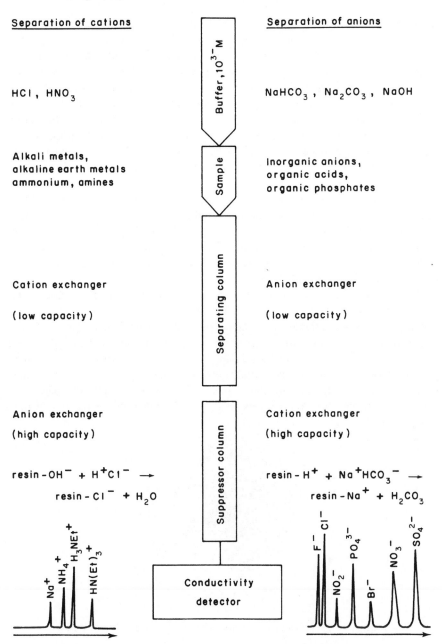

Figure 14.1 Ion chromatography with supression column.

for cations. Chemical suppression is used almost exclusively in anion analysis where eluents with carbonate/hydrogen carbonate, sodium hydroxide or potassium hydroxide are used. Mobile phases with potassium hydroxide can be generated electrochemically from water and commercially available cartridges with potassium electrolyte, thus eliminating the need to prepare and monitor the eluent at the workplace.

The mobile phase should be of highest purity with regard to the respective analyte ion or ions. The detection limit of a given ion depends on its concentration in the eluent; the higher this concentration is, the worse is the limit. Water of 'HPLC quality' can have rather high concentrations of iron and of other ions even if its conductivity is as low as $1 \, \mu\mathrm{S} \, \mathrm{cm}^{-1}$. It may be necessary to use a water quality designed for inorganic trace analysis.[4]

If a carbonate mobile phase is used a system peak (see Section 19.9) is always observed, the position of which depending on the column in use; it can be close to or even under the chloride peak (therefore the sum of Cl^- and CO_3^{2-} is often determined in water analysis) or it appears much later. The equilibrium of carbonate and hydrogen carbonate is influenced by the CO_2 in air which makes it necessary to equip the eluent bottle with a CO_2 absorber.

Materials for anion separations are based on styrene–divinylbenzene, polymethacrylate (with limited pressure stability) or poly(vinyl alcohol). In most cases the ion exchange group is a derivative of trimethylamine or dimethylethanol amine. Cation exchangers are made from silica, polyethylene or polybutadiene maleic acid. The ion exchange group can be chemically bonded. However, the functionality of many phases is based on small latex beads with quaternary amino groups. These beads have a diameter of ca. $0.1 \, \mu\mathrm{m}$ and are held in place on the sulfonated packing material by electrostatic or van der Waals bonds (Figure 14.2). Such phases have good mass

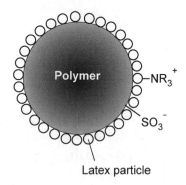

Figure 14.2   Anion exchanger with aggregated latex beads.

[4] I. Kano *et al.*, *J. Chromatogr. A*, **1039**, 27 (2004).

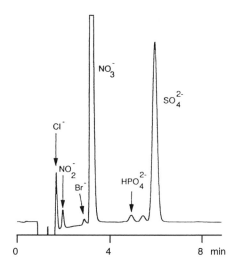

Figure 14.3 Separation of inorganic anions in rain water [after W. Shotyk, *J. Chromatogr.*, **640**, 309 (1993)]. Conditions: stationary phase, AS4A; mobile phase, 1.8 mM Na_2CO_3 + 1.7 mM $NaHCO_3$; conductivity detector after membrane suppressor. Concentrations (in ng g^{-1}): chloride, 17; nitrite, 51; bromide, 5; nitrate, 1329; hydrogenephosphate, 50; sulfate, 519.

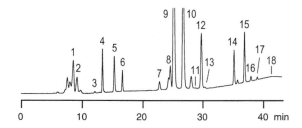

Figure 14.4 Separation of organic acids (as anions) and inorganic anions in wine (reproduced with permission of Dionex). Conditions: column, 25 cm × 4 mm i.d. and precolumn; stationary phase, IonPac AS11-HC; mobile phase, 1.5 ml min^{-1}, nonlinear gradient from 1 mM to 60 mM NaOH and from 0 to 20% methanol; temperature, 30 °C; conductivity detector after packed suppressor. Peaks: 1 = lactate; 2 = acetate; 3 = formate; 4 = pyruvate; 5 = galacturonate; 6 = chloride; 7 = nitrate; 8 = succinate; 9 = malate; 10 = tartrate; 11 = fumarate; 12 = sulfate; 13 = oxalate; 14 = phosphate; 15 = citrate; 16 = isocitrate; 17 = *cis*-aconitate; 18 = *trans*-aconitate.

transfer properties because the diffusion paths are short, therefore the number of theoretical plates of these columns is high but the chemical stability is poorer than with chemically bonded phases.

14.4   APPLICATIONS

Figure 14.3 shows the ion chromatographic separation of anions in rain water and Figure 14.4 represents the analysis of organic acids and inorganic anions in wine.

15 Size-Exclusion Chromatography[1]

15.1 PRINCIPLE

Size-exclusion chromatography is basically different from all other chromatographic methods in that a simple molecule size classification process rather than any interaction phenomena forms the basis of separation.

A porous material is used as the column packing. The only space available to sample molecules that are too large to diffuse into the pores is that between the individual stationary phase particles; hence they become *excluded*. The column then appears to be filled with massive impenetrable beads and, as no attractive forces are apparent, i.e. there is no interaction between beads and sample molecules, the latter are transported by the mobile phase through the column within the shortest possible time.

However, if molecules that are small enough to penetrate all the pores are present, then the whole of the mobile phase volume becomes available to them. As the mobile phase stagnates in the pores, diffusion is the only way in which molecules can escape and as a result they move more slowly than the excluded molecules. They are the last to appear in the detector, namely at the breakthrough time known from all the other column chromatographic methods. The solvent itself does not interact with the stationary phase but passes through all the pores, and solvent molecules are the first to appear in the detector in all other modes of

---

[1] A.M. Striegel, W.W. Yau, J.J. Kirkland and D.D. Bly, *Modern Size-Exclusion Chromatography*, John Wiley & Sons, Ltd, Chichester, 2nd ed., 2009.

Practical High-Performance Liquid Chromatography, Fifth edition   Veronika R. Meyer
© 2010 John Wiley & Sons, Ltd

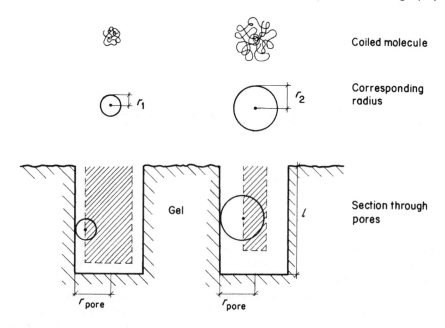

Figure 15.1 Simple cylinder pore model of size-exclusion chromatography. In reality, not all pores are equal in size.

liquid chromatography (provided that there is no exclusion). In size-exclusion chromatography, molecules as small as the solvent molecules are eluted last. They have fully penetrated the stationary phase (*total permeation*). This means that each sample component is flushed from the column at the latest at the breakthrough time.

Medium-sized molecules can only utilize part of the available pore volume. A small coiled molecule with a statistical mean radius of r_1 can remain in a larger pore volume than a larger molecule with a mean radius r_2. The available pore volume is represented by the narrowly shaded area in Figure 15.1. The relevant equation is:

$$\text{accessible pore volume} = \left(r_{\text{pore}} - r_{\text{molecule}}\right)^2 \times \pi \times \left(l - r_{\text{molecule}}\right)$$

for cylindrically shaped pores with radius r_{pore} and length l. The smaller a molecule the greater is the amount of pore volume available and the longer is its journey through the column; the gel separates according to molecule size:

$r_{molecule} > r_{pore}$	No pore volume accessible	First peak
$r_{molecule} < r_{pore}$	Available pore volume is a function of molecular radius	Separation according to molecular size
$r_{molecule} \ll r_{pore}$	All pore volume available	Final peak

The situation can be summed up as follows:

(a) All injected molecules are eluted (as long as no unwanted adsorption effects occur).
(b) The breakthrough time represents the completion of the chromatogram; hence the separation time can be predicted.
(c) The elution volume or time is a function of molecular size alone and hence indirectly of molar mass. Size-exclusion chromatography can be used for molar mass determinations.
(d) As the separation is completed after only a small amount of mobile phase has passed through, the peak capacity (i.e. the number of peaks that can be separated from each other with a specific resolution) is limited.

Moreover:

(e) Two types of molecules can be separated, provided that there is at least a 10% difference in their molar masses.
(f) Peaks do not broaden as the retention time increases.
(g) In contrast to other methods, molecules are promoted to reside in the stagnant mobile phase; hence the plate numbers are generally lower than normally found in HPLC.
(h) The van Deemter curve does not follow the normal pattern of passing through a minimum and having a steep gradient when flow rates are low. The low rate of macromolecule diffusion means that the B term (see Section 8.6) is negligible and the separation performance increases the more slowly the chromatography is carried out.[2]
(i) There are no interaction equilibria; hence relatively large amounts can be separated. However, the viscosity of the sample solution should not be more than double that of the mobile phase.
(j) As the elution volumes are small, a reduction of the extra-column volumes in all the equipment used becomes of vital importance.

---

[2] Possibilities for fast size-exclusion chromatography: S.T. Popovici and P.J. Schoenmakers, *J. Chromatogr. A*, **1099**, 92 (2005).

Problem 30

What is the size-exclusion chromatographic separation sequence for the sugars shown in Figure 12.5 (ion-exchange chromatogram)?

Solution

1. Starch	Molar mass large
2. Maltoheptaose	1153
3. Maltohexaose	991
4. Maltopentaose	829
5. Maltotetraose	667
6. Maltotriose	504
7. Maltose	342
8. Mannitol, sorbitol	182
9. Glucose	180
10. Rhamnose	164
11. Xylose, arabinose	150
12. Erythritol	122

Starch is a macromolecule and would be excluded. It is followed by the oligo-saccharides in decreasing degree of polymerization. Maltose is a disaccharide; the other compounds are monosaccharides or small sugar alcohols. Fractions 8 and 9 are impossible to resolve because their masses are too similar. The elution order partly differs from the ion-exchange separation, and some compounds cannot be separated at all due to equal size.

15.2 THE CALIBRATION CHROMATOGRAM

The column can be calibrated with a test mixture of compounds of accurately defined molecular mass, as a means of finding the elution volume, V_E, for a specific molecular size.[3] The sizes of test molecules must be selected to ensure that:

(a) one component is excluded;
(b) several components partly penetrate the pores; and
(c) one component completely penetrates the stationary phase.

---

[3] Monodisperse samples (a single type of molecule per peak) are best for calibration but polydisperse ones can also be considered: M. Kubin, *J. Liq. Chromatogr.*, **7** Suppl., 41 (1984).

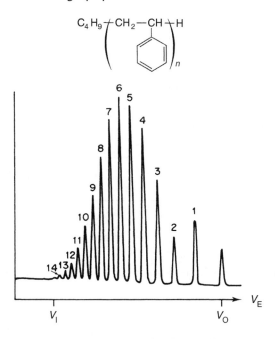

Figure 15.2 Size-exclusion chromatogram of styrene oligomers, with $n = 1$–14 [reproduced with permission from W. Heitz, *Fresenius Z. Anal. Chem.*, **277**, 324 (1975)]. Conditions: column, 100 cm × 2 mm i.d. (coiled); stationary phase, Merckogel 6000; mobile phase, dimethylformamide.

A chromatogram for styrene oligomers, with $n = 1$–14, may be taken as an example (Figure 15.2). The peak on the far right may be derived from styrene. This is the smallest molecule (next to those of the mobile phase) and is eluted last. The void volume, V_0, is its relevant elution volume. The first small peak on the left is produced by excluded molecules, the relevant volume being the interstitial volume, V_I (liquid found *between* the individual particles of the stationary phase).[4]

Molecules with a degree of polymerization n of between 14 and 1 are eluted between V_I and V_0. The volume $V_0 - V_I$ corresponds to the pore volume, V_P, of the

---

[4] The void volume V_0 defined here corresponds to the elution volume of breakthrough time as presented in Section 2.3. However, in size-exclusion literature often V_I, as defined here, is termed V_0 because it is the volume where the first peak can occur.

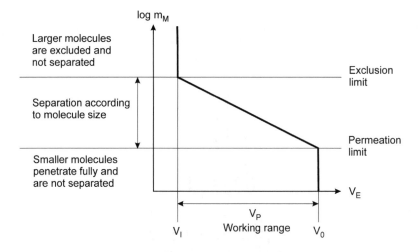

Figure 15.3   Calibration graph for a size-exclusion column.

stationary phase (liquid in the pores). V_p is the only effective separation volume and should, therefore, be as large as possible. The peak capacity is a function of V_P. It is clear that peaks follow on more closely, the smaller is the difference in molecule size (mass).

In an ideal case, a graph of log (molar mass) against elution volume is a straight line that characterizes the column (Figure 15.3). However, there may also be some sort of curve in the calibration graph.

Problem 31

Figure 15.14 presents a separation of milk proteins. Draw a calibration curve which corresponds to this chromatogram.

Solution

Figure 15.4 shows the theoretical calibration graph. The white dot corresponds to the first peak of high molecular weight proteins. Its position is only known with regard to the retention volume which can be estimated from the chromatogram to be 8.8 ml; it must be somewhere on the vertical branch at this volume but its log m_M position is unknown. The black dots represent from left to right: peak 2 with

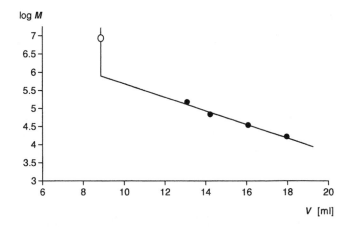

Figure 15.4 Calibration graph of the milk proteins of Figure 15.14.

retention volume 13.1 ml and mass 150 000 (log $m_M = 5.18$); peak 3 with 14.2 ml and 69 000 (log $m_M = 4.84$); peak 4 with 16.1 ml and 35 000 (log $m_M = 4.54$); peak 5 with 17.9 ml and 16 500 (log $m_M = 4.22$). Note that true calibration graphs are curved, not straight.

Manufacturers of stationary phases or columns for size-exclusion chromatography always give details of the molecular mass range covered by each product. Figure 15.5 depicts μStyragel (Waters) as an example, the values in ångstroms representing the mean pore width of the respective gel ($10 \text{ Å} = 1$ nm).

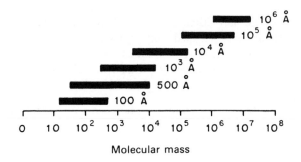

Figure 15.5 Pore width and separating range of a typical commercial column set (reproduced with permission of Waters).

Problem 32

Which column would be suitable for separating the mixture described in Problem 30?

Solution

The molar mass varies between 122 and 1153. The '500 Å' column, which has an exclusion limit of around 10^4 and a permeation molecular mass of about 50, can separate the sugars. The first peak, starch, should be clearly resolved from the other compounds.

15.3 MOLECULAR MASS DETERMINATION BY MEANS OF SIZE-EXCLUSION CHROMATOGRAPHY[5]

An unknown fraction can be characterized or the molecular mass distribution of a polymer material determined by comparing the sample chromatogram (Figure 15.6, bottom) with the calibration graph (Figure 15.6, centre).

As elution volume is a function of molecular size and not directly of its mass, this comparison is only valid for the same type of molecules (e.g. homologues or polystyrenes in this particular case) in the same solvent:

(a) Different types of molecule (e.g. polyamide instead of polystyrene) may have a different mass for a specific elution volume (i.e. one that is the same size) or, to put it more clearly, the density may be different to that of the standard molecules. Polystyrene standards can be converted for other products but it is much better to calibrate the column with a standard that is as similar as possible in nature to that of the sample, if molecular mass determinations are required.

(b) The same molecule may differ in size in the various solvents (be this real or apparent) and hence may be eluted at different rates. A coiled molecule may swell or it may shrink (Figure 15.7). A molecule may be solvated in one solvent, thus appearing bigger than it actually is, and remain unchanged in another (Figure 15.8).

Even the concentration of the sample solution can strongly influence both peak width and retention time.[6]

---

[5] Chapter 8 of Striegel *et al.*, see first page of this chapter.
[6] Anonymous, *J. Chromatogr. Sci.*, **30**, 66 (1992).

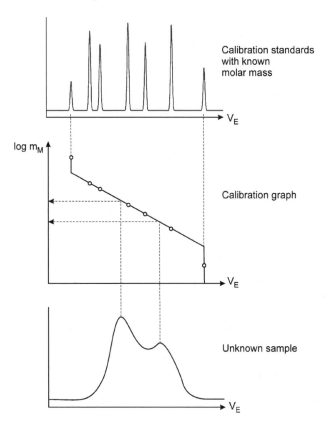

Figure 15.6   Calibration graph as an aid in the determination of molecular mass distribution.

The determination of molecular mass and molecular mass distribution by size-exclusion chromatography is rapid and easy.[7] However, as it is based exclusively on elution volume, this latter must be determined extremely accurately. Elution volume is a function of log(molar mass) and hence even small errors in the determination of V_E have a dramatic effect. For this reason, either a pump which is both reproducible and precise plus thermostatic control of the system should be

[7] For data evaluation in size-exclusion chromatography, see: D. Held and P. Kilz, in *Quantification in LC and GC*, S. Kromidas and H.J. Kuss, eds, Wiley-VCH, Weinheim, 2009, chapter 13, pp. 271–302.

Figure 15.7   Coiled molecule in different swelling states.

considered or, alternatively, a system of continuous volume measurement must be established.

The second alternative is simple and cheap. After passing through the detector, the eluate is channelled into a 1 or 5 ml siphon which tips over when full and triggers a signal which is marked accordingly on the chromatogram. The elution volume is then determined by signal number. Elution time can be used for direct characterization if the pump is good (and expensive!) enough. Thermostating is vital in view of the different coefficients of expansion attributed to the various components (mobile phase, steel, glass, PTFE). Eluent, pump head and column must be thermostated to $\pm 0.1\,°C$ if retention volumes have to be more accurate than 1%.

A fully penetrating material should be chosen for the determination of theoretical plate number. Manufacturers recommend methanol for $100\,\mathring{A}$ μStyragel and o-dichlorobenzene (not more than 1 μl) for $500{-}10^6\,\mathring{A}$ columns. At least 3000–4000 plates should be recorded in this way for a 30 cm column.

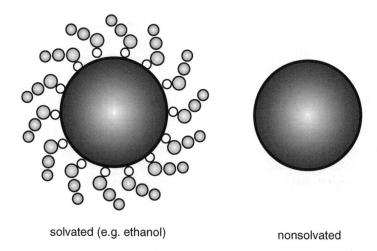

solvated (e.g. ethanol)   nonsolvated

Figure 15.8   Different apparent molecular sizes caused by solvation.

However, the specific resolution, R_{sp}, provides a better basis for characterizing a size-exclusion column:

$$R_{sp} = \frac{0.576}{b\sigma}$$

where b is a constant related to the slope of the calibration graph and σ is the standard deviation of a monodisperse sample peak ($\sigma = w/4$). The accuracy of molecular mass determination is a function of b and σ.[8]

15.4 COUPLED SIZE-EXCLUSION COLUMNS

There are two basic possibilities:

(a) If component resolution is not good enough, then it may be improved by adding one or more *columns of the same type*, thus providing extra pore volume. Figure 15.9 shows how two columns can be used to improve resolution over a single column. The elution volume is doubled, but so is analysis time.

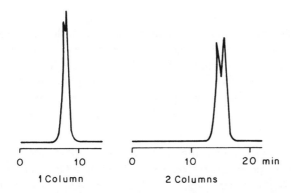

Figure 15.9 Improved resolution due to double-pore volume. Conditions: sample, dodecane and hexadecane; column(s), 30 cm × 7.8 mm i.d.; gel, µStyragel 100 Å; mobile phase, 1 ml min^{-1} toluene; refractive index detector.

---

[8] W.W. Yau, J.J. Kirkland, D.D. Bly and H.J. Stoklosa, *J. Chromatogr.*, **125**, 219 (1976).

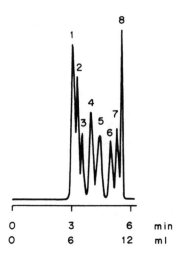

Figure 15.10   Separation of a polystyrene mixture with a broad molecular mass range [reproduced with permission from R.V. Vivilecchia, B.G. Lightbody, N.Z. Thimot and H.M. Quinn, *J. Chromatogr. Sci.*, **15**, 424 (1977)]. Mobile phase, 2 ml min^{-1} dichloromethane. Molecular masses: $1 = 2\,145\,000$; $2 = 411\,000$; $3 = 170\,000$; $4 = 51\,000$; $5 = 20\,000$; $6 = 4000$; $7 = 600$; $8 = 78$ (benzene).

(b) A 10^4 Å column, for example, separates molecules with molar masses of about 4000–200 000. However, if the molecular mass range is even greater, then a series of *different types of column* may be used to advantage. The full set as shown in Figure 15.5 can separate samples with a molar mass ranging from 20 to 2×10^6, i.e. the total spectrum. The separation of polystyrene standards using four μBondagel columns (125, 300, 500, 1000 Å) can be used as an example (Figure 15.10).

A set of *bimodal columns* is useful in this case.[9] These consist of two or more columns with two specific pore sizes which differ by a factor of 10 (e.g. 100 and 1000 Å columns). If the pore volume of both gel types is the same, then a calibration graph with optimum linearity is obtained.

Note that a sample component which migrates nonselectively through a column, i.e. is subject to total exclusion or total permeation, inevitably undergoes unnecessary band broadening and hence impairs the resolution. The mobile phase should pass through the columns according to increasing pore diameter.

---

[9] W.W. Yau, C.R. Ginnard, and J.J. Kirkland, *J. Chromatogr.*, **149**, 465 (1978).

15.5 PHASE SYSTEMS

Mobile and stationary phases have to satisfy only three conditions:

(a) The sample must be readily soluble in the mobile phase.
(b) The sample must not interact with the stationary phase at all, e.g. by adsorption.
(c) The mobile phase must not damage the stationary phase.

If these three conditions are satisfied, then size-exclusion chromatography is relatively trouble-free, simple, versatile and fast.

The mobile phase must be a solvent in which the sample dissolves well and must also be compatible with the stationary phase. For example, μStyragel columns must not come into contact with water, alcohols, acetone, ethylmethyl ketone or dimethyl sulfoxide.

If the sample does not dissolve well in the mobile phase, then an unwanted interaction between sample and stationary phase may ensue. Interactions may be manifest in the form of tailing and, more especially, be followed by delayed elution. If a component appears later than V_0, then this is a sign that it has been retained by the stationary phase in some way. Adsorption effects are by no means a rare occurrence with porous glass and silica gel. Modified silica (e.g. diol or glyceropropyl) is often chosen over the nonderivatized form for separating biopolymers that are known to be highly polar, ionic strength and mobile phase pH being very significant aspects under these circumstances; optimum parameters may be determined empirically.

Adsorption effects can be reduced by using a mobile phase which is chemically related to the stationary phase, e.g. toluene with styrene–divinylbenzene columns.[10]

A detergent may be added to aqueous systems to prevent tailing and sometimes size-exclusion chromatography may require a fungicide to prevent fungal growth. Gases and suspended matter can render a gel column completely useless.

The smaller the particle size of the stationary phase the greater is the peak capacity of the column[11] and the better and faster are the separations.[12]

Size-exclusion chromatography is generally subdivided further into mobile phase categories. The mobile phase is aqueous in GFC (gel filtration chromatography) and is an organic solvent in GPC (gel permeation chromatography).

---

[10] A discussion of possible effects that are not exclusion phenomena as well as remedies were given by S. Mori, in *Steric Exclusion Liquid Chromatography of Polymers*, J. Janca, ed., Dekker, New York, 1984, pp. 161–211.
[11] H. Engelhardt and G. Ahr, *J. Chromatogr.*, **282**, 385 (1983).
[12] G. Guiochon and M. Martin, *J. Chromatogr.*, **326**, 3 (1985).

15.6 APPLICATIONS

Size-exclusion chromatography is an important separation method. It provides an easy way of solving many types of problem and as such should form an integral part of any HPLC laboratory set-up. There are four main areas of application:

(a) As a supplement to other HPLC methods. As columns are available in which a molar mass of 20 represents the permeation threshold, size- exclusion chromatography is now regarded as an effective means of separating small molecules.[13] The expenditure involved is roughly as low as for gas chromatography or even lower for preparative work. Size-exclusion chromatography should be considered for all samples in which the molar mass of the components differs by at least 10%. The determination of methanol and water in tetrahydrofuran (Figure 15.11) shows how effective the method is.

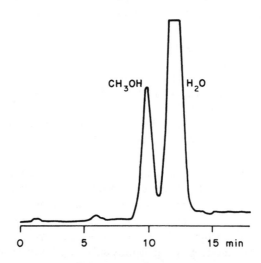

Figure 15.11 Analysis of methanol and water in tetrahydrofuran [reproduced with permission from R.K. Bade, L.V. Benningfield, R.A. Mowery and E.N. Fuller, *Int. Lab.*, **11**(8), 40 (1981)]. Conditions: sample, 40 µl of THF with 1% each of methanol and water; column, 25 cm × 7.7 mm i.d.; stationary phase, OR-PVA-500; mobile phase, 1 ml min^{-1} THF; temperature, 50 °C; dielectric constant detector.

---

[13] See, for example, R.A. Grohs, F.V. Warren and B.A. Bidlingmeyer, *J. Liquid Chromatogr.*, **14**, 327 (1991).

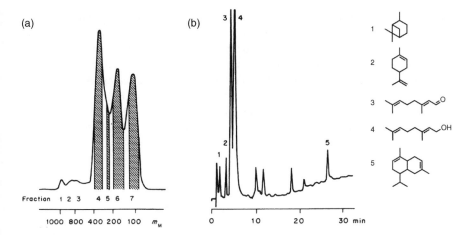

Figure 15.12 (a) Size-exclusion chromatogram of fruit juice [after J.A. Schmit, R.C. Williams and R.A. Henry, *J. Agric. Food Chem.*, **21**, 551 (1973)]. Conditions: column, 1 m × 7.9 mm i.d.; stationary phase, Bio-Beads SX-2; mobile phase, 0.8 ml min^{-1} chloroform; UV detector, 254 nm. (b) Reversed-phase chromatogram of fraction 6. Conditions: column, 1 m × 2.1 mm i.d.; stationary phase, Permaphase ODS; mobile phase, gradient from 5 to 100% methanol in water at 3% min^{-1}, UV detector, 254 nm. Peaks: 1 = pinene; 2 = limonene; 3 = neral; 4 = geranial; 5 = cadinene.

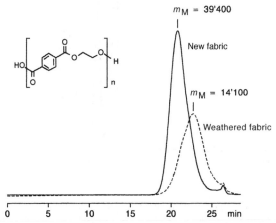

Figure 15.13 Analysis of synthetic polymers, using new and aged polyester fabric (after Swiss Federal Laboratories for Materials Testing and Research, St. Gallen). Conditions: sample, polyethylene terephthalate dissolved in hexafluoroisopropanol; columns, four columns 25 cm × 7 mm i.d.; stationary phases, Hibar LiChrogel PS 1, PS 20, PS 400, and PS 4000 in series, 10 μm; mobile phase, 1 ml min^{-1} chloroform–hexafluoroisopropanol (98:2); temperature, 35 °C; UV detector, 254 nm.

(b) Separation of complex mixtures to replace the processing or cleaning stage. Examples are removal of salts and other lower molecular mass components from biological material; removal of plasticizers from synthetic polymers; analysis of components in chewing gum following prior separation into polymers (rubber), plasticizer, stabilizer and flavouring. Fruit juice analysis provides a good example of this, seven fractions being picked up by size-exclusion chromatography (Figure 15.12a). Fraction 6 was orange oil and subjected to further analysis using a reversed-phase column (Figure 15.12b).

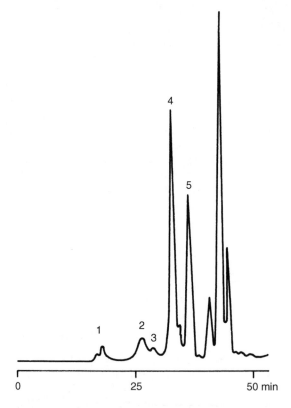

Figure 15.14 Separation of milk proteins [reproduced with permission from B.B. Gupta, *J. Chromatogr.*, **282**, 463 (1983)]. Conditions: sample, 100 μl of whey from raw skimmed milk (casein precipitated at pH 4.6); column, 60 cm × 7.5 mm i.d.; stationary phase, TSK 3000 SW (silica, 10 μm); mobile phase, 0.5 ml min^{-1} buffer containing 0.1 M NaH$_2$PO$_4$, 0.05 M NaCl and 0.02% NaN$_3$ (pH 6.8); UV detector, 280 nm. Peaks (with molecular masses): 1 = high molecular weight proteins; 2 = γ-globulin (150 000); 3 = bovine serum albumin (69 000); 4 = β-lactoglobulin (35 000); 5 = α-lactalbumin (16 500); other components not identified.

(c) Analysis of synthetic polymers (plastics). With size-exclusion chromatography it is possible to judge, e.g., if a polymerization reaction is running as expected, if raw materials fulfil the requirements, or if end products have the desired properties of usage due to their molecular mass distribution. Figure 15.13 compares a new polyethylene terephthalate fabric (Trevira®) with an aged fabric (weathered for one year). The average length of the polymer chains has decreased markedly, a fact which manifests itself by a drastically reduced tensile strength (only 18% of the original value). Together with appropriate software it is easily possible to determine the polydispersity from weight-average M_w and number-average M_n.

(d) Characterization of samples of biological origin (qualitative and quantitative analysis, preparative isolation or individual proteins) (Figure 15.14).

16 Affinity Chromatography

16.1 PRINCIPLE[1]

Affinity chromatography is the most specific chromatographic method. The interaction is biochemical in nature, e.g.:

antigen	$\longleftrightarrow$	antibody
enzyme	$\longleftrightarrow$	inhibitor
hormone	$\longleftrightarrow$	carrier

The highly specific nature of these interactions is due to the fact that the two
participating compounds are ideally suited to each other both spatially and electro-
statically. One component (ligand) is bonded to a support (in a similar way to a phase
chemically bonded to silica) and the other (sample) is adsorbed from solution, the
process being reversible (Figure 16.1).

If other sample components are present in the solution, e.g.:

then they do not match the ligand and are not adsorbed.

The sample is specifically retained by the stationary phase, whereas the
other molecules (proteins, enzymes, etc.) are removed by the mobile phase. The

---

[1] R.R. Walters, *Anal. Chem.*, **57**, 1099A (1985); N. Cooke, *LC GC Mag.*, **5**, 866 (1987); K. Jones, *LC GC Int.*, **4**(9), 32 (1991).

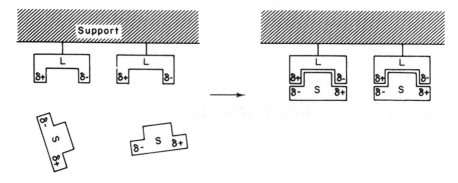

Figure 16.1 Principle of affinity chromatography. L = ligand; S = sample. $\delta +$ and $\delta -$ represent partial charges (less than one elementary charge).

sample is now freed from all other impurities. It cannot be isolated until it has been separated from the stationary phase, this being done by elution with a solution containing a product with a great affinity to the ligand, or even by a change in pH or ionic strength. As any biochemical activity can take place in just one specifically defined medium, it is clear that a pH or concentration gradient may break the specific interaction.

The sample [S] can be eluted as shown in Figure 16.2.

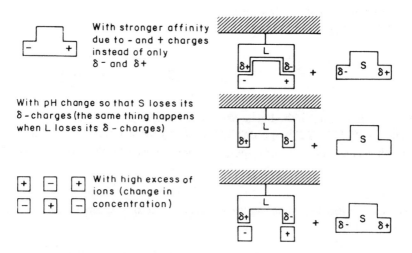

Figure 16.2 Elution alternatives in affinity chromatography.

16.2 AFFINITY CHROMATOGRAPHY AS A SPECIAL CASE OF HPLC[2]

Affinity chromatography differs from other chromatographic modes in that a suitable stationary phase can specifically 'catch' either a single or several components out of a random mix of products owing to a naturally occurring biospecific bond. A suitable elution process then provides the pure compounds(s).

In some cases it is not necessary to use a columm for isolation purposes; hence affinity chromatography is also carried out with membranes or disks.[3] Classical columns with a hydrostatic eluent feed offer a further possibility. This chapter, however, is confined to a description of separations with high-performance stationary phases (10 µm and below) with which rapid chromatography can be achieved. Very small columns may be used.

Stationary phases[4] for affinity chromatography can be prepared by the user, often starting from diol- or aminosilica. However, it is much simpler to buy an 'activated' gel which allows the desired ligand to be bound according to well known methods. A possible example for activation and binding is shown in Figure 16.3. For separation problems that require common ligands, ready-to-use stationary phases

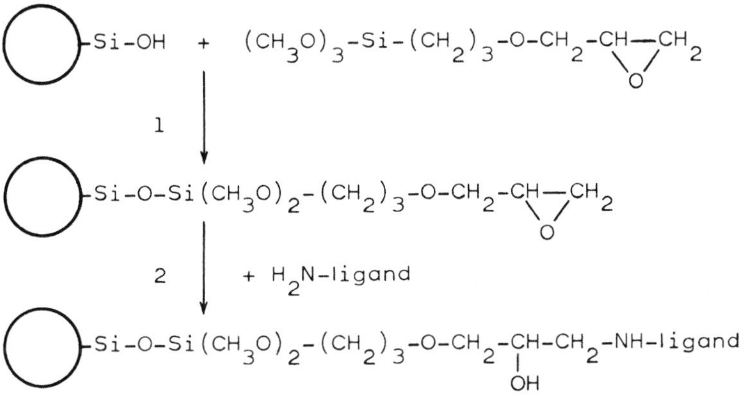

Figure 16.3 Reaction of silica with γ-glycidoxypropyl trimethoxysilane for the preparation of 'activated' epoxy silica (step 1) and ligand binding (step 2). The ligand needs to be an amine. The reactions take place at room temperature.

---

[2] P.O. Larsson, *Methods Enzymol.*, **104**, 212 (1984); G. Fassina and I.M. Chaiken, *Adv. Chromatogr.*, **27**, 247 (1987); K. Ernst-Cabrera and M. Wilchek, *TrAC*, **7**, 58 (1988); A.F. Bergold, A.J. Muller, D.A. Hanggi and P.W. Carr, in *HPLC Advances and Perspectives*, vol. 5, C. Horváth (ed.), Academic Press, New York, 1988, pp. 95–209.

[3] M. Peterka *et al.*, *J. Chromatogr. A*, **1109**, 80 (2006).

[4] N.E. Labrou, *J. Chromatogr. B*, **790**, 67 (2003); J.E. Schiel *et al.*, *J. Sep. Sci.*, **29**, 719 (2006).

are commercially available. Most phases are built up with a long-chain group between the silica and the ligand, the so-called *spacer*, to guarantee free accessibility of the sample molecules to the bonding site at the ligand.

The bonded sample can be eluted by a gradient (pH, ionic strength or competing sample) or by a 'pulse', the latter being a typical affinity chromatography characteristic. The bond-breaking compound is injected whilst the mobile phase passes through the column unchanged.

The sample size is restricted only by the bonding capacity of the column. The yields of biologically active protein frequently reach 100%, which means that denaturation and irreversible adsorption are negligibly low in many cases.

16.3 APPLICATIONS

Anti-IgG[5] ligands bond all antibodies of the IgG class specifically. The stationary phase synthesis required for the separation of IgG shown in Figure 16.4 was described

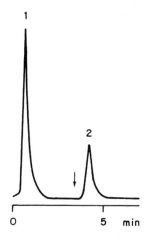

Figure 16.4 Affinity chromatography of IgG [reproduced with permission from J.R. Sportsman and G.S. Wilson, *Anal. Chem.*, **52**, 2013 (1980) © American Chemical Society]. Conditions: sample, 10 μl of solution with 14 μg of human IgG; column, 4 cm × 2 mm i.d.; stationary phase, antihuman-IgG on LiChrospher Si 1000, 10 μm; mobile phase, initially 0.5 ml min⁻¹ PBS (pH 7.4) (phosphate-buffered saline); ↓ shows change to 0.01 M phosphate buffer pH 2.2; fluorescence detector, 283/335 nm. Peaks: 1 = nonretained components, unspecified; 2 = IgG.

---

[5] IgG = immunoglobulin G.

by the authors in the reference cited in the caption. The mobile phase starts off with a pH of 7.4; a switch to pH 2.2 breaks the antigen–immunoadsorbent bond and the IgG is eluted. As the characteristic IgG fluorescence is quenched at pH 2.2, the pH of the column effluent had to be raised by adding a buffer of pH 8. The yield of active IgG was in excess of 97%. The stationary phase remains active for a long time provided that the column is stored at 4 °C when not in use.

Lectins are plant proteins that can specifically bond hexoses and hexosamines. Concanavalin A is a lectin widely used in the affinity chromatography and can isolate glycoproteins, glycopeptides and glycolipids. Elution is triggered by a pulse of sugar solution. Separation of the enzyme peroxidase (a glycoprotein) is shown in Figure 16.5. Detection is effected at two wavelengths simultaneously: 280 nm (nonspecific for all proteins) and 405 nm (peroxidase-specific owing to the proto-haemin group).

Cibacron Blue is a popular ligand as it can bind a large number of enzymes and also some blood proteins.[6] It is often referred to as a 'pseudo-affinity ligand' because it is a

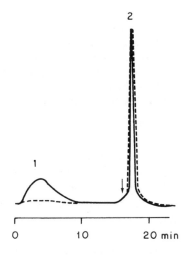

Figure 16.5 Affinity chromatography of peroxidase (after A. Borchert, P.O. Larsson and K. Mosbach, *J. Chromatogr.*, **244**, 49 (1982)]. Conditions: sample, 4.1 ml of solution with 1 μg ml^{-1} of protein; column, 5 cm × 5 mm i.d.; stationary phase, concanavalin A on LiChrospher Si 1000, 10 μm; mobile phase, 0.05 M sodium acetate, 0.5 M sodium chloride, 1 mM calcium chloride, 1 mM manganese chloride, pH 5.1;↓ shows pulse of 4.1 ml of 25 mM α-methyl-D-glycoside; detector, solid line, UV, 280 nm; broken line, visible, 405 nm. Peaks: = nonretained proteins; this peak contains less than 2% peroxidase; 2 = peroxidase.

[6] M.D. Scawen, *Anal. Proc. (Lond.)*, **28**, 143 (1991).

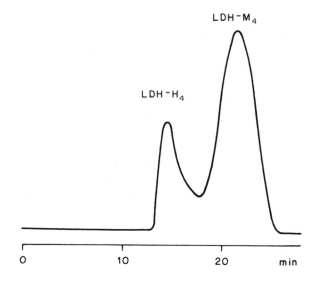

Figure 16.6 Separation of isoenzymes H_4 and M_4 of LDH [after C.R. Lowe, M. Glad, P.O. Larsson, S. Ohlson, D.A.P. Small, T. Atkinson and K. Mosbach, *J. Chromatogr.*, **215**, 303 (1981)]. Conditions: sample, 1.3 µg of LDH-H_4 and 11.8 µg of LDH-M_4; column, 10 cm × 5 mm i.d.; stationary phase, Cibacron Blue F3G-A on LiChrosorb Si 60, 5 µm; mobile phase, 1 ml min^{-1} potassium phosphate buffer (pH 7.5); after 3 min a 0–4 mM NADH gradient starts; detector, enzyme activity with UV, 340 nm.

synthetic triazine dye and not a naturally occurring biomolecule. This greatly reduces the activity of the eluted enzyme in most cases. One advantage of Cibacron Blue is its low price. The separation of isoenzymes H_4 and M_4 of lactate dehydrogenase (LDH), eluted in sequence by a gradient of reduced nicotinamide adenine dinucleotide (NADH), is shown in Figure 16.6. A post-column reactor for measuring enzyme activity was used as a detector (see Section 19.8).

Antibodies, as well as antigens, can be used as ligands. This case is often referred to as immuno affinity chromatography.[7]

[7] T.M. Phillips, in *High Performance, Liquid Chromatography of Peptides and Proteins*, C.T. Mant and R.S. Hodges, eds, CRC Press, Boca Raton, 1991, p. 507; G.W. Jack, *Mol. Biotechnol.*, **1**, 59 (1994).

17   Choice of Method

17.1   THE VARIOUS POSSIBILITIES

It should not be too difficult to select a suitable phase pair from all the background information given hitherto.

Small, nonionic molecules can be separated by adsorption, reversed-phase or chemically bonded phase. Under certain circumstances it may be difficult to choose between the various alternatives (in general terms, between normal- and reversed-phase methods), although the specific features of each individual system have been discussed in the relevant chapters. Should any doubt persist as to the success of a particular method (e.g. as could be the case if both silica and octadecylsilane were potential separation materials), then any column which is to hand should be tried. It is better if the mobile phase is too strong to begin with. Most components are eluted by a gradient of 10–90% methanol or acetonitrile in water over octadecylsilane.[1] The actual composition of the mobile phase at the moment of elution provides information on a suitable eluent mixture under isocratic[2] conditions.[3] As already mentioned in Section 4.7, the sample should not be completely insoluble in the mobile phase.

Ionic samples are separated by ion exchange, ion or ion pair chromatography. As explained in Section 10.5, reversed-phase chromatography may be considered as an alternative.

If the sample components are sufficiently different in *size*, then it is worth trying size-exclusion chromatography (gel filtration in the case of polar samples, gel permeation for apolar samples). Molecules with a molar mass in excess of 2000 are

---

[1] A. Ceccato et al., *Organic Process Res. Dev.*, **11**, 223 (2007).

[2] Isocratic = under constant separation conditions with no gradient (especially solvent gradient).

[3] A.C.J.H. Drouen, H.A.H. Billiet, P.J. Schoenmakers and L. de Galan, *Chromatographia*, **16**, 48 (1982).

Practical High-Performance Liquid Chromatography, Fifth edition   Veronika R. Meyer
© 2010 John Wiley & Sons, Ltd

often impossible to separate by a method other than size-exclusion chromatography (for biomolecules affinity chromatography may work).

Affinity chromatography is suitable for special biochemical problems. A prior knowledge of the molecules to which the relevant components have a strong affinity is an essential prerequisite of ligand selection.

Samples that are sufficiently *volatile* (boiling point below 350 °C) are best separated by gas chromatography, provided that no decomposition occurs at elevated temperatures.

The more physical and chemical data there are available for each sample, the more chance there is of selecting the right phase system (and the most suitable detector), thus ensuring success right from the start.[4] Every successful separation adds to the analyst's wealth of experience. To a limited extent the choice of a method can be done by an expert system.[5]

An HPLC laboratory which has to deal with a range of different problems must be stocked with the following stationary phases:

(a) silica;
(b) octadecylsilane or octylsilane;
(c) strong cation exchanger;
(d) strong anion exchanger;
(e) gel filtration columns to cover the molecular size range of interest;
(f) gel permeation columns to cover the molecular size range of interest;
(g) possibly a few chemically bonded phases such as nitrile and amine.

For a number of important classes of compound specially designed phases are available.

The tables in this chapter are intended to help in making a decision, yet a phase system selected as a result does not have to provide the optimum separation. There are problems that can be solved by methods that are not actually recommended for them.

Many stationary phases can be obtained in thin layer form, in which case the TLC test mentioned in Section 9.4 may be very useful.[6]

Note the principal fortes of the various types of stationary phases:

(a) silica: isomers;
(b) nonpolar chemically bonded phases: homologues;
(c) polar chemically bonded phases: functional group separation.

Proposal for Separating Mixtures of Unknown Composition

Table 17.1 shows a scheme according to Hewlett-Packard. The sample should be injected on to a reversed-phase column and eluted with a gradient of 10–90% acetonitrile in water.

---

[4] B. Serkiz *et al.*, *LC GC Int.*, **10**, 310 (1997).

[5] M. Peris, *Crit. Rev. Anal. Chem.*, **26**, 219 (1996).

[6] P. Renold, E. Madero and T. Maetzke, *J. Chromatogr. A*, **908**, 143 (2001).

TABLE 17.1 Scheme for separating mixtures of unknown composition (after Hewlett-Packard). Run a gradient of 10–90% acetonitrile in water on an octadecylsilane column

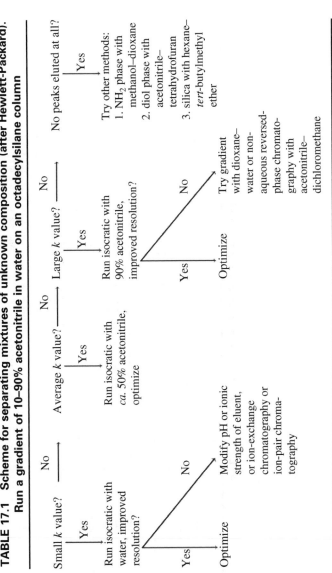

TABLE 17.2 Determination of separation method (after Varian)

Molar mass of sample				
< 2000			**> 2000**	
Water soluble		Water insoluble	Water soluble	Water insoluble
Non ions	Ions			
Reversed phase (for preference)	Basic compounds: cation exchange	Gel permeation	Gel filtration	Gel permeation
Gel filtration	Acidic compounds: anion exchange	Adsorption (silica): for isomers, class separations, preferably for nonpolar compounds		
	Gel filtration	Polar bonded phase (nitrile, amino, diol): for homologues, labile compounds, preferably polar compounds		
		Reversed phase (octadecyl, octyl): for homologues, labile compounds, preferably nonpolar compounds		

| Glycols, oligosaccharides | Fatty acids, alcohols, phenothiazines, surfactants | Metal cations, carbamates, vitamins, trisulfapyrimidines, catechol amines, nucleosides, purines, aromatic amines, glycosides | Organic acids, nucleotides, inorganic anions, sugars, analgesics, sulfonamides | Oligomers, lipids | Antioxidants, amines, steroids, prostaglandins, dyes, vitamins, barbiturates, hydrocarbons, phenols, alkaloids, amides, carotenoids, aflatoxins, anthraquinones, lipids, PTH-amino acids, nitrophenols, quinones, acids, nucleotides | Plasticizers, glycols, alcohols, steroids, phenols, anilines, alkaloids, benzodiazepines, pesticides, dyes, aromatics, metal complexes, polyethylene glycol | Alcohols, aromatics, antibiotics, barbiturates, carbamates, chlorinated pesticides, steroids, vitamins A, D, E, sulfonamides, anthraquinones, alkaloids, oligomers | Biopolymers, proteins, nucleic acids, oligosaccharides, peptides, poly(vinyl alcohol), poly(acrylic acid), pectins, carboxyethyl cellulose, polyvinyl-pyrrolidone | Organic polymers: polyolefins, polystyrene, polyvinyls, polyacrylates, polycarbonates, polyamides, natural rubber, silicones |

Determination of Separation Method

Table 17.2 shows a scheme due to Varian.

After the choice of the method its *optimization* will be necessary; especially for routine analysis this is indispensable. Figure 17.1 depicts the steps involved (note also the retention times of this hypothetical chromatogram!). Presumably the first separation will not be very successful. In the example shown only peak 5 is suited for quantitative analysis and no statement about the number of compounds present is possible (peak 0 is the breakthrough time). The first goal is the optimization of retention time. In most cases this is not difficult; the more complex a chromatogram is the higher k_{max} will be. The most difficult step is selectivity optimization.[7] This is dealt with in Chapter 18. In Figure 17.1 the elution order has now changed in comparison to the first chromatogram. (If no satisfactory selectivity can be obtained the choice of method was wrong.) For a single analysis the separation problem is now solved, otherwise system optimization is recommended. Here it was impossible to reduce the analysis time from 600 to 30 s, yet the peak resolution is markedly lower. Probably also method ruggedness has decreased, which means that the separation is more prone to disturbances (see Section 18.6). Finally, validation is necessary if the method is to be used in quality control (see Section 20.3).

17.2   METHOD TRANSFER[8]

Method transfer from one laboratory to another one (from development to routine, from manufacturer to customer and so on) can be difficult because HPLC separations are influenced by many parameters. At the new place the resolution of a critical peak pair can be worse than required or the whole chromatogram looks different. In order to prevent such surprises, whenever possible, it is necessary to describe every detail of the method: column dimensions, stationary phase (maybe even the batch number), preparation of the mobile phase (the order the individual components are mixed can be critical), temperature, volume flow rate, extra-column volumes of the instrument, the dwell volume in the case of gradient separations (see Section 4.3) as well as detection and integration parameters. It can be useful to designate alternative stationary phases, i.e. materials which are located close to each other in representations such as Figure 10.9. The true temperature in a column oven must be verified because it can differ from the requested one! Method transfer also includes the detailed description of sampling, storage and sample preparation.

[7] How to find hidden peaks: J. Pellett *et al.*, *J. Chromatogr. A*, **1101**, 122 (2006).

[8] J.J. Kirschbaum, *J. Pharm. Biomed. Anal.*, **7**, 813 (1989); D. Guillarme, J. L. Veuthey and V. R. Meyer, *LC GC Eur.*, **21**, 322 (2008); F. Bernardoni et al., *Chromatographia*, **70** 1561 (2009).

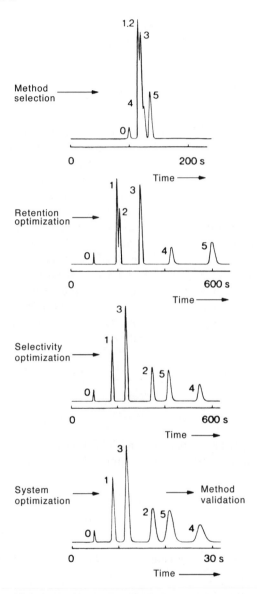

Figure 17.1 Method optimization [reproduced with permission from P.J. Schoenmakers, A. Peeters and R.J. Lynch, *J. Chromatogr.*, **506**, 169 (1990)].

When a separation is transferred to a column with a different inner diameter d_c, the flow rate F needs to be adapted in order to get identical retention times:

$$F_{new} = \frac{F_{old} \times d_{c,new}^2}{d_{c,old}^2}$$

The only parameter to be considered is the ratio of the squared diameters. If the particle diameter d_p is altered too, then the flow rate should also be corrected with regard to the van Deemter curve because the transferred separation should run at the same reduced velocity ν as before (preferably at the optimum). Smaller particles allow faster runs because ν is proportional to their diameter; in Section 8.5 we saw that $\nu = (d_p \times u)/D_m$. For a constant ν it is therefore necessary to work with a constant product $d_p \times u$. Thereby the pressure increases markedly. In Figure 7.3 the particle diameter was decreased from 10 µm to 3 µm. For an identical number of theoretical plates it was possible to reduce the column length to 30% of the original value (note that all the extra-column volumes of the HPLC instrument need to be adapted!). Despite the shorter column the pressure increases by a factor of ten when the condition $\nu = const$ is fulfilled as demonstrated in Figure 7.3. The benefit of these changes lies in the separation time which decreased to one tenth of its original value.

If column and particle diameters are changed simultaneously the following equation yields the necessary change in flow rate:

$$F_{new} = \frac{F_{old} \times d_{c,new}^2 \times d_{p,old}}{d_{c,old}^2 \times d_{p,new}}$$

Problem 33

The old method required a 4.6 mm column with 5 µm packing, used at a flow rate of 1.5 ml min^{-1}. Now a 2.0 mm column with 3 µm packing is planned to be used. Which flow rate is needed?

Solution

$$F_{new} = \frac{1.5 \times 2.0^2 \times 5}{4.6^2 \times 3} \text{ ml min}^{-1} = 0.47 \text{ ml min}^{-1} \approx 0.5 \text{ ml min}^{-1}$$

With such a method transfer it is usually not necessary to perform a re-validation as described in Section 20.3 (e.g., a re-determination of linearity is not needed). A documentation which shows that the Performance Qualification (Section 20.6) ist still fulfilled and that the chromatographic numbers of merit did not deteriorate will do the job.

18 Solving the Elution Problem

18.1 THE ELUTION PROBLEM

Complex mixtures, i.e. those that contain 20 or more components, in most cases present separation problems. Under isocratic conditions (Figure 18.1) the initial peaks are likely to be poorly resolved and the final peaks will probably be broad and flat and may be swamped by background noise. If a weaker solvent is used, the initial peaks show improved resolution but the final ones are not eluted at all. A stronger solvent compresses the early peaks together more, so that some components can no longer be distinguished.

This is then the 'general elution problem' and there are several ways of dealing with it:

(a) solvent gradient;
(b) column switching;
(c) temperature and flow rate gradients;[1]
(d) comprehensive two-dimensional HPLC.

All the above alternatives aim to create a greater distance between the early peaks and to speed up the elution of the latter ones so that they become closer together.

The elution problem does not occur in size-exclusion chromatography, which is one of its major advantages. Instead, elution of all components is guaranteed, the final ones coinciding with the void volume at the very latest. The only exception to this is in the case of adsorption effects, peaks being eluted later than V_0, but this can be prevented by a careful choice of stationary and mobile phases.

---

[1] T. Greibrokk and T. Anderson, *J. Sep. Sci.*, **24**, 899 (2001); B.A. Jones, *J. Liq. Chromatogr. Rel. Techn.*, **27**, 1331 (2005).

Practical High-Performance Liquid Chromatography, Fifth edition   Veronika R. Meyer
© 2010 John Wiley & Sons, Ltd

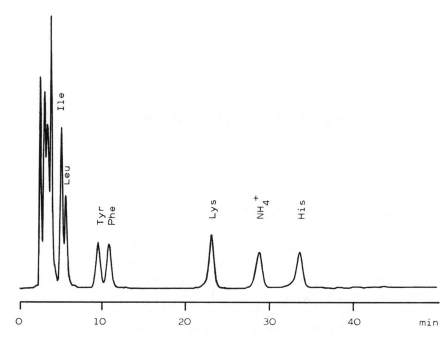

Figure 18.1 Isocratic separation of 17 amino acids, the early appearing ones being insufficiently resolved and the last one not eluted (reproduced by permission of A. Serban, Isotope Department, Weizmann Institute of Science, Rehovot). Conditions: sample, 50 µl containing 5 nmol of each amino acid; column, 15 cm × 4 mm i.d.; stationary phase, Amino Pac Na-2 (cation exchanger) 7 µm; mobile phase, 0.4 ml min^{-1} sodium citrate 0.2 N pH 3.15–sodium phosphate 1 N pH 7.4 (1:1); temperature, 55 °C; VIS detector 520 nm after derivatization with ninhydrine.

18.2 SOLVENT GRADIENTS[2]

In reversed-phase and ion-exchange chromatography it is a common procedure to run a gradient scouting run if the conditions for a successful separation are unknown. Such a run is performed from 10 to 100% B solvent (stronger solvent) with a linear profile. (As already explained, a 100% A mobile phase is often not recommended in reversed-phase chromatography because the alkyl chains are collapsed and equilibration with

───────────────

[2] P. Jandera, in: *Advances in Chromatography*, vol. 43, P. R. Brown et al., eds., Dekker, New York, 2005, pp. 1–108; L.R. Snyder and J.W. Dolan, *High-Performance Gradient Elution*, Wiley-Interscience, Hoboken, 2007.

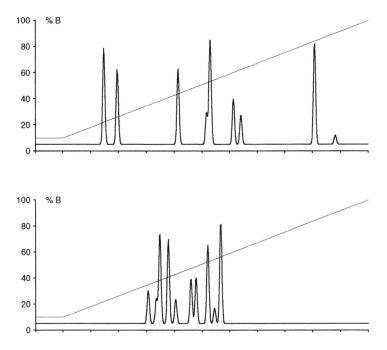

Figure 18.2 The decision for gradient (top) or isocratic (bottom) separation with a gradient scouting run from 10 to 100% B. The initial delay in the gradient profile comes from the dwell volume of the system. Hypothetical chromatograms are from a computer simulation.

an organic solvent is slow.) The decision whether it is better to run the separation in isocratic or gradient mode then proceeds as follows:[3]

If more than 25% of the gradient runtime is occupied with peaks, a gradient separation is better or even the only possibility; however, the %B span can be less than from 10 to 100%. A hypothetical peaks pattern is shown in the upper half of Figure 18.2.

If less than 25% of the gradient runtime is occupied with peaks (Figure 18.2 bottom) an isocratic separation should be tried. The favourable percentage of B solvent can be estimated from the chromatogram.

In both cases the separation often needs to be optimized by searching optimum selectivity (type of stationary phase, B solvent, mobile phase additives, temperature). The gradient itself must be selected to give a mobile phase that is initially just strong

[3] J.W. Dolan, *LC GC Int.*, **9**, 130 (1996) or *LC GC Mag.*, **14**, 98 (1996); J.W. Dolan, *LC GC Eur.*, **13**, 388 (2000) or *LC GC Mag.*, **18**, 478 (2000).

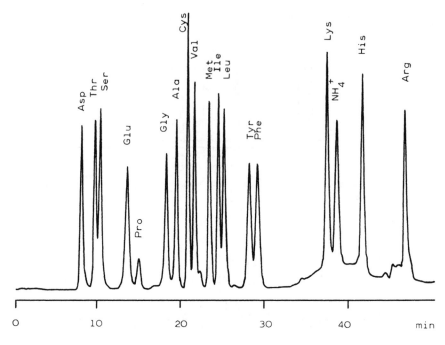

Figure 18.3 Separation of the amino acid mixture of Figure 18.1 with a solvent gradient (reproduced by permission of A. Serban, Isotope Department, Weizmann Institute of Science, Rehovot). Mobile phases: A, sodium citrate 0.2 N pH 3.15; B, sodium phosphate 1 N pH 7.4; C, sodium hydroxide 0.2 N. Gradient: 9 min with 100% A, then linear gradient to 100% B for 25 min, finally linear gradient to 15% C for 18 min. Other conditions as in Figure 18.1.

enough to elute the fastest peaks. Later, its composition is changed to elute the more retained components by increasing the amount of the strong solvent. Figure 18.3 shows again the separation of the 17 amino acids, now with an appropriate gradient program.

One possibility for gradient elution is the use of *step gradients*. Various solvents and a multiport valve are needed. The switching can be performed manually (e.g. during method development) or by means of a computer (Figure 18.4). Optimization is no more or less difficult than with a continuous gradient, nor is the quality of the final separation inferior in any way. However, ghost peaks may appear if solvent fronts build up in the column as a result of solvent demixing. This effect is more likely to arise the greater the difference in polarity between the individual eluents involved. In any event, a reference 'chromatogram' should

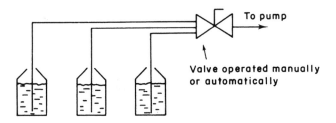

Figure 18.4 Step gradient system.

be set up for both step and continuous gradients, under identical conditions but without sample injection, and in this way the ghost peaks can be clearly identified.

Section 4.3 described the possibilities of producing *continuous solvent gradients*. The change in solvent composition may be linear, composed linear, concave or convex (Figure 18.5).

Concave gradients are useful if the second component elutes much more strongly than the first. In this instance, a small addition of B produces a considerable change in the mobile phase elution strength. Ghost peaks can also be found with continuous gradients. All solvents used for gradient elution must be extremely pure otherwise trace impurities from eluent A (or B) may be adsorbed on top of the column and eluted later by eluent B, giving rise once again to ghost peaks. Figure 18.6 shows an example

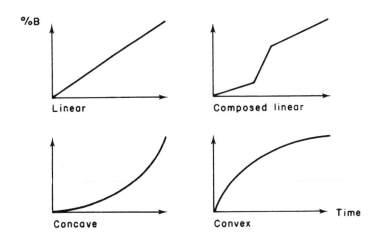

Figure 18.5 Gradient elution alternatives.

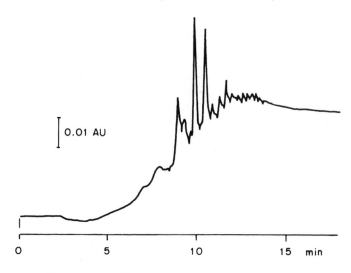

Figure 18.6   Ghost peaks with solvent gradient.

of this. A gradient of 20–100% B has been run in 10 min without sample injection (subsequently 100% B isocratic). The conditions were as follows: solvent A, double-distilled water; solvent B, acetonitrile (Merck, pro analysi); flow, $1\,\text{ml}\,\text{min}^{-1}$; stationary phase, Spherisorb ODS, $5\,\mu\text{m}$; column, $25\,\text{cm} \times 3.2\,\text{mm}$ i.d.; detector, UV, 254 nm. Conclusion: the acetonitrile or water used was not pure enough for this purpose!

A suitable gradient must be determined empirically (solvents involved, period between steps, shape of gradient profile), possibly following certain rules[4] or by computer aid.[5]

The disadvantage of solvent gradients is that the column has to be regenerated afterwards. However, this is relatively simple in the case of chemically bonded phases which are ready for injection again after about five times the column volume of the first eluent (A) has been pumped through. Successive solvents must, of course, be miscible! Pay attention to the *dwell volume* of the instrument when working with gradients, see Section 4.3.

Macromolecules behave differently to small molecules during gradient elution.[6] As can be seen from Figure 18.7, a protein (or another macromolecule) has a steep

---

[4] P. Jandera, *J. Chromatogr.*, **485** 113 (1989).
[5] B.F.D. Ghrist and L.R. Snyder, *J. Chromatogr.*, **459**, 43 (1988).
[6] D.W. Armstrong and R. E. Boehm, *J. Chromatogr. Sci.*, **22**, 378 (1984).

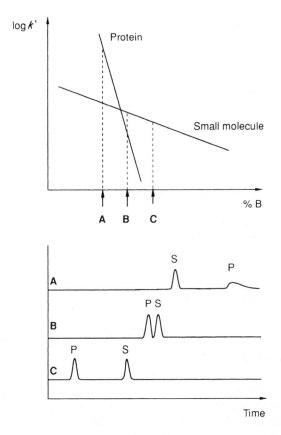

Figure 18.7 Elution behaviour of proteins P and small molecules S as a function of modifier concentration in the mobile phase, here presented as %B (stronger solvent, but could also be something else). The slope of the linear relationship is known as the *S* value; it is high for P and low for S. Small changes in %B have a distinct effect on *k* of the protein.

slope or a high so-called S value in a log $k = f(\%B)$ diagram whereas the S values of small molecules are much lower. Therefore it is not possible to separate a mixture of proteins without gradient in reversed-phase or ion-exchange chromatography.

As already mentioned, gradient elution is usually not recommended for adsorption columns and is unsuitable for refractive index and conductivity detection. Applications may be found everywhere in this book.

18.3 COLUMN SWITCHING[7]

Column switching provides an alternative to gradients elution, the various parts of the chromatogram being separated over two or more columns. The less significant parts of the chromatogram are often discarded; the *front, heart* or *end cuts* are obtained and separated further.

Columns and valves can be arranged in whatever manner the analyst cares to devise, provided that the main objective of maximum selectivity is kept in mind. Some possible column combinations are as follows:

(a) adsorbents with different specific surface areas;
(b) reversed phases with different chain lengths;
(c) ion exchangers of different strength;
(d) combination of anion and cation exchangers;
(e) combination of various methods such as ion-exchange and reversed-phase chromatography, size-exclusion and adsorption chromatography, affinity and reversed-phase chromatography, etc.[8]

The last alternative, which generally involves more than one mobile phase, is known as *multidimensional chromatography* and appears to be the most promising. All arrangements should attempt to provide a peak compression effect so that unavoidable extra-column volumes become insignificant and the highest possible separation performance is achieved.

Equipment for column switching is available commercially but equally good results can be obtained from a do it yourself system. Six- or ten-port valves can be used, one ten-port valve being capable of replacing two six-port valves.

There is a certain risk of system peaks (Section 19.9) with column-switching methods. System peaks can be invisible but they alter the shape of co-eluted peaks.[9]

One example can serve to illustrate the enormous potential of column switching,[10] *viz.* the determination of the aminoglycoside antibiotic tobramycin in serum. Patients on this drug require careful control to prevent any risk of harmful overdose. Figure 18.8 shows the column-switching arrangement required.

The proteins in 100 µl of serum are precipitated with 100 µl of 0.078 N sulfosalicyclic acid. Following centrifugation, 50 µl of the supernatant solution are injected into the sample loop (step A). The antibiotic is in protonated form. The injection valve is then turned so that the sample is transferred by pump 1 with acidic mobile phase to

---

[7] R.E. Majors *et al.*, *LC GC Mag.*, **14**, 554 (1996); E. Hogendoorn, P. van Zoonen and F. Hernández, *LC GC Eur.*, **16** (12a), 44 (2003).

[8] Ion-exchange and size-exclusion chromatography of proteins: M.W. Bushey and J.W. Jorgenson, *Anal. Chem.*, **62**, 161 (1990).

[9] T. Arvidsson, *J. Chromatogr.*, **407**, 49 (1987).

[10] G.J. Schmidt and W. Slavin, *Chromatogr. Newsl. Perkin-Elmer*, **9**, 21 (1981).

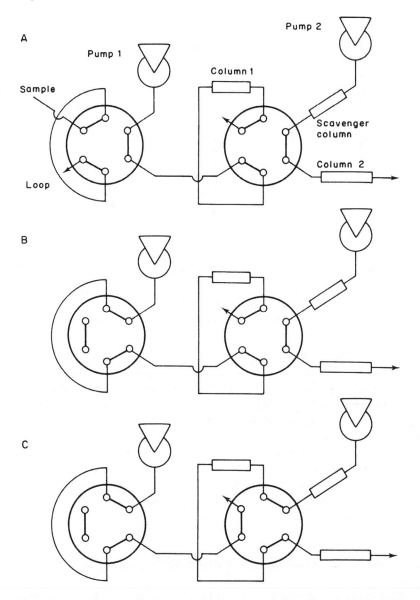

Figure 18.8 Individual stages in the determination of tobramycin. Conditions: column 1, 3 cm × 4.6 mm i.d. with 10 μm cation exchanger; column 2, 12.5 cm × 4.6 mm i.d. with 5 μm RP-18; scavenger column, 2.5 cm × 2.2 cm i.d. with 37–54 μm silica; mobile phase 1, 1.5 ml min⁻¹ of 10 mM sodium phosphate buffer (pH 5.2); mobile phase 2, 1.5 ml min⁻¹ of 50 mM EDTA (pH 8.8).

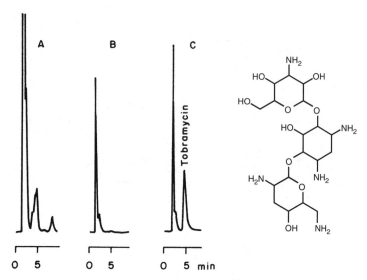

Figure 18.9 Determination of tobramycin (reproduced by permission of Perkin-Elmer). (A) Serum blank, direct on column 2; (B) serum blank, on columns 1 and 2; (C) serum from a patient treated with tobramycin, on columns 1 and 2.

column 1 (step B). As this short column contains a cation exchanger, tobramycin is retained whilst most of the impurities are flushed to waste. Following this, the switching valve is turned (step C), starting the back-flushing of the antibiotic with basic mobile phase from pump 2 and separation commences on column 2. (The scavenger column at pump 2 saturates the mobile phase with SiO_2.) Postcolumn derivatization with *o*-phthaladehyde is followed by fluorescence detection at 340–440 nm.

Figure 18.9 shows how successful this two-dimensional process actually is. A is the chromatogram for 'blank' serum with direct injection on to column 2 which is accompanied by too many impurities. B represents a switching between columns 1 and 2; the 4–6 min range contains no peaks. C is the determination of tobramycin in a patients's serum.

18.4 COMPREHENSIVE TWO-DIMENSIONAL HPLC[11]

The word says it: comprehensive chromatography is an all-including technique, a method which allows to investigate the whole eluate in greater detail. In the case of

---

[11] S.A. Cohen and M.R. Schure, eds., *Multidimensional Liquid Chromatography*, John Wiley & Sons, Ltd, Chichester, 2008; D.R. Stoll *et al.*, *J. Chromatogr. A*, **1168**, 3 (2007); P. Dugo *et al.*, *J. Chromatogr. A*, **1184**, 353 (2008).

column switching, as described in Section 18.3, only a part of the eluted fractions is transferred to and analysed on a second column. With the comprehensive approach, all of the eluate is separated on-line a second time on another short column with different selectivity. This allows the separation of peak clusters which are resolved only poorly or not at all in the first dimension. However, a success can only be expected if the two columns involved differ markedly in their properties.[12] The mobile phases can be identical or different. Usually they are completely miscible; obvious separation pairs are aqueous size-exclusion chromatography (gel filtration) and reversed phase, or ion exchange and reversed phase. It is even possible to combine two eluents with only partial miscibility if they are selected cleverly. The first dimension can run in the normal phase mode and the second one in reversed phase.[13] This approach guarantees for the maximum difference of the phase systems, thus the maximum width of the separation procedure.

The first dimension can be a gradient run with long separation time. The eluting mobile phase is fed to the second dimension in short intervals of 1 to 5 min, realized by loops of appropriate size (to store the fractions during the cycle time), by short 'storage columns', or by using two identical columns of the second-dimension type which are used alternately.[14] The duration of the second separation must not be longer than the desired pace of the second series of chromatograms, i.e. 1, 2 or 5 min to the maximum. The second dimension can be operated in the gradient mode as well although the instrumentation for isocratic separations is simpler.

Such chromatograms are often represented in two dimensions, namely as spots with contour lines or with coloured intensity gradation. We look at them as we look on maps, i.e. from 'above'.

With comprehensive two-dimensional chromatography the attainable peak capacity increases enormously,[15] thus also the chance for the separation of difficult and complex separation problems. Figure 18.10 shows the two-dimensional separation of antioxidants on a polar (polyethylene glycol silica) and a nonpolar (C_{18} silica) stationary phase.

18.5 OPTIMIZATION OF AN ISOCRATIC CHROMATOGRAM USING FOUR SOLVENTS

Many separation problems require no gradient at all if the mobile phase is optimized. The disadvantage of all gradients is that they require reconditioning, which takes a certain amount of time.

[12] P. Jandera, *J. Sep. Sci.*, **29**, 1763 (2006); P. Jandera, *LC GC Eur.*, **20**, 510 (2007) or *LC GC North Am.*, **26**, 72 (2008).
[13] P. Dugo et al., *Anal. Chem.*, **76**, 2525 (2004).
[14] I. François et al., *J. Chromatogr. A*, **1178**, 33 (2008).
[15] X. Li, D.R. Stoll and P.W. Carr, *Anal. Chem.*, **81**, 845 (2009).

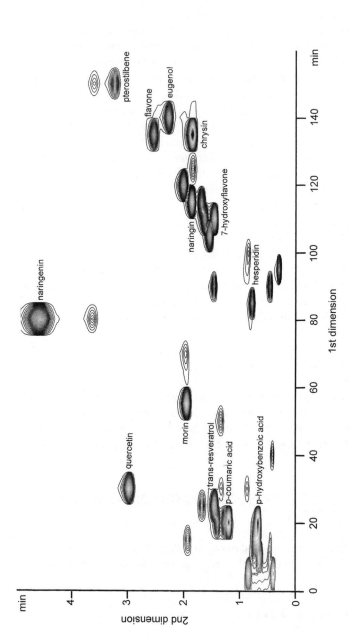

Figure 18.10 Two-dimensional separation of a mixture of phenolic and flavone antioxidants [after P. Jandera, University of Pardubice, Czech Republic; see also F. Cacciola *et al.*, *J. Chromatogr. A*, **1149**, 73 (2007)]. Conditions in the first dimension: column, 15 cm × 4.6 mm i.d.; stationary phase, PEG silica 5 μm; mobile phase, 0.3 ml min^{-1} water–acetonitrile, gradient 1–55% acetonitrile in 200 min. Interface: ten-port valve with two storage columns X-Terra C18 2.5 μm, 3 cm × 4.6 mm i.d. which concentrate the eluate. Cycle time, 5 min. Conditions in the second dimension: column, 10 cm × 4.6 mm i.d.; stationary phase, SpeedROD RP-18e (monolith); mobile phase, 2 ml min^{-1} water–acetonitrile, gradient 1–40% acetonitrile in 5 min. Diode array detector with 254 + 260 + 280 + 320 nm. Of the numerous identified analytes only the most important ones are specified in the figure.

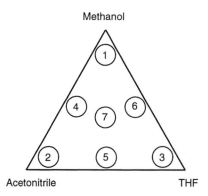

Figure 18.11   Position of seven mixtures in the selectivity triangle.

Sections 9.4 and 10.3 have already provided the basis for optimization by attempting to work with three different solvent mixtures: hexane–ether, hexane–dichloromethane and hexane–ethyl acatate for adsorption chromatography and water–methanol, water–acetonitrile, water–tetrahydrofuran for reversed-phase systems. However, this concept is not restricted to binary mixtures but a third or even a fourth component may be added in an attempt to improve the separation. An arrangement of seven different mixtures (Figure 18.11) provides the best basis for systematic evaluation.[16] An example is outlined below.

The best mixture (with respect to retention) for separating ten phenols by reversed-phase chromatography was found to be water–methanol (60 : 40), represented by chromatogram ①. The mobile phase polarity can be calculated as follows:[17]

$$P'_{mixture} = P'_1 \varphi_1 + P'_2 \varphi_2$$
$$= (0 \times 0.6) + (2.6 \times 0.4) = 1.04$$

P' is then kept constant for all seven chromatograms.
Chromatogram ② with acetonitrile:

$$\varphi_{ACN} = \frac{P'_{mixture}}{P'_{ACN}} = \frac{1.04}{3.2} = 0.33$$

Water: acetonitrile $= 67 : 33$.

---

[16] J.L. Glajch, J.J. Kirkland and J.M. Minor, *J. Chromatogr.*, **199**, 57 (1980).
[17] Compare with Section 10.3. The P' values used here are different, namely: $P'_{MeOH} = 2.6$, $P'_{ACN} = 3.2$, $P'_{THF} = 4.5$.

Chromatogram ③ with THF:

$$\varphi_{THF} = \frac{1.04}{4.5} = 0.23$$

Water: THF $= 77 : 23$.

Chromatogram ④ with methanol and acetonitrile. The mobile phase is a 1 : 1 mixture of the eluents used in ① and ②.

Water: $(0.5 \times 60) + (0.5 \times 67) = 63.5$ parts
Methanol: $0.5 \times 40 = 20$ parts
Acetonitrile: $0.5 \times 33 = 16.5$ parts
[control: $P' = (0.635 \times 0) + (0.2 \times 2.6) + (0.165 \times 3.2) = 1.05$]

Chromatogram ⑤ with acetonitrile and THF. The 1 : 1 mixture of 2 and 3 consists of water–acetonitrile–THF in the ratio $72 : 16.5 : 11.5$.

Chromatogram ⑥ with methanol and THF. The 1 : 1 mixture of 1 and 3 consists of water–methanol–THF in the ratio $68.5 : 20 : 11.5$.

Chromatogram ⑦ with methanol, acetonitrile and THF. The mobile phase is a 1 : 1 : 1 mixture of the eluents used in ①, ② and ③.

Water: $(0.33 \times 60) + (0.33 \times 67) + (0.33 \times 77) = 68$ parts
Methanol: $0.33 \times 40 = 13$ parts
Acetonitrile: $0.33 \times 33 = 11$ parts
THF: $0.33 \times 23 = 8$ parts

One of these separations presumably lies very close to the optimum, i.e. has a neighbouring peak resolution of 1 at the very least. Should none of the separations be entirely satisfactory, it is obvious now which other mixtures appear to show promise.

This type of optimization is tiresome to carry out in practice unless the chromatograph is provided with four solvent reservoirs and a suitable computer control facility. The latter can also be used to calculate and control the seven eluent mixtures.

Figure 18.12 shows chromatograms obtained for the seven mobile phases. Phenols must be separated in weakly acidic (acetic acid) medium, this also having its own effect on selectivity. Hence, the computer calculates slightly different compositions to those described above. The final chromatogram ⑧ represents the optimum separation obtained out of all the solvent combinations tested and requires only three components (63% water with 1% acetic acid–34% methanol–3% acetonitrile).

18.6 OPTIMIZATION OF THE OTHER PARAMETERS

The mobile phase can be optimized following the method outlined above, as a matter of routine. The completeness of the method is particularly striking: mobile phase

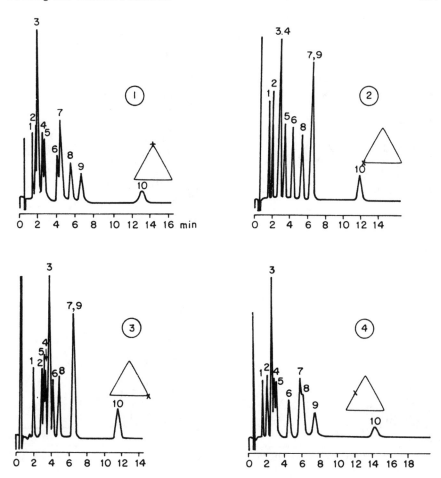

Figure 18.12 Systematic optimization of the separation of ten phenols (reproduced by permission of Du Pont). Peaks: 1 = phenol; 2 = *p*-nitrophenol; 3 = 2,4-dinitrophenol; 4 = *o*-chlorophenol; 5 = *o*-nitrophenol; 6 = 2,4-dimethylphenol; 7 = 4,6-dinitro-*o*-cresol; 8 = *p*-chloro-*m*-cresol; 9 = 2,4-dichlorophenol; 10 = 2,4,6-trichlorophenol.

components are not chosen at random but according to their position in the selectivity triangle and the sum of all components always totals 100%.

Then there are other parameters whose choice is not so compelling: stationary phase, temperature, pH[18] (or other ionic effects) and secondary chemical equilibria,

---

[18] P.J. Schoenmakers, S. van Molle, C.M.G. Hayes and L.G.M. Uunk, *Anal. Chim. Acta*, **250**, 1 (1991).

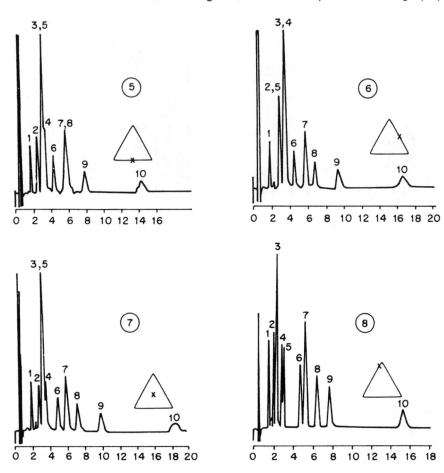

Figure 18.12 (*Continued*).

such as ion pairs. Optimization is much more difficult because the processes involved are open; hence conditions are often chosen by chance. However, systematic variation of some of the more promising parameters can be very successful, as the following example shows.[19]

A mixture of six compounds with different acid and base properties was separated isocratically on a reversed C_{18} phase. The influencing parameters that needed to be optimized simultaneously were the methanol content (%B) and the pH of the mobile phase. For systematic optimization it was necessary to perform a limited number of

[19] Y. Hu and D.L. Massart, *J. Chromatogr.*, **485**, 311 (1989).

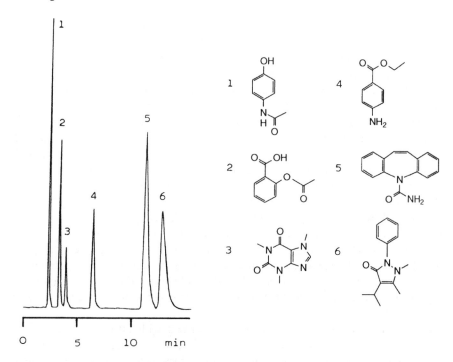

Figure 18.13 Optimized chromatogram of a six-component mixture with acids and bases [reproduced with permission from Y. Hu and D.L. Massart, *J. Chromatogr.*, **485**, 311 (1989)]. Conditions: column, 25 cm × 4 mm i.d.; stationary phase, LiChrosorb RP18 5 μm; mobile phase, 1 ml min^{-1} methanol-phosphate buffer pH 3.0 (1 : 1); UV diode array detector. Peaks: 1 = paracetamol; 2 = acetylsalicylic acid; 3 = caffeine; 4 = benzocaine; 5 = carbamazepine; 6 = propyphenazone.

experiments within the boundaries 45–65% B and pH 3.0–6.0. The results, i.e. the minimum resolution of the critical peak pair as a function of these parameters, were evaluated mathematically and represented as a surface in three-dimensional space (similar to Figure 18.16). It could be predicted that all components should be separated with sufficient resolution at 50% B and pH 3.0 or 4.0. This does, in fact, mirror reality and the separation of the mixture is shown in Figure 18.13.

Optimization may be systematic, as in the example shown, or directed and both processes can be automated.[20] *Systematic optimization* means that the interesting

[20] S.N. Deming and S.L. Morgan, *Experimental Design: A Chemometric Approach*, Elsevier, Amsterdam, 1991; B. Dejaegher and Y. Vander Heyden, *LG GC Eur.*, **20**, 526 (2007); S.L.C. Ferreira *et al.*, *J. Chromatogr. A*, **1177**, 1 (2008); P.F. Vanbel and P.J. Schoenmakers, *Anal. Bioanal. Chem.*, **394**, 1283 (2009).

parameters are varied according to a well designed pattern within certain limits. Boundaries and step widths must be selected by the analyst according to previous experience. Statistics may help in optimizing experimental design, thereby keeping the necessary number of chromatograms surprisingly low. *Directed optimization* involves deciding after each chromatogram whether optimization is proceeding in the right direction or not and continuing towards the optimum or changing direction accordingly. The simplex method is the best known example of this type of process.[21]

Automatic optimization requires an efficient computer system to establish whether one separation is better than another. A so-called 'resolution function', for which various difinitions have been proposed, is used as an objective criterion, covering resolution between neighbouring peaks and the analysis time involved. Identification by the computer of individual peaks, whether by the injection of standards or by spectroscopic means, is a distinct advantage. The simplex process may be used to optimize automatically the separation of complex acid–base mixtures by simultaneous variation of pH, ion pair reagent concentration, composition of ternary mobile phase, flow rate and temperature. Some automatic optimization systems are available commercially and, although they do leave some room for improvement, there is no doubt that they provide satisfactory answers to a great many problems.

Computer simulation of a separation performed with different experimental conditions can be very helpful in saving time and chemicals.[22] For this it is necessary first to run two chromatograms with differing composition of the mobile phase and to feed the results (the k values) to the computer. Several programs of this type are commercially available. Often the results are presented as *window diagrams*. Figure 18.14 depicts a window diagram with the resolution of the critical peak pair as a function of gradient runtime. Other influence parameters can be plotted similarly. This type of diagram allows optimum conditions to be found at a glance.

Simultaneous Optimization of Gradient Runtime and Temperature

Figure 18.15 represents an example of the computer-assisted optimization of linear gradient runtime and temperature. It was necessary to perform four preliminary experiments which included two different temperatures (50 and 60 °C) and two gradient runtimes (17 and 51 min). From the data of these four runs the optimum conditions were calculated by the computer to be 57 °C and 80 min. The simulated chromatogram of this proposal is shown at the top of Figure 18.15. Some peak pairs are not resolved better than with $R = 0.7$. Since not all peaks are of equal

[21] J.C. Berridge, *J. Chromatogr.*, **485**, 3 (1989).
[22] I. Molnar, *J. Chromatogr. A*, **965**, 175 (2002).

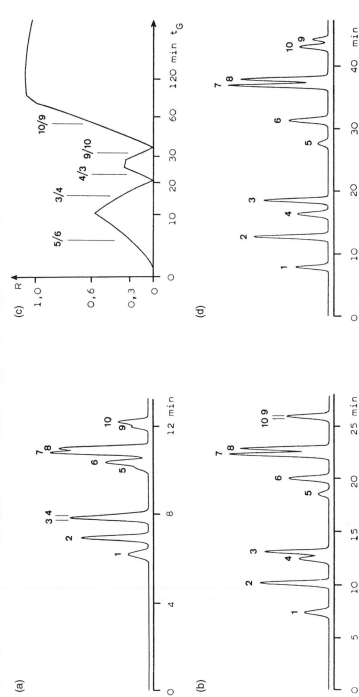

Figure 18.14 Window diagram. The separation was performed with a linear gradient from 0 to 45% B and the optimum runtime needs to be found out. (a) Gradient in 15 min; (b) gradient in 45 min with some elution orders reversed; (c) window diagram calculated from the initial two experiments with a linear relationship between retention time and %B assumed; the plot shows the resolution *R* of the peak pair which is critical under the respective conditions; it is necessary to use long gradient runtimes to obtain a good resolution; (d) optimized chromatogram with 0–45% B in 80 min, but separation is already finished after 45 min and 25% B.

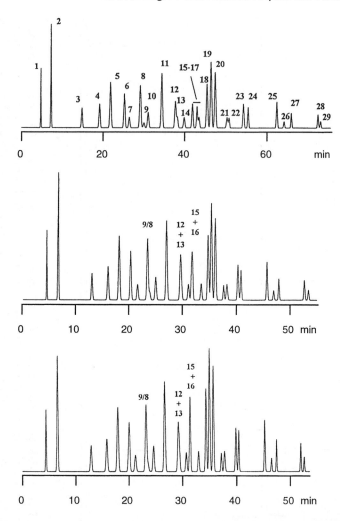

Figure 18.15 Optimization of gradient runtime and temperature [after J.W. Dolan *et al.*, *J. Chromatogr. A*, **803**, 1 (1998)]. Conditions: sample, algal pigments; column, 25 cm × 3.2 mm i.d.; stationary phase, Vydac 201tp C_{18}, 5 μm; mobile phase, 0.65 ml min^{-1}, gradient from 70 to 100% methanol in 28 mM tetrabutyl-ammonium acetate buffer pH 7.1. Top: computer simulation of a separation at 57 °C and 80 min gradient runtime; middle: computer simulation with three fused peak pairs at 55 °C and 54 min; bottom: experimental chromatogram under these conditions.

scientific importance, it was decided that peaks 8, 9, 12, 13, 15, and 16 did not need to be resolved. With this facilitation the computer proposed 55 °C and 54 min (middle). The real chromatogram is almost identical with the simulated one (bottom).

Besides separation quality (resolution of the critical peak pair) one needs to consider the *ruggedness* of a method (see also Section 20.3). It is not favourable at all to work at conditions where small fluctuations of the method, such as an unwanted change in pH, strongly influence the quality criterion. For routine separations it is much better to work under rugged conditions, even if this means a certain decrease in resolution. A hypothetical example is shown in Figure 18.16 where the lower quality conditions at point B have markedly higher ruggedness than a method located at the maximum point A.

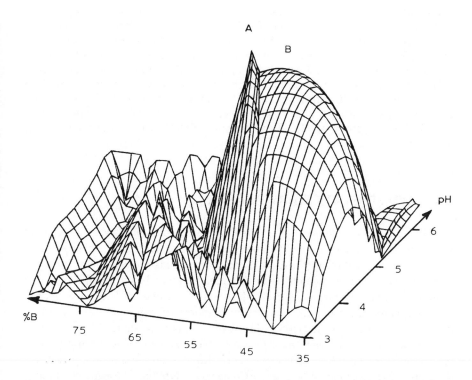

Figure 18.16 Ruggedness of a method. Here the quality criterion, e.g. resolution, is drawn as a surface in a space governed by modifier content and pH. Point A gives maximum quality but is highly prone to fluctuations in method conditions, especially towards higher %B or lower pH. Point B is more rugged although slightly lower in quality.

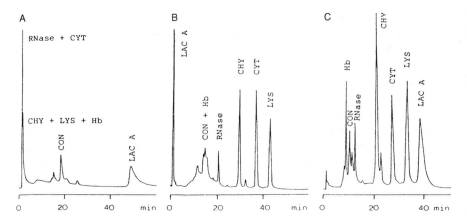

Figure 18.17 Separation of proteins on a mixed stationary phase [reproduced with permission of Elsevier Science Publishers BV from Z. el Rassi and C. Horvath, *J. Chromatogr.*, **359**, 255 (1986)]. Stationary phases: (A) strong anion exchanger (Zorbax SAX-300); (B) strong cation exchanger (Zorbax SCX-300); (C) 1:1 mixture of A and B. Particle size, 7 µm. Column, 8 cm × 6.2 mm i.d. (A and B), 10 cm × 4.6 mm i.d. (C); mobile phase, 20 mM tris-HCl (pH 7.0), gradient from 0 to 0.3 M sodium chloride in 40 min, 1.5 ml min^{-1} (A and B), 1.0 ml min^{-1} (C); UV detector, 280 nm. RNase = ribonuclease A; CYT = cytochrome c; CHY = α-chymotrypsinogen A; LYS = lysozyme; Hb = haemoglobin; CON = conalbumin; LAC A = β-mactoglobulin A.

18.7 MIXED STATIONARY PHASES

In optimization the stationary phase also has to be considered. Various materials should be tested, despite the expenditure necessary. It may even be advantageous to use two different stationary phases at the same time. This can be done with *coupled columns,* two or more columns with different packings being arranged in series (without column switching).[23] A useful device is the use of *mixed stationary phases*: a column is packed with a mixture of two or more different phases after a careful determination of promising types and ratio of components. Figure 18.17 shows the separation of acidic proteins (e.g. β-lactoglobulin A with an isoelectric point of pH 5.1) and basic proteins (e.g. lysozyme, pH 11.0). It is not possible to separate all of these proteins on a single cation or anion exchanger as is the case on a mixed stationary phase.

[23] S. Nyiredi, Z. Szücs and L. Szepesy, *Chromatographia Suppl.*, **63**, S-3 (2006); S. Louw *et al.*, *J. Chromatogr. A*, **1208**, 90 (2008).

19 Analytical HPLC

19.1 QUALITATIVE ANALYSIS[1]

The aim of qualitative analysis is to identify the peaks on the chromatogram or a specific component in the eluate. If the chemical nature of the peak is totally unknown, the special techniques described in Section 6.10 may be used as a basis or, alternatively, enough material must be collected from preparative or semi-preparative HPLC (Chapter 21) for the peaks to be identified using various analytical methods.

If one or several components in the sample mixture are presumed, then a comparison of k values of reference and sample under identical chromatographic conditions is the easiest means of analysis. If a reference compound has the same retention time as a peak in the chromatogram, then the two substances could be identical. However, it is necessary to make some more tests in order to obtain a higher degree of certainty:

(a) The reference compound is mixed with part of the sample and the mixture is injected. The suspect peak should become higher without producing any shoulders or broadening phenomena. The greater the number of theoretical plates of the column, the better is the chance of separating at least partially two different components with similar chromatographic behaviour.

(b) The same test should also be carried out with another mobile phase and preferably with another stationary phase (plus a suitable eluent). Phase pairs that behave entirely differently, i.e. a normal and a reversed phase, are preferred. If the sample-reference mixture continues to show no sign of separation, then the identity is fairly certain.

As mentioned above, the suspect peak must be analysed before absolute certainty can be established. However, qualitative analysis can be improved without

---

[1] M. Valcárcel et al., J. Chromatogr. A, **1158**, 234 (2007).

Practical High-Performance Liquid Chromatography, Fifth edition   Veronika R. Meyer
© 2010 John Wiley & Sons, Ltd

much expenditure:

(a) The more specific the detector is, the more it contributes to peak identification. An RI detector can pick up all components but a UV detector is only suitable for UV-absorbing materials and the fluorescence detector is specifically for fluorescing substances. The high specificity (that a compound fluoresces) can be further increased by the position of the excitation and emission wavelengths.

(b) A change in UV detector wavelength must be accompanied by an identical change in the signal of the sample and the reference, i.e. the increase or decrease should be of the same degree. The absorbance ratio between two random wavelengths is compound-specific.[2]

(c) The signal ratio in a UV and a subsequent RI detector (or any other detector pair that may be used) is characteristic for each compound, under specified conditions.

By far more evident but also more demanding are the special methods mentioned in Section 6.10; in particular, the diode array detector and the coupling with mass spectroscopy are instruments for qualitative analysis.

Any refractive index peaks due to the sample solvent should not be confused with analyte peaks (Figure 19.1). The sample is best dissolved in the mobile phase in order

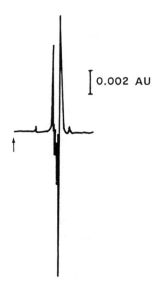

0.002 AU

Figure 19.1 Refractive index peak. Injection of 10 μl of methanol–water (1 : 1) (serves as sample solvent); mobile phase, dioxane–water (10 : 90); UV detector, 254 nm.

---

[2] A.C.J.H. Drouen, H.A.H. Billiet and L. De Galan, *Anal. Chem.*, **56**, 971 (1984).

to avoid being deceived by these ghost peaks. However, should this not be possible, a blind test can be carried out with the solvent used.

Retention time comparisons depend greatly on the constancy of the solvent delivery system and are less sensitive than the sample-reference mixture injection alternative.

19.2 TRACE ANALYSIS

If the analytical task is the qualitative or quantitative determination of an analyte at low concentration it is first necessary to optimize the chromatographic system. Under isocratic conditions the concentration at the peak maximum c_{max} is lower than in the injection solution c_i:[3]

$$c_{max} = \frac{c_i V_i}{V_R} \sqrt{\frac{N}{2\pi}}$$

where:

V_i = injected sample volume
V_R = retention volume, $V_R = t_R F$
N = theoretical plate number of the column.
(With gradient elution c_{max} is higher than when calculated with this formula.)

Problem 34

$10\,\mu l$ of a solution with 1 ppm $(10^{-6}\,g\,ml^{-1})$ of the analyte of interest are injected. The retention value of its peak is 14 ml. With this peak a plate number of $10\,000$ can be calculated. What is the peak maximum concentration?

Solution

$$c_{max} = \frac{1 \times 10}{14 \times 10^3} \sqrt{\frac{10\,000}{2\pi}}\,ppm = 0.028\ ppm$$

For trace analysis it is necessary to obtain as high a signal, i.e. a maximum concentration, as possible. How can this be done?

---

[3] B.L. Karger, M. Martin and G. Guiochon, *Anal. Chem.*, **46**, 1640 (1974).

Problem 35

For the separation of problem 34, a column of 15 cm length and 4.6 mm i.d. with 5 μm packing was used. The maximum concentration of 0.028 ppm gave a signal of 1 mV in the UV detector. Calculate the signal intensity under the following conditions, all other parameters remaining unchanged:

(a) column length 25 cm;
(b) column inner diameter 3 mm;
(c) column packing 3.5 μm.

Solution

(a) If the packing of the new column is of identical quality, its theoretical plate height H is also identical. Therefore, the plate number of the 25 cm column is 1.7 times higher and its retention volume is 1.7 times higher as well ($25 : 15 = 1.7$).

$$c_{max} = \frac{1 \times 10}{1.7 \times 14 \times 10^3} \sqrt{\frac{1.7 \times 10\,000}{2\pi}} \text{ppm} = 0.022 \text{ ppm}$$

corresponding to a signal of 0.8 mV.
(b)

$$V_R = \frac{14 \times 3^2}{4.6^2} \text{ml} = 6.0 \text{ ml}$$

$$c_{max} = \frac{1 \times 10}{6 \times 10^3} \sqrt{\frac{10\,000}{2\pi}} \text{ppm} = 0.066 \text{ ppm}$$

corresponding to a signal of 2.4 mV.
(c) With identical packing quality the column with 3.5 μm particles will give a plate number which is 1.4 times higher ($5.0 : 3.5 = 1.4$).

$$c_{max} = \frac{1 \times 10}{14 \times 10^3} \sqrt{\frac{1.4 \times 10\,000}{2\pi}} \text{ppm} = 0.034 \text{ ppm}$$

corresponding to a signal of 1.2 mV.

It follows that the lowest dilution or the highest peaks are obtained if a short and, most importantly, thin column with as high a theoretical plate number as possible, is used. The origin of the separation efficiency of the column must come from its fine, high-quality packing and not from its length. A longer column gives a lower maximum concentration because V_R as well as N is proportional to column length L_c; therefore c_{max} depends on $1/\sqrt{L_c}$. With a given sample concentration c_i, the peak dilution is less pronounced with a large injection volume, a high plate number, and a small retention volume.

The equation given above can also be written in a different manner (because $V_R = L_c d_c^2 \pi \varepsilon (k+1)/4$ and $N = L_c/H$):

$$c_{max} = 0.5 \frac{c_i V_i}{d_c^2 \varepsilon (k+1) \sqrt{L_c H}}$$

where:

d_c = column inner diameter
ε = total porosity
k = retention factor
L_c = column length
H = height of a theoretical plate.

Now the best approach is obvious:

— Use a column with small inner diameter. A twofold reduction gives a fourfold increase of c_{max}.
— Use a short column.
— Use a column packing with small theoretical plate height. This is obtained by a small particle diameter, excellent packing and good mass transfer properties of the stationary phase. The column must be run at its van Deemer optimum.
— Elute the analyte with a small retention factor.

Although a short column is to be preferred, its separation performance must be high enough to solve the separation problem. The method needs to be optimized with regard to the trace component.

The equations also make clear that the sample mass, i.e. the product $c_i V_i$, should be large. The column may be overloaded with regard to the main component as long as the resolution of the trace peak is not affected (Figure 19.2). However, the injection volume is restricted if band broadening needs to be avoided. The maximum allowed injection volume which restricts band broadening to 9% is given by:[4]

$$V_{i\,max} = 0.6 t_R F / \sqrt{N}$$

---

[4] M. Martin, C. Eon and G. Guiochon, *J. Chromatogr.*, **108**, 229 (1975).

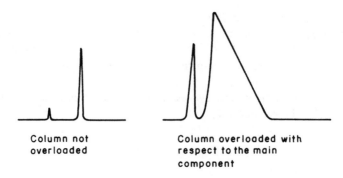

Column not
overloaded

Column overloaded with
respect to the main
component

Figure 19.2 Column overloading in trace analysis.

An alternative way of writing this equation is as follows [because retention time $t_R = t_0(k+1)$ and flow rate $F = L_c d_c^2 \pi \varepsilon / 4 t_0$]:

$$V_{i\,max} = 0.48 \frac{(k+1)L_c d_c^2 \varepsilon}{\sqrt{N}}$$

If enough sample is available (e.g. in food analysis) it is possible to inject $V_{i\,max}$ even if early eluted peaks will be broadened or the main component will be overloaded (as long as the resolution of the trace component is not affected).

Then the peak maximum concentration will be defined by the combined equations for c_{max} and $V_{i\,max}$ giving:

$$c_{max} = 0.24 c_i$$

This is the most favourable case with isocratic elution (and 9% peak broadening). Even then the fourfold analyte dilution is remarkably high.

Note that under these circumstances (enough sample available and injection of $V_{i\,max}$) all the above-mentioned requirements such as thin and short column, low plate height, and short retention time, come to nothing. If resolution is not affected at $V_{i\,max}$ the equation $c_{max} = 0.24 c_i$ is also valid for thick, long, and poorly packed columns. Nevertheless, it is always a good recommendation to optimize the separation system because the expenditure with regard to time and solvent will be lower.

In trace analysis some other parameters also need to be considered:

(a) Avoid tailing. An asymmetric peak is broader, thus less high than a symmetrical one.
(b) Peaks are compressed by gradients, therefore they are higher. Gradient elution is recommended if it is not possible to elute the trace compound at low retention factor. A step gradient at the right moment can be simple and effective.

(c) The detector should have as low a detection limit as possible. A bulk-property detector such as a refractive index detector is not suitable for trace analysis. Fluorescence and electrochemical detectors are most useful. With UV detection it is possible to conceal interfering impurities by choosing a suitable wavelength (see Figure 6.7); the same is possible with MS and single-ion monitoring. Derivatization (see Section 19.8) may increase the detection limit by several orders of magnitude. Quantitation by peak height may be more accurate than by peak area.

Although it is necessary to run a quantitative analysis at a signal-to-noise ratio of better than 10 (Section 6.1) an S/N ratio of 3 can be adequate for qualitative evaluation.

For trace analysis, the sample preparation (if possible combined with analyte enrichment) is of the utmost importance. HPLC itself can provide a highly effective method of concentration. Samples are concentrated at the column inlet if the elution strength of the solvent is too low.[5] If a mobile phase of high elution strength is used for elution, band broadening is small and the concentration in the eluate is high. This procedure is most effective if elution can be done with $k = 0$, i.e. with the solvent front. This is referred to as *displacement chromatography*. As an example, it was possible to enrich chlorophenols in water by a factor of 4000 on a styrene-divinylbenzene stationary phase.[6] *Column switching* (Section 18.3) may be used very effectively to separate the main components and to obtain trace analytes.

For the limits of quantification and of detection see the next section.

19.3 QUANTITATIVE ANALYSIS[7]

The detector signal is much more dependent on specific compound properties in liquid chromatography than it is in gas chromatography. For example, the UV signal is a function of the molar absorptivity, which varies between 0 and $10\,000\,l\,mol^{-1}\,cm^{-1}$ depending on the compound used. The molar absorptivity and absorbance maximum also vary between homologues. Hence at least one calibration chromatogram must be obtained for each quantitative analysis.[8]

---

[5] M. Hutta *et al., J. Sep. Sci.*, **29**, 1977 (2006): triazines in soil, injection volume $= 20\,ml$.

[6] F.A. Maris, J.A. Stab, G.J. De Jong and U.A.T. Brinkman, *J. Chromatogr.*, **445**, 129 (1988).

[7] J. Asshauer and H. Ullner, in *Practice of High Performance Liquid Chromatography*, H. Engelhardt, ed., Springer, Berlin, 1986, pp. 65–108; E. Katz (ed.) *Quantitative Analysis Using Chromatographic Techniques*, John Wiley & Sons, Inc., New York, 1987, 31–98; S. Lindsay, *High Performance Liquid Chromatography*, ACOL Series, John Wiley & Sons, Ltd, Chichester, 2nd ed., 1992., pp. 229–250 (with problems).

[8] Two-component mixtures can be quantitatively analysed without a calibration chromatogram or calibration graph. Solutions of identical concentration are prepared of each pure compound and the UV spectra are recorded. If the wavelength at which the spectra lines intersect is chosen for detection (isosbestic point), then the peak area is identical with the mixing ratio of the sample, on condition that the UV detector used shows high stability and excellent reproducibility of the selected wavelength (at least to $\pm\,0.2\,nm$).

Calibration can be performed with three different methods: external standard, internal standard and standard addition (see Figure 19.3). The procedures are explained here for single-point calibration although it is much better to establish a calibration graph with at least three data points in any case. The calculations are then not performed with simple rules of three, as below, but by means of the calibration graph slope. Calibration graphs should be straight and run through the origin. It is necessary to do a new calibration for each set of quantitative analyses. If older data are used it is quite possible that the actual chromatographic conditions are slightly different which may give rise to systematic deviations. Since quantitative analysis can be based on either peak area or peak height, the term 'signal' means both possibilities, see Section 19.5.

As an example, the problem concerns quantitative determination of glucose in soft drink. The content can be expected to be in the $5 \, \mathrm{g \, l^{-1}}$ range. The three procedures are as follows:

(a) Procedure for *external standard.* A standard solution of $6 \, \mathrm{g \, l^{-1}}$ (not necessarily $6.000 \, \mathrm{g \, l^{-1}}$ but exactly known) is prepared. It gives a peak of area 5400. The sample gives a peak of area 3600.

$$\text{Calibration factor CF} = \frac{6}{5400} = 1.11 \times 10^{-3}$$

$$\begin{aligned} \text{Content} &= \text{sample peak area} \times \text{CF} = 1.11 \times 10^{-3} \\ &= 4 \, \mathrm{g \, l^{-1}} \end{aligned}$$

(b) Procedure for *internal standard.* To both sample and standard solution (as above) a further component, which is not present in the sample, is added in equal amounts. Here this was done with $10 \, \mathrm{g \, l^{-1}}$ of fructose, with peak areas: glucose in sample, 3600; glucose in standard, 5400; fructose in both solutions, 6300. Now an amount ratio is obtained by comparison with the signal ratio:

$$\text{CF} = \frac{\text{standard amount ratio}}{\text{standard signal ratio}} = \frac{6 \times 6300}{10 \times 5400}$$

$$= \frac{0.6}{0.86} = 0.7$$

$$\text{Amount ratio of sample} = \text{signal ratio} \times \text{CF} = \frac{3600}{6300} \times 0.7 = 0.4$$

$$\begin{aligned} \text{Content} &= \text{internal standard amount} \times \text{amount ratio} \\ &= 10 \times 0.4 = 4 \, \mathrm{g \, l^{-1}} \end{aligned}$$

(c) Procedure for *standard addition.*[9] The sample is fortified with a known amount of the compound to be determined. Here this was done by addition of

---

[9] DIN 32633: Chemical analysis – Methods of standard addition – Procedure evaluation, Beuth, Berlin, 1998.

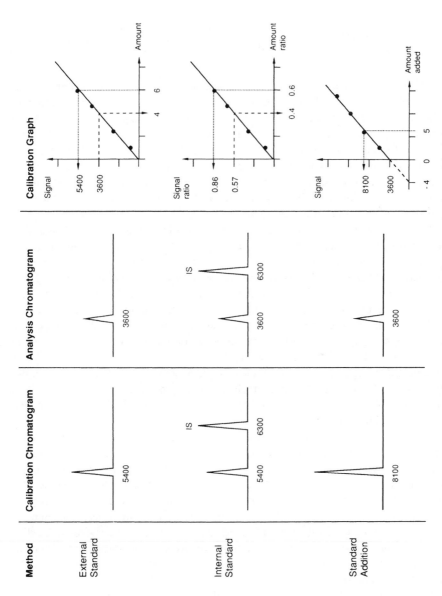

Figure 19.3   Calibration methods for quantitative analysis.

$5\,\mathrm{g\,l^{-1}}$ of glucose. The peak areas are 3600 for the sample itself and 8100 for the fortified sample.

$$\text{Content} = \frac{\text{addition} \times \text{sample net signal}}{\text{signal difference}} = \frac{5 \times 3600}{8100 - 3600} = 4\,\mathrm{g\,l^{-1}}$$

The external standard method is the simplest one and should therefore only be used for simple analytical problems. The injection must be performed with good reproducibility; thus it is recommended that the complete loop filling method be used (see Section 4.6). With multiple-point calibration it is not recommended to inject different volumes from a reference stock solution (e.g. 10, 20, 30, 40 and 50 µl) because it is well possible that neither accuracy nor precision of these injections are high enough. It is better to prepare a number of calibration solutions with different concentrations and to inject equal volumes of them (with complete loop filling or with exactly the same procedure as is used also for the sample).

In the case of internal standard calibration this is not necessary and small variations in injection volume cease to be important. If sample preparation is complicated or demanding, the internal standard approach is strongly recommended. In this case the standard is added before the first preparation step has begun. The choice of a suitable internal standard may not be easy. It must be a pure, clearly defined compound with similar properties with respect to sample preparation, chromatographic separation and detection to the compound(s) of interest. If possible it should be eluted in a chromatogram gap and not at the very beginning or end. Examples can be found in Figures 6.11, 11.2, 11.6, 13.1, 13.2 and 22.4.

Standard addition is elegant if the sample amount is not limited. It allows calibration of the analysis under realistic conditions, i.e. not with a standard chromatogram which is free from interferences. The standard addition and internal standard methods can be combined.

It is part of the validation process (see Section 20.3) to define how often a calibration chromatogram or calibration graph needs to be determined again. It is always necessary to consider the possible fundamental errors of calibration curves. They are shown in Figure 19.4. There is a difference between *constant-systematic deviations* and *proportional-systematic deviations*. In the first case the calibration curve does not run through the origin. The deviation can be positive, as in the figure, or negative. With the standard addition method this type of error cannot be recognized! A curve with proportional error has a false slope which can be too low, as in the figure, or too high. Both types of deviation can occur simultaneously. They can be detected by the determination of the recovery function (Section 19.4).

Various methods have been proposed to find the *Limit of Quantification* (LOQ). What needs to be defined is the lowest amount or concentration of an analyte whose quantitative determination can be done with a defined accuracy and repeatability. The most pragmatic approach (although some work is necessary) is to define the upper

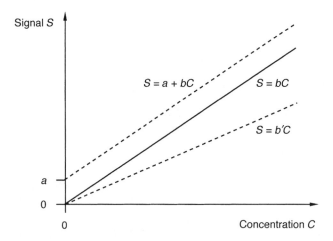

Figure 19.4 Erroneous calibration curves. The accurate curve with solid line runs through the origin and its slope is *b*. The lower broken curve has a proportional-systematic bias because its slope b' is wrong. The upper broken curve has the accurate slope but it runs through the *y* axis at point *a*, therefore it has a constant-systematic bias.

limit of the allowed repeatability standard deviation and to find it experimentally.[10] At least six different concentrations in the vicinity of the expected LOQ must be investigated with at least six determinations each.

Problem 36

Six different concentrations *c* of the calibration solution with matrix were analysed ten times each. The maximum allowed repeatability as relative standard deviation is 10%. The following data were found:

$c = 5$ ppm: 5.38 / 4.06 / 5.48 / 5.52 / 5.70 / 6.12 / 4.30 / 5.18 / 5.44 / 4.17 ppm.

$c = 10$ ppm: 8.09 / 9.36 / 11.29 / 9.92 / 10.76 / 10.50 / 9.35 / 11.38 / 10.21 / 12.78 ppm.

$c = 15$ ppm: 17.05 / 13.21 / 18.76 / 17.45 / 16.05 / 14.68 / 14.19 / 14.03 / 17.59 / 15.82 ppm.

$c = 20$ ppm: 17.86 / 19.36 / 20.87 / 18.97 / 18.37 / 19.82 / 17.90 / 18.25 / 23.42 / 22.79 ppm.

$c = 25$ ppm: 21.59 / 24.22 / 20.33 / 27.41 / 26.54 / 25.99 / 24.63 / 26.90 / 23.55 / 22.76 ppm.

$c = 30$ ppm: 31.03 / 28.82 / 32.32 / 27.84 / 31.86 / 31.37 / 26.96 / 25.47 / 30.91 / 33.09 ppm.

Where is the LOQ?

---

[10] J. Vial, K. Le Mapihan and A. Jardy, *Chromatographia Suppl.*, **57**, S-303 (2003).

Solution

$c = 5$ ppm: mean $\bar{x} = 5.14$ ppm, standard deviation $s(x) = 0.71$ ppm, RSD = 13.8%.
$c = 10$ ppm: $\bar{x} = 10.4$ ppm, $s(x) = 1.31$ ppm, RSD = 12.6%.
$c = 15$ ppm: $\bar{x} = 15.9$ ppm, $s(x) = 1.82$ ppm, RSD = 11.5%.
$c = 20$ ppm: $\bar{x} = 19.8$ ppm, $s(x) = 2.00$ ppm, RSD = **10.1%**.
$c = 25$ ppm: $\bar{x} = 24.4$ ppm, $s(x) = 2.37$ ppm, RSD = 9.7%.
$c = 30$ ppm: $\bar{x} = 30.0$ ppm, $s(x) = 2.54$ ppm, RSD 8.5%.

The LOQ is at 20 ppm.

The LOQ depends on the analyte due to the different detector properties and the k value. The *Limit of Detection* (LOD) is 30% of the LOQ, i.e. 7 ppm in the example given above.

Gradient elution can be source of additional error; therefore quantitative analysis should be performed in isocratic mode if possible. It is also necessary to keep the flow rate (for peak-area determinations) or mobile phase composition (for peak height determinations) constant (as can be seen later from Figure 19.5).

The sample should be dissolved in the mobile phase. If the sample solvent and mobile phase are not identical, then the analysis can become less accurate;[11] it is also possible that detector response is strongly dependent on sample solvent[12] (see also Section 4.7).

Quantitative analysis is strictly conditional on the column not being overloaded and the detector operating in the linear range. Experiments must be carried out to check these points. Accuracy may be increased by thermostating the column. Asymmetric peaks should be avoided. The analytical results can be verified in a number of ways.[13]

19.4   RECOVERY[14]

It is necessary to determine the recovery function and the recovery rate in order to detect a possible error in the analytical result by the process of sample preparation or even by the sample matrix itself.

---

[11] M. Tsimidou and R. Macrae, *J. Chromatogr. Sci.*, **23**, 155 (1985); S. Perlman and J. Kirschbaum, *J. Chromatogr.*, **357**, 39 (1986); F. Khachik, G.R. Beecher, J.T. Vanderslice and G. Furrow, *Anal. Chem.*, **60**, 807 (1988).
[12] J. Kirschbaum, J. Noroski, A. Cosey, D. Mayo and J. Adamovics, *J. Chromatogr.*, **507**, 165 (1990).
[13] J. Kirschbaum, S. Perlman, J. Joseph and J. Adamovics, *J. Chromatogr. Sci.*, **22**, 27 (1984).
[14] W. Funk, V. Dammann and G. Donnevert, *Quality Assurance in Analytical Chemistry*, Wiley-VCH, Weinheim, 2nd ed., 2006.

Procedure (Recovery of Sample Preparation)

(a) The pure reference analyte is analysed without sample preparation procedure (or without a certain sample preparation step). The calibration function is calculated:

$$y = a_0 + b_0 x_0$$

where:

y = signal
x_0 = mass or concentration of reference analyte
a_0 = y axis intercept
b_0 = slope .

(b) The pure reference analyte is analyzed using the sample preparation procedure. The analytical result is calculated with the calibration function:

$$x_p = \frac{y - a_0}{b_0}$$

where x_p = mass or concentration of reference analyte after sample preparation.

(c) The data x_p are plotted against the respective masses or concentrations x_0 and the recovery function is calculated:

$$x_p = a_p + b_p x_0$$

where the index p denotes the parameters found with sample preparation.

(d) The recovery rate is calculated:

$$RR = \left(\frac{a_p}{x_0} + b_p\right) \times 100\%$$

The recovery function should be linear with y axis intercept $a_p = 0$ and slope $b_p = 1$. If $a_p \neq 0$ a constant-systematic error is present, if $b_p \neq 1$ there is a proportional-systematic error of sample preparation.

The recovery function and rate can be determined for each individual step of sample preparation. This allows the specific improvement of the critical steps. Afterwards the same procedure can be performed with spiked samples for the determination of matrix effects.

Problem 37

The following peaks areas were found for aflatoxin solutions without sample preparation:

10 ng	2432 counts
20 ng	4829 counts
30 ng	7231 counts
40 ng	9628 counts

After sample preparation with solid phase extraction the following data were found:

10 ng	1763 counts
20 ng	4191 counts
30 ng	6617 counts
40 ng	9050 counts

Calculate the recovery function and the recovery rate. Judge the recovery function.

Solution

(a) Calibration curve with linear regression:

$$y = 32.5 + 240\, x_0$$

This function has a y axis intercept (a constant-systematic error) which is not zero. The reason probably comes from the HPLC separation and needs to be found and improved later.

(b)

$$x_p(10) = \frac{1763 - 32.5}{240} = 7.2 \text{ ng}$$

Analogously, the following data are found: $x_p(20) = 17.3$ ng, $x_p(30) = 27.4$ ng, $x_p(40) = 37.6$ ng.

(c) The recovery function is found by linear regression of the data pairs (7.2; 10), (17.3; 20), (27.4; 30) and (37.6; 40):

$$x_p = -2.91 + 1.01\, x_0$$

The slope is almost prefect but there is a small (1%) proportional-systematic error. The y axis intercept has a constant-systematic bias; one finds an aflatoxin content which is too low by 2.9 mg. The solid phase extraction must be improved.

(d) Due of this error the recovery rate depends on the absolute sample mass:

$$RR(10) = \left(\frac{-2.91}{10} + 1.01 \right) 100\% = 71.9\%$$

Analogously, $RR(20) = 86.5\%$, $RR(30) = 91.3\%$ and $RR(40) = 93.7\%$.

19.5 PEAK-HEIGHT AND PEAK-AREA DETERMINATION FOR QUANTITATIVE ANALYSIS[15]

For a well resolved peak, its area as well as its height is proportional to the amount of analyte. Therefore, one can choose between these two possibilities. Perhaps area determination is more precise[16] but this should be tested during the validation of the method (Section 20.3). If the signal-to-noise ratio is small, then area determination is a difficult task for the integrator and so it is better to use quantitation by height for trace analysis.

Two problems of peak quantitation need special attention: the influence of external parameters on area and height and the errors coming from peak overlap.

The Influence of Retention and Flow Rate on Peak Height and Area

Figure 19.5 shows the influence of these parameters on quantitation.

(a) In isocratic separations, peak height is strongly influenced by the retention, i.e. the %B solvent. Early eluted peaks (high %B) are narrow and high, whereas late eluted ones (low %B) are broad and small. For quantitation by height it is necessary that the composition of the mobile phase is not altered between calibration and analysis runs (e.g. by evaporation of the component with lower boiling point from a premixed mobile phase during a day). Height is much less influenced by a change in the flow rate and is governed by the van Deemter equation.

(b) When concentration-sensitive detectors are used the peak area is strongly influenced by the flow rate of the mobile phase. If the peak flows slowly through the detector cell, it will be broad but its height is only determined by the concentration at the peak maximum; therefore its area will be large. If the peak is running fast, its area will be small. As a consequence, for area quantitation it is necessary to use a pump with highly constant flow delivery, even if the back pressure is changing. The peak area is barely influenced by a change in %B.

---

[15] S.T. Balke, *Quantitative Column Liquid Chromatography*, Elsevier, Amsterdam, 1984, pp. 147–162; N. Dyson, *Chromatographic Integration Methods*, Royal Society of Chemistry, London, 2nd ed., 1998, pp. 83–85; H.J. Kuss, and S. Kromidas, eds. *Quantification in LC and GC*, Wiley-VCH, Weinheim, 2009.
[16] R.E. Pauls *et al.*, *J. Chromatogr. Sci.*, **24**, 273 (1986).

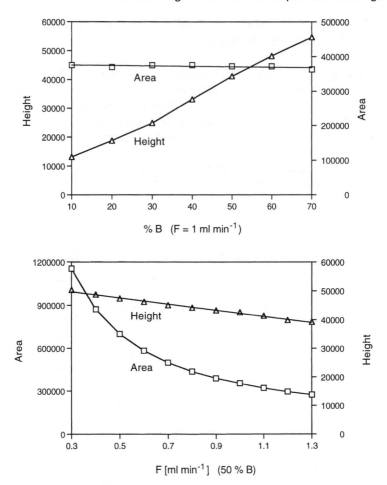

Figure 19.5 Influence of %B solvent and eluent flow rate on peak area and height. Conditions: sample, 10 µg of phenol dissolved in water–methanol (1:1); column, 25 cm × 3.2 mm i.d.; stationary phase, Spherisorb ODS, 5 µm; mobile phase, water–methanol; UV detector, 254 nm. The numbers at the vertical axes are height and area counts of the integrator used.

Overlapped Peaks[17]

Peak overlap is an important cause of erroneous quantitation. The reason becomes clear when looking at Figure 19.6. If the integrator divides the two peaks by a vertical

[17] V.R. Meyer, *J. Chromatogr. Sci.*, **33**, 26 (1995); V.R. Meyer, *Chromatographia*, **40**, 15 (1995); V.R. Meyer, *LC GC Int.*, **7**, 94 (1994) or *LC GC Mag.*, **13**, 252 (1995).

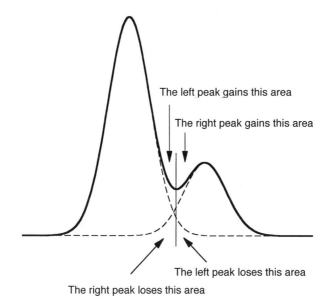

The left peak gains this area

The right peak gains this area

The left peak loses this area

The right peak loses this area

Figure 19.6   Overlapped peak pair and area partition by a vertical drop.

drop some fractions of their areas become a part of the wrong peak. Similarly, area partition is not accurate if the division is done by a tangent instead of a vertical line (this method is often used for small 'rider' peaks). Figure 19.7 shows the effects for various resolutions, asymmetries, and peak orders. The effects with regard to areas can be summarized as follows:

(a) With Gaussian (symmetrical) peaks the large peak is too large and the small one is too small, irrespective of the order of elution. Only in the case of two symmetrical peaks of identical size is there no error when they are cut apart by a vertical line.
(b) With tailed peaks the first one is too small, the second one is too large, irrespective of the size ratio.
(c) The ratio of relative errors is inversely proportional to the peak area ratio. Small peaks are affected more severely.

Some guidelines to avoid this type of error can be given:

(a) Resolve the peaks! The necessary resolution depends on the peak size ratio. A resolution of 1.5 may be adequate for peaks of similar size but it is not large enough for extreme peak area ratios such as 20:1 and larger.
(b) The determination of peak heights is often less inaccurate than area integration. Quantitation by peak height is almost error-free, even in the case of tailing, if a small peak is eluted in front of a large one.

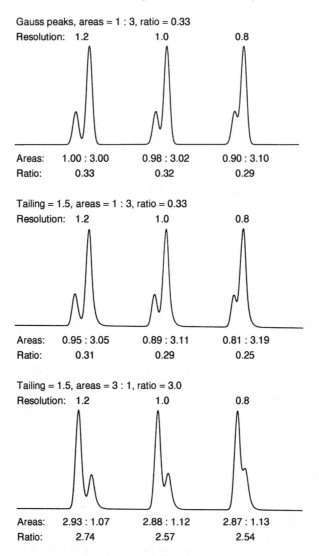

Figure 19.7 Overlapped peaks and their areas as integrated when a vertical drop is used. The true areas are always 1 : 3 or 3 : 1, respectively.

(c) Avoid tailing. Tailing reduces resolution and has a very detrimental effect on the integration of small peaks behind large ones. Look for minimum extra-column volumes in the instrument, excellent column packing, suitable phase systems which allow rapid mass transfer between mobile and stationary phase, suitable sample solvent, and no column overload.

19.6 INTEGRATION ERRORS

Electronic integrators and data systems are of such indispensable importance that possible errors as a result of their use can be overlooked. In order to become familiar with these problems it is helpful to read the easy to understand book by Dyson[18] or some review papers.[19] Different types of error can occur, the most important ones being presented in Figure 19.8.

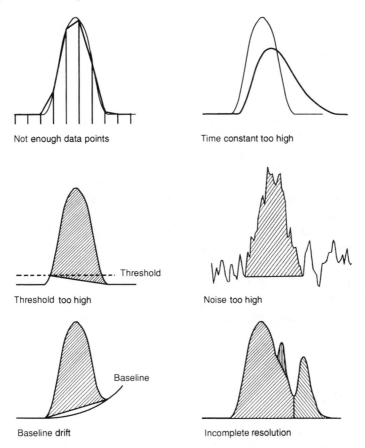

Figure 19.8 Integration errors. Areas counted by the integrator are shaded.

[18] N. Dyson, *Chromatographic Integration Methods*, Royal Society of Chemistry, London, 2nd edn, 1998.
[19] K. Ogan, in *Quantitative Analysis using Chromatographic Techniques*, E. Katz, ed., John Wiley & Sons, Ltd, Chichester, 1987, p. 31; D.T. Rossi, *J. Chromatogr. Sci.*, **26**, 101 (1998); E. Grushka and I. Zamir, in *High Performance Liquid Chromatography*, P.R. Brown and R.A. Hartwick, eds, John Wiley & Sons, Inc., New York, 1989, pp. 529–561.

Types of Error

(a) *Not enough data points.* It is necessary to register a peak with at least ten data points in order to obtain its true shape. If data acquisition is too slow its area and probably also its height will not be correct. Raw data smoothing by the integrator is not a remedy in this case. The acquisition rate can be chosen and perhaps it is necessary to alter it during the registration of a chromatogram.

(b) *Time constant too high.* Too slow an electronics, located in the detector or integrator, distorts the signal. In principle, peak area remains unaffected, peak height decreases, and tailing increases. It will probably be found that peak area is detracted and resolution to neighbouring peaks will be decreased.

(c) *Threshold too high.* Integrators recognize a peak if its signal is higher than the baseline. The threshold can be set and needs to be higher than the noise. In any case peak area and height are decreased.

(d) *Noise too high.* Obviously it is not possible to do a correct integration if the signal-to-noise ratio is too low. The result will be random, even with data systems that allow the baseline to be defined by estimation after the run. For peak-area determination the signal-to-noise ratio should be at least 10.

(e) *Baseline drift.* The integrator usually draws a straight baseline. If the true baseline is bent the calculated area will be too small (as shown) or too high.

(f) *Incomplete resolution.* The drawing only shows two typical problems. Is it best to use a tangent or a perpendicular drop? With the exception of really ideal cases, incomplete resolution always gives rise to erroneous integration. Peak asymmetry is even more detrimental. See Section 19.5.

In many cases integration errors cannot be compensated for by injection of standards and determination of calibration factors because a standard mixture is usually less complex than the sample.

19.7   THE DETECTION WAVELENGTH

If a conventional UV detector is used it is necessary to choose a certain wavelength for detection (this is not the case with diode array detectors). Several points need to be considered:

(a) The UV spectra of the different sample components, especially the wavelengths of the absorption maxima;

(b) the molar absorptivities at the maximum and at the wavelength which will be chosen, respectively;

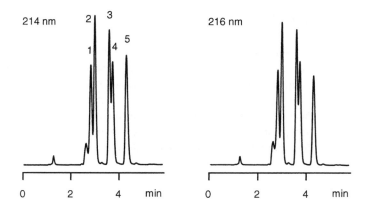

Figure 19.9 Analysis of explosives at a UV wavelength change of 2 nm. Conditions: column, 25 cm × 4.0 mm i.d.; stationary phase, LiChrospher 60 RP-Select B, 5 µm; mobile phase, 1 ml min^{-1} water–acetonitrile (30 : 70). Peaks: 1 = octogen; 2 = hexogen; 3 = tetryl; 4 = trinitrotoluene; 5 = nitropenta.

(c) the retention factors of the different sample components if the separation is in the isocratic mode (the peaks are broadening gradually and thus their height decreases);
(d) the importance of the different sample components (in some cases only a single compound needs to be quantified).

The smaller a peak is, the more affected is its signal-to-noise ratio. The precision and probably also the accuracy of a quantitative determination decreases. Therefore, important peaks should be eluted early and detected at their absorption maximum. As a general rule, at low detection wavelengths the analysis becomes less rugged, system peaks (Section 19.9) have a greater chance to show up and the baseline of a gradient separation is more prone to drift.

It is obvious that it is necessary to verify the chosen wavelength at the instrument and to perform the apparatus test (Chapter 25) regularly in order to guarantee accuracy over months and years. Even a minor deviation from the correct wavelength can strongly influence the analytical result. In Figure 19.9 the proper wavelength for the determination of nitropenta (the last peak) would be 214 nm which is the absorption maximum of this compound. If the wavelength changes to 216 nm due to an improper setting or an unperceived change of instrument optics, the area of the nitropenta peak drops to 85% of its true value, whereas the other peaks are affected to a much lower degree.

19.8 DERIVATIZATION[20]

Detection is the greatest problem associated with HPLC. If highly sensitive detectors such as the fluorescence type cannot be used, then the detection limit is much higher than in gas chromatography with a flame ionization detector. Derivatization reactions provide a useful way of increasing the detection limit in liquid chromatography.[21]

Derivatization means changing the sample in a specific way be chemical reaction and can be carried out prior to chromatographic injection (precolumn derivatization) or between the column and detector (postcolumn derivatization).

Precolumn Derivatization

Advantages

(a) Reaction conditions can be freely chosen.
(b) Reaction may be slow.
(c) Derivatization can serve as a purification and processing step.
(d) Excess of reagent can be removed.
(e) This type of derivatization may also improve the chromatographic properties of the sample (e.g. faster elution, less tailing).

Disadvantages

(a) Byproducts may be formed as a result of the reaction period, which is often long.
(b) Reaction should be quantitative.
(c) The various sample compounds become more similar by derivatization, so the chromatographic selectivity may be lower.

Postcolumn Derivatization: Reaction Detectors[22]

In this case, a reagent is added to the column effluent. In general, a high concentration of reagent solution is preferred as this reduces dilution effects to a minimum. The reactor is often heated, as chemical reactions are generally faster at higher temperatures.

---

[20] I.S. Krull, Z. Deyl and H. Lingeman, *J. Chromatogr. B*, **659**, 1 (1994); Derivatization for fluorescence detection: H. Lingeman *et al., J. Liquid Chromatogr.*, **8**, 789 (1985).

[21] In contrast, the aim of derivatization in GC is to improve the volatility of the sample and to increase its temperature stability.

[22] U.A.T. Brinkman, *Chromatographia*, **24**, 190 (1987); W.J. Bachman and J.T. Stewart, *LC GC Mag.*, **7**, 38 (1989); U.A.T. Brinkman, R.W. Frei and H. Lingeman, *J. Chromatogr.*, **492**, 251 (1989). Practical problems of post-column derivatization are discussed by M.V. Pickering, *LC GC Int.*, **2**(1), 29 (1989). Biochemical reaction detectors: J. Emnéus and G. Marko-Varga, *J. Chromatogr. A*, **703**, 191 (1995).

Advantages

(a) The reaction does not have to be quatitative.
(b) An additional detector may be added prior to derivatization, e.g. column → UV detector → reaction with fluorescence reagent → fluorescence detector.
(c) Highly specific detection systems such as immunoassays[23] can be installed.
(d) It is possible to analyse compounds which give identical reaction products (e.g. formaldehyde) because separation is prior to detection.

Disadvantages

(a) The reaction must be in the mobile phase used, which restricts possibilities.
(b) The derivatization agent itself should not be detected. Post-column derivatization with the aim of improving UV absorption is virtually impossible as all suitable reagents are strongly absorbing in the UV region.

The three most important types of reaction detectors are:

(a) *Open capillaries.* The smaller the capillary diameter, the less band broadening occurs; a 0.3 mm capillary is a good size to choose. A specific capillary bend ('knitted tube')[24] in which the liquid is well mixed, thus minimizing band broadening, is a great advantage. Capillaries are suitable for reaction times of up to 1 min (knitted tube versions up to 5 min).
(b) *Bed reactors.* These are columns packed with nonporous material such as glass beads. (A nonporous packing is essential as it enables crossflow.) They are suitable for reactions lasting between 0.5 and 5.0 min. A catalyst or immobilized enzyme[25] packing may play an active part in the reaction.
(c) *Segmented systems.* The liquid flux is segmented, i.e. separated into smaller portions, by air bubbles or a nonmiscible solvent for slow reactions (up to 30 min), thus preventing band broadening. Phases are generally separated again prior to the detector but noise can also be suppressed electronically.

The following figures provide examples of separation problems in which derivatization was carried out:

Figure 12.7: lanthanides were converted into coloured complexes by post-column derivatization with 4-(2-pyridylazo)resorcinol. There is no information on the reaction detector.

---

[23] I.S. Krull *et al.*, *LC GC Int.*, **10**, 278 (1997) or *LC GC Mag.*, **15**, 620 (1997); A.M. Girelli and E. Mattei, *J. Chromatogr. B*, **819**, 3 (2005).
[24] H. Engelhardt, *Eur. Chromatogr. News*, **2**(2), 20 (1988); B. Lillig and H. Engelhardt, in *Reaction Detection in Liquid Chromatography*, I. S. Krull, ed., Dekker, New York, 1986.
[25] Y.I. Nie and W.H. Wang, *Chromatographia Suppl.*, **69**, S-5 (2009).

Figure 13.2: α-keto acids were pre-column derivatized for 2 h at 80 °C with o-phenylenediamine, producing fluorescent quinoxalinol compounds.

Figure 16.6: lactate dehydrogenase isoenzymes were detected by their activity. Either lactate + NAD$^+$ or pyruvate + NADH was a post-column addition. The reaction detector was a 10 cm × 5 mm i.d. column packed with 150 µm glass beads (as diol derivative) at 40 °C. Detection principle: NADH absorbs at 340 nm whereas NAD$^+$ does not.

Figures 18.1 and 18.3: the amino acids were transformed into coloured products by means of ninhydrine added after the column. The reaction took place in a knitted tube reactor within 2 min at 130 °C.

Figure 18.9: tobramycin is an aminoglycoside. It was derivatized with o-phthalaldehyde in a 1.83 m × 0.38 mm i.d. capillary after passing through the column. Fluorescent products result from a combination of this reagent and amines after a reaction period of 6 s.

Figure 22.4: for the successful separation of the propranolol enantiomers on a chiral stationary phase the molecule should have a rigid structure. This was obtained by a precolumn derivatization with phosgene. This reagent gives an oxazolidone ring from the alcohol and secondary amino groups. The reaction is fast at 0 °C.

Figure 22.5: one possibility of enantiomer resolution consists of the reaction of the compounds of interest with enantiomerically pure, chiral reagents to obtain diastereomers. With a proper choice of reagent the diastereomers can be separated on a reversed phase (or another nonchiral stationary phase). For the separation presented here the quantitative reaction with chloroformate was obtained within 30 min.

Figure 23.1: a 2 cm × 0.34 mm i.d. microbed reactor containing immobilized 3α-hydroxy steroid dehydrogenase was set up after the column. The NAD$^+$ required for the reaction was added to the mobile phase; hence no second pump and no mixing tee between the column and reactor were required. The 3α-hydroxy groups of the bile acids were oxidized to ketones in the reactor, the resulting NADH being registered by a fluorescence detector.

19.9 UNEXPECTED PEAKS: GHOST AND SYSTEM PEAKS

At any time it is possible that unexpected positive or negative peaks will occur in a chromatogram. If these peaks cannot be reproduced they are called *ghost peaks*.[26] It is necessary to eliminate them, although this can be tedious and time-consuming. Reasons for ghost peaks include bubbles in the detector, impurities in the mobile

---

[26] S. Williams, *J. Chromatogr. A*, **1052**, 1 (2004).

phase, demixing phenomena in the mobile phase, bleeding of the column, inadequate adjustment of equilibrium in gradient operation, carry-over from previous injection[27] and many others.

Truly reproducible unexpected peaks are *system peaks*,[28] as shown in Figures 13.3 and 22.2. They always occur when the mobile phase is a mixture of several

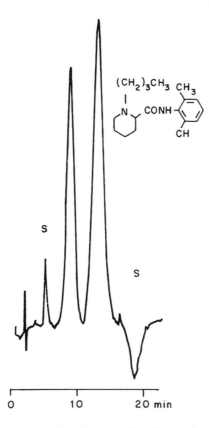

Figure 19.10 Peak areas and system peaks. (Reproduced by permission of Elsevier Science Publishers BV from G. Schill and J. Crommen, *Trends Anal. Chem.*, **6**, 111 (1987).) The two peaks indicated by *S* are system peaks. Sample, racemate of bupivacaine (equal amounts of both isomers); stationary phase, EnantioPac (α_1-acid glycoprotein on silica); mobile phase, phosphate buffer (pH 7.2)–isopropanol (92 : 8), UV detector, 215 nm (the mobile phase shows some absorbance at this wavelength).

[27] J.W. Dolan, *LC GC Int.*, **7**, 74 (1994).
[28] J. Srbek *et al.*, *J. Sep. Sci.*, **28**, 1263 (2005).

components, provided that the detector used is sensitive to one or more of these components, e.g. if the eluent contains a trace amount of an aromatic and if detection is carried out at 254 nm. In indirect detection (cf. Section 6.9) system peaks always have to be taken into account; indeed, the transition from direct to indirect detection may be continuous. Surprises are always possible if in UV detection a lower wavelength than usual is chosen in order to obtain a lower detection limit. System peaks can interfere considerably in qualitative and quantitative analysis if they are not identified (this can be difficult because system peaks may even be invisible!).[29] They may cause peak compression and double peaks.[30] In phase systems with system peaks, the areas of the individual peaks depend on their positions relative to the system peak! Therefore, the peaks of the two enantiomers of bupivacaine, which had been injected as a racemate, in Figure 19.10 are not equal in area.

---

[29] T. Arvidsson. *J. Chromatogr.*, **407**, 49 (1987).
[30] T. Fornstedt, D. Westerlund and A. Sokolowski, *J. Liquid Chromatogr.*, **11**, 2645 (1988).

20   Quality Assurance[1]

20.1   IS IT WORTH THE EFFORT?

Laboratory analyses only make sense if their results are reliable. In scientific terms this statement requires that the analyses and their results are reasonable, accurate and precise.

Reasonable means: the analytical problem can be solved with a suitable method. Imagine an ion-sensitive electrode which allows to find out quickly and cheaply that the amount of cadmium in a drinking water sample is far below the legal maximum limit. For this very purpose it is not necessary to perform further investigations. However, there are situations where a trace concentration needs to be known precisely, calling for a rather big effort in instrumentation, personnel, time and money.

Accurate means in principle: the analysis yields the true result. This statement is less trivial than it seems because the 'true value' of an amount or a concentration is unknown as a matter of principle. We can only try to obtain as small a difference between the found and true values. This can be verified by using alternative analytical methods (Section 20.2) and by the participation in interlaboratory tests.

Precise means: the scatter of the results is low if the analysis of the same material is repeated, i.e. the standard deviation is small. In most cases a high precision is only attainable with high effort as described above. High precision is only meaningful if it is accompanied by high accuracy.

---

[1] W. Funk, V. Dammann, G. Donnevert and S. Ianelli, *Quality Assurance in Analytical Chemistry*, Wiley-VCH, Weinheim, 2nd edn, 2007; E. Prichard and V. Barwick, *Quality Assurance in Analytical Chemistry*, John Wiley & Sons, Ltd, Chichester, 2007; P. Konieczka and J. Namiesnik, *Quality Assurance and Quality Control in the Analytical Chemical Laboratory*, CRC Press, Boca Raton, 2009; I.N. Papadoyannis and V.F. Samaridou, *J. Liq. Chromatogr. Rel. Tech.*, **27**, 753 (2004).

Practical High-Performance Liquid Chromatography, Fifth edition   Veronika R. Meyer
© 2010 John Wiley & Sons, Ltd

This chapter explains some techniques and instruments of quality assurance but an exhaustive presentation is not possible within this book. Numerous textbooks and other material is available which allows you to familiarize yourself quickly with the topic if you are the new quality assurance manager of your company.

20.2   VERIFICATION WITH A SECOND METHOD

A good approach for the verification of quantitative results is the investigation of the sample with a second method. The latter should be as different from the original method as possible, e.g. by using gas or thin layer chromatography instead of HPLC. Since sample preparation is often the main reason for poor accuracy it is highly recommended to use different sample preparation techniques for the two methods. If identical quantitative results are found in both cases it may be assumed that they are accurate. If the results differ it is necessary to find out the reason (or the multiple reasons) although such efforts can be tedious and time-consuming.

As an example, Urakova et al. compared the analysis of chlorogenic acid in green coffee bean extracts by thin layer chromatography on silica (i.e. normal-phase liquid chromatography) and by reversed-phase HPLC.[2] The validation data (LOD, LOQ, repeatability, and various precision parameters), the recoveries and the quantitative results were totally comparable. It can be assumed that both methods find the 'true' value. Either method can be used, depending on the preference or instrumentation of a laboratory.

This kind of verification is virtually mandatory for completely new procedures and for samples with difficult matrix. It is obvious that the verification is performed previously to the validation (Section 20.3) of the method which will be chosen finally.

20.3   METHOD VALIDATION[3]

Validation is a procedure which shows that a method is able to yield the needed results in a reliable manner, with appropriate precision and equally appropriate accuracy. It is a matter of course that it is not possible to set up generally valid recipes of how to

---

[2] I.N. Urakova *et al., J. Sep. Sci.*, **31**, 237 (2008).

[3] J. Ermer and J.H.M. Miller, eds., *Method Validation in Pharmaceutical Analysis,* Wiley-VCH, Weinheim 2005; D.M. Bliesner, *Validating Chromatographic Methods*, Wiley, Chichester, 2006, A.G. González and M.A. Herrador, *Trends Anal. Chem.*, **26**, 227 (2007); G.A. Shabir *et al., J. Liq. Chromatogr. Rel. Tech.*, **30**, 311 (2007).

perform a validation. However, several or all of the following points need to be covered:

(a) *Selectivity* (or specifity). The ability to find and quantify the compound of interest also in the presence of other compounds. This means for chromatographic methods that the analyte can be separated with sufficient resolution from all accompanying peaks and that it can be detected with a suitable instrument.

(b) *Linearity*. The properties of the calibration curve (Section 19.3). It should be straight and run through the origin. If deviations occur they need to be known; it is desirable that they can be explained.

(c) *Working range*. The well defined range of amount or concentration which has been validated. It can be smaller than the linear range of the calibration curve. Analyses which are performed outside the working range are not validated, thus their results cannot become effective.

(d) *Precision*. The ability to re-run an analysis with low standard deviation. It is necessary to distinguish between *repeatability* (the analysis is repeated after a short time interval, with the same instrument, by the same person, in the same laboratory) and *reproducibility* (the analysis is repeated after long time, by other people, with other instruments and/or in another laboratory).

(e) *Accuracy*. The ability to run an analysis with low difference between the 'true' and found value. (Note: the true value is unknown.)

(f) *Recovery* (Section 19.4). The degree to which the analyte is found in the sample. The recovery and its repeatability need to be known. A 100% recovery is the goal but quite often it is impossible to reach it. In many cases recovery is the most important influence factor of accuracy.

(g) *Lower limits* in the case of trace analysis. The ability to analyse samples with low analyte content; it has a lower end. Note the difference between the limit of quantification (LOQ) and the limit of detection (LOD) which is approximately one third of the LOQ (Section 19.3). Sometimes the LOD is defined as the threefold height of the noise.

(h) *Measurement uncertainty* in the case of quantitative analyses, see Section 20.5.

(i) *Ruggedness*.[4] The insensitivity of the method against external parameters (see the end of Section 18.6). It depends on the particular method and the analytical task how many of the parameters need to be controlled more or less strictly: temperature, composition of the mobile phase, detector properties, sample preparation, personnel (can the analysis be performed by anybody or by specialists only?) and laboratory (is it possible to run it also elsewhere?). This list is not complete. Concerning the ruggedness of a method, unexpected experiences can never be excluded; this is especially the case if the analysis needs to be performed

---

[4] J.W. Dolan, *LC GC Eur.*, **19**, 268 (2006) or *LC GC North Am.*, **24**, 374 (2006); B. Dejaegher and Y. Vander Heyden, *LC GC Eur.*, **19**, 418 (2006).

by another laboratory (see also the considerations to method transfer in Section 17.2). Ruggedness should be determined with a statistical design of experiments in order to get valid results with a limited number of experiments.[5] As a positive side effect, a rational plan of experiments demands for completeness and results in a deeper understanding of the method.

Problem 38

Starting with a separation at 30 °C and pH 4.5 (middle) these two parameters were studied further. What can be said about their ruggedness?

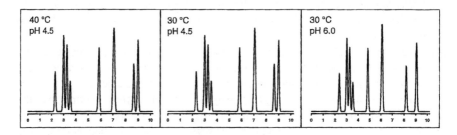

Solution

The method is rugged with regard to temperature (at least within the range investigated) but not with regard to pH. At pH 6 three peaks are eluted at shorter retention times. It is absolutely necessary to note this fact in the Standard Operating Procedure (Section 20.4). In addition to temperature and pH, other influence parameters should be studied, too, as mentioned above.

It is recommended to create a detailed instruction (in fact, a Standard Operating Procedure) about how to perform a validation, matching the needs of the company or the laboratory. It is then obligatory in all cases where a new method is established or an existing one is altered.

20.4 STANDARD OPERATING PROCEDURES

Standard operating procedures (SOPs) are detailed descriptions of analytical methods (or of other laboratory tasks such as weighing or document archiving).

[5] S.N. Deming and S.L. Morgan, *Experimental Design: A Chemometric Approach*, Elsevier, Amsterdam, 2nd edn, 1993; R.B. Waters and A. Dovletoglou, *J. Liquid Chromatogr. Rel. Tech.*, **26**, 2975 (2003); Y. Vander Heyden, *LC GC Eur.*, **19**, 469 (2006); B. Dejaegher and Y. Vander Heyden, *LC GC Eur.*, **20**, 526 (2007) and **21**, 96 (2008). An application example: M. Enrique *et al.*, *J. Chromatogr. Sci.*, **46**, 828 (2008).

Following such a guideline, an educated and trained person must be able to perform the described activity, therefore some detailedness is needed. It is best to let write an SOP by the person who is actually doing the respective analysis and then to ask another person for a critical review. Afterwards the SOP is released and is mandatory. It can be revised at any time; then only the new version is valid but the old one must be kept in the archive.

SOPs for quantitative analyses shall include the complete equation of the measurand as well as an example which shows how the calculation of the result is performed.

SOPs are kept in the laboratory in order to have them at hand whenever they are needed.

20.5 MEASUREMENT UNCERTAINTY[6]

The result of a quantitative analysis consists not only of the measured value but also of its measurement uncertainty. The latter means the bandwidth of the possible fluctuations of the value within the 68, 95 or 99.7% confidence range. As is well known the values found with repeated analyses are not identical if enough digits are considered. For the assessment of such deviations an established procedure is needed[7] because without a budget of measurement uncertainty it is difficult to decide if a limit was exceeded or if both of two different laboratories are able to perform trustworthy analyses.

The measurement uncertainty can be determined by different methods. One possibility is to add the standard uncertainties of all parameters which have an influence on the final value ('bottom up' procedure). However, concerning chromatographic separations it is difficult to determine the number of parameters which, by their interplay, yield the area of a peak, and their quantification would be even more intricate.[8] Figure 20.1 illustrates the problem of a comprehensive presentation; some question marks at the separation itself (the HPLC branch) indicate that probably not all parameters are known. Therefore a pure 'bottom up' approach cannot be used in HPLC.

It is possible to take up the problem from the other end and to declare the reproducibility of the procedure, i.e. the interlaboratory standard deviation s_R, as the measurement uncertainty ('top down' procedure). If the same analytical procedure is performed in numerous laboratories at the occasion of a collaborative study it

[6] V.R. Meyer, *J. Chromatogr. A*, **1158**, 15 (2007).
[7] S.L.R. Ellison, M. Rösslein and A. Williams (eds), *EURACHEM/CITAC Guide Quantifying Uncertainty in Analytical Measurement*, 2nd edn, 2000, ISBN 0-948926-15-5. Free download at www.measurementuncertainty.org.
[8] V.R. Meyer, *J. Chromatogr. Sci.*, **41**, 439 (2003).

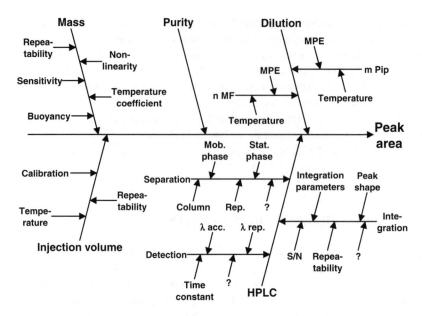

Figure 20.1 Ishikawa diagram of the parameters which lead to the area of an HPLC peak. 'Mass' means the weighed sample or analyte. For its dilution *n* measuring flasks (MF) and *m* pipettes (Pip) are needed. 'MPE' is the maximum permissible error of a volumetric operation, i.e. the combined effects of calibration uncertainty and repeatability. Other abbreviations: λ = wavelength; acc. = accuracy; rep. = repeatability; S/N = signal-to-noise ratio.

can be assumed that all possible influences add to the interlaboratory standard deviation which finally is calculated from all data. If no interlaboratory test was performed, it is only the own intra-laboratory standard deviation s_r which is known, a value that may be too small and thus too optimistic, especially when it is necessary to compare the own results with the ones from other laboratories (suppliers, customers, authorities, competitors).

The measurement uncertainty can also be determined from validation data.[9]

A possible approach for many problems is the long-term registration of the peak area or height of a reference material (with a control chart) and the calculation of its standard deviation; this yields the repeatability Rep_{Ref}. This is not yet a measurement uncertainty because what must also be considered are the uncertainty of the purity of the reference material and the repeatability of the recovery (as well, strictly speaking, as the mass uncertainties of sample and reference, values which are very low if the weighing goods are not critical).

[9] E. Compos Giménez and D. Populaire, *J. Liq. Chromatogr. Rel. Tech.*, **28**, 3005 (2005).

The measurement uncertainty of the analyte concentration in the sample $u(c_{Sample})$ is then calculated as follows:

$$u(c_{Sample}) = c_{Sample} \sqrt{\left[\frac{u(Pur_{Ref})}{Pur_{Ref}}\right]^2 + \left[\frac{s(Rec_{Sample})}{Rec_{Sample}}\right]^2 + s_{rel}^2(Rep_{Ref})}$$

where u stands for standard uncertainty and s stands for standard deviation.

A problem is not set here because the determination of some standard uncertainties and standard deviations is not trivial.

20.6   QUALIFICATIONS, INSTRUMENT TEST AND SYSTEM SUITABILITY TEST

It is not enough to use suitable and validated analytical methods. The proper condition of the instruments, i.e. their "fitness for purpose", is a necessary prerequisite as well to obtain accurate and precise results. A new instrument needs to pass four different qualifications:

(a) *Design qualification* (DQ), prior to the purchase, i.e. the definition of the requirements and specifications an instrument needs to fulfil. A task for the laboratory management.
(b) *Installation qualification* (IQ), during the installation of the instrument, i.e. the documentation and verification of the installation. A task for both the service technician and the laboratory management.
(c) *Operational qualification* (OQ), during the installation, i.e. the first test which shows that the instrument is working in accordance to the DQ. A task for both the service technician and the laboratory staff.
(d) *Performance qualification* (PQ),[10] after the installation, i.e. the test with a typical analytical task. The PQ will be repeated later regularly and is then called the *System Suitability Test*, see below. A task for the laboratory staff.

After the successful completion of these four steps it will be possible to perform reliable analyses. However, it is necessary to perform the *instrument test*[11] at regular time intervals, at least once a year. The test for HPLC instrumentation is described in detail in Chapter 25.

In addition to all these qualifications, the laboratories of the pharmaceutical industry need to perform *System Suitability Tests* SSTs.[12] Any given method must only be used after a successful SST; as a consequence, the SST has to be performed regularly, usually daily and in any case at the beginning of a new series of analyses. It

---

[10] J. Crowther *et al.*, *LC GC North Am.*, **26**, 464 (2008); L. Kaminski et al., *J. Pharm. Biomed. Anal.*, **51**, 557 (2010).

[11] G. Maldener, *Chromatographia*, **28**, 85 (1989).

[12] J.W. Dolan, *LC GC Eur.*, **17**, 328 (2004) or *LC GC North Am.*, **22**, 430 (2004).

is done with a mixture which is typical for the samples to follow and not with any test mixture whatever. The United States Pharmacopoeia USP requires five replicate injections which allow to obtain the following data:

(a) the calibration function (slope, intercept),
(b) the repeatability of the peak area(s) or the peak height(s), calculated as standard deviation,
(c) the repeatability of the retention time(s),
(d) the peak tailing,
(e) the resolution (to the next peak or of the critical peak pair),
(f) the baseline noise,
(g) the baseline drift.

The USP demands the following maximum relative standard deviations of the peak area:

(a) less than 2% for standard solutions,
(b) less than 15% for bioanalytical samples,
(c) less than 20% for bioanalytical samples at the limit of quantitation.

20.7 THE QUEST FOR QUALITY

An analytical laboratory (or a company as a whole) can strive for ISO certification or EN accreditation. The certification in accordance with ISO 9001[13] refers to formal and technical processes only and not to quality issues. For an institution which sells goods or services an accreditation in accordance to EN 17025[14] is to be preferred. This norm claims:

'Qualified personnel perform well defined tests in accordance with validated methods and with tested measurement equipment. All procedures are clearly regulated and adequate documentation is guaranteed.'

An accreditation needs a long and intense preparation; it is only granted after a thorough inspection by external experts who visit the laboratory or company for several days. The list of requirements to be fulfilled is long and detailed.

Even without certification or accreditation it is highly recommended to implement a number of quality assurance actions (they are mandatory for an accreditation):

(a) use only chemicals of well defined quality,
(b) note the opening and expiration dates on the bottles and packages of chemicals,

---

[13] *ISO 9001:2000, Quality Management Systems – Requirements*, International Organization for Standardization, Geneva, 2000.
[14] *EN ISO/IEC 17025:2005, General Requirements for the Competence of Testing and Calibration Laboratories*, International Organization for Standardization, Geneva 2005.

(c)   define column specifications,
(d)   perform column tests,
(e)   repeat column tests in regular time intervals,
(f)   keep log books for the columns,
(g)   define instrument specifications,
(h)   perform instrument tests,
(i)   repeat instrument tests in regular time intervals,
(j)   keep log books for the instruments,
(k)   perform system suitability tests (perhaps with a simplified protocol),
(l)   use control charts[15] to depict trends and outliers (Figure 20.2),
(m)   participate in interlaboratory tests[16] to get a feedback of the laboratory's competence.

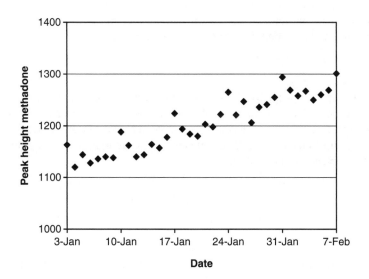

Figure 20.2   Simple control chart. The daily test analysis is plotted, in this case the peak height of a methadone standard. The chart shows two critical facts at a glance: The first analysis of a week yields a markedly higher peak than the next ones; in addition, there is a trend to increasing peak heights. The reasons for these incidents need to be found out and actions for constant results must be implemented.

[15] E. Mullins, *Analyst*, **119**, 369 (1994) and **124**, 433 (1999).
[16] J. Vial and A. Jardy, *Chromatographia Suppl.*, **53**, S-141 (2001).

Problem 39

You become the new quality manager of your company. So far no formal quality measures exist and money is short. You decide to implement five actions taken from the list below. How do you start? What is your long-term goal?

(a) Appropriate software makes it impossible to alter electronic data without authority.
(b) Control charts are established.
(c) Date and signature are noted on every written or electronic document.
(d) Dates of opening and of expiration are noted on the labels of chemicals.
(e) Instrument log books are kept.
(f) Seminars and discussions are established in order to convince the personnel of the new actions.
(g) System suitability tests are established.
(h) The company aspires for certification or accreditation.
(i) The company participates successfully in interlaboratory tests.
(j) Written Standard Operating Procedures are established.

Solution

There is no general true answer. However, your efforts will be of no avail if you cannot convince the staff. Other quickly established actions, which are free or cheap, are the date and signature regulation, the date regulation for chemicals, the control charts and the protection of electronic data. Certification or accreditation are long-term goals with a time frame of one to three years.

21 Preparative HPLC[1]

21.1 PROBLEM

Preparative chromatography is used to separate and purify substances for further use, the amounts obtained varying greatly according to application. A few micrograms, i.e. one analytical peak, are enough for registration of a UV spectrum, the eluate generally being collected directly in the UV cell. At least 10 mg of pure compound are required for analysis based on ^{1}H NMR, ^{13}C NMR and MS and a further 5 mg for determination of the empirical formula by elemental analysis. Amounts of 10–100 mg are required for the synthesis of derivatives from the components or investigating reactivities. These amounts can just about be provided by an analytical set-up, although the process is often a lengthy one, sometimes referred to as 'scaling up' (i.e. extending the range of application to a previously unplanned level). Amounts of 1–100 g of product can also be purified by chromatography, this being the most important area of application of classical column chromatography since a very long time. However, separation can be greatly speeded up and also improved by using modified HPLC equipment. Industrial plants have been built incorporating 1.6 m diameter columns for separating between 1 and 3 kg h^{-1}.

This chapter describes the method for separating amounts varying from 10 mg up to several grams, there being no essential differences for variations above or below these limits.

Note: Every analytical separation can be performed in preparative scale!

---

[1] G. Guiochon, A. Felinger, D.G. Shiraz and A.M. Katti, *Fundamentals of Preparative and Nonlinear Chromatography*, Elsevier Academic Press, New York, 2nd edn, 2006; H. Schmidt-Traub, ed., *Preparative Chromatography*, Wiley-VCH, Weinheim, 2005; G. Guiochon, *J. Chromatogr. A*, **965**, 129 (2002).

Practical High-Performance Liquid Chromatography, Fifth edition Veronika R. Meyer
© 2010 John Wiley & Sons, Ltd

21.2   PREPARATIVE HPLC IN PRACTICE[2]

An analytical chromatogram generally provides the starting point for preparative separation, and conditions favourable for separating the sample mixture isocratically with good resolution have to be established:

(a) Gradients are not recommended for preparative separations as they involve considerable expense. If necessary, the sample may be pre-treated in a suitable manner.
(b) The better the resolution on the analytical chromatogram, the more heavily loaded the preparative column can be.

Figure 21.1 shows an analytical separation of reaction products from thermolysis of *trans*-dihydrocarvone oxime:

The structures of the products were not known, nor was there any certainty that each peak represented one component only.

A preparative high-performance column packed with 5, 7 or 10 μm material and available from various manufactures is used after the analytical separation optimization stage (columns with coarser particles of around 40 μm are more suitable for easy separations requiring less than 100 theoretical plates). A 10 mm i.d. column is designed for 10–100 mg samples, rising to 21.5 mm (o.d. 1 in) for samples between 100 mg and 1 g. The transfer from analytical to preparative conditions works best if both columns are packed with the same stationary phase.

Thicker columns require a pump that ensures a high flow rate, an identical linear mobile phase throughput and hence comparative retention times, being dependent on the flow rate, increasing with the square of the column diameter: 1 ml min^{-1} for a 4 mm i.d. column rising to 6.25 ml min^{-1} for a 10 mm i.d. column!

The detector chosen must be insensitive. A refractive index detector (with a preparative cell if required) or a UV detector with an optical path length of 0.1–0.5 mm

---

[2] V.R. Meyer, *J. Chromatogr.*, **316**, 113 (1984); B. Porsch, *J. Chromatogr. A*, **658**, 179 (1994); P. Renold, E. Madero and T. Maetzke, *J. Chromatogr. A*, **908**, 143 (2001); P. López-Soto-Yarritu *et al.*, *LC GC Eur.*, **18**, 669 (2005).

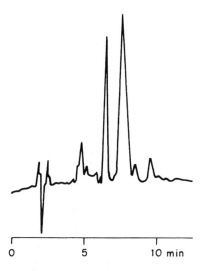

Figure 21.1 Analytical separation of the reaction mixture. Conditions: sample, 20 μl solution in mobile phase; column, 25 cm × 3.2 mm i.d.; stationary phase, LiChrosorb Si 60, 5 μm; mobile phase, 1 ml min^{-1} pentane-diethyl ether (95 : 5);[3] refractive index detector.

is the most suitable. Minimal path length is obtained with a film detector whereas the eluate flows over an inclined glass plate.[4]

The sample mixture should contain no strongly retained components and thin-layer chromatography can be used to check this (no trace should remain at the starting point). The sample may be cleaned by adsorptive filtration (suction filter with a coarse stationary phase) or column chromatography if required. This is of special importance in the case of reaction mixtures or samples of natural origin. A guard column is essential if impurities or decomposition in the column cannot be fully ruled out. Losses can be avoided by ensuring that the sample loop is not filled to capacity (see Section 4.6).

The mobile phase should be extremely pure and volatile so that the pure form of each separated component is easy to obtain: see Problem 43 in Section 21.4. Volatile buffer reagents are acetic acid, trifluoroacetic acid, formic acid, ammonia, ammonium hydrogen carbonate, triethyl amine, trimethyl amine, ethanolamine and pyridine (see also Section 5.4).

---

[3] As the reaction products are relatively volatile, the mobile phase had to have a low boiling point, otherwise hexane-*tert*-butylmethyl ether would have been tried.

[4] T. Leutert and E. von Arx, *J. Chromatogr.*, **292**, 333 (1984).

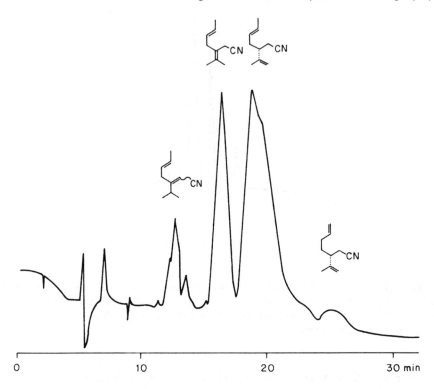

Figure 21.2 Preparative separation of the reaction mixture. Conditions: sample, 1 ml of solution containing 250 mg; guard column, 5 cm × 4.5 mm i.d., dry-packed with silica (40 µm); column, 25 cm × 21.5 mm i.d.; stationary phase, silica, 7 µm; mobile phase, 14 ml min^{-1} pentane-diethyl ether (95:5); refractive index detector.

Preparative separation of the reaction mixture described above, from which four pure components were obtained, is shown in Figure 21.2. NMR was used to determine structures. Some peaks are poorly shaped owing to column overload.

In the optimization of routine or industrial separations, economic requirements (cost of mobile and stationary phases) and the throughout per unit time have also to be taken into consideration in addition to peak purity.[5]

---

[5] M. Kaminski, B. Sledzinska and J. Klawiter, *J. Chromatogr.*, **367**, 45 (1986); R. M. Nicoud and H. Colin, *LC GC Int.*, **3**(2), 28 (1990).

Problem 40

A compound needs to be prepared in pure form by means of preparative HPLC. The separation is done fully automated with an autoinjector and computer-controlled fraction collection. The following systems and columns can be used:

(a) Analytical column 4.6 mm × 25 cm with 5 μm stationary phase. Sample amount (of the compound of interest) 1.2 mg per injection, complete resolution. Mobile phase flow rate 2 ml min^{-1}, duration of a chromatogram 8 min.
(b) Preparative column 22 mm × 25 cm with 5 μm stationary phase. Sample amount 120 mg, complete resolution. Mobile phase flow rate 14 ml min^{-1}, duration 30 min.
(c) Preparative column 22 mm × 25 cm with 40 μm stationary phase. Sample amount 120 mg, but only 80% can be obtained in pure form because the resolution is incomplete. Mobile phase flow rate 14 ml min^{-1}, duration 37 min.

Calculate the throughput (mass/time unit) of pure compound for the three methods. Calculate the production costs (€ g^{-1}) of the three methods (without amortization, the purchase costs are high in any case). The price of the mobile phase is 24 € 1^{-1}, but 90% of it can be reused after distillation, the cost of which is 5 € 1^{-1}. In all three cases the column needs to be refilled after 200 injctions; the price of the stationary phase is: (a) 60 €, (b) 1400 €, (c) 110 €.

Solution

Throughput: (a) 9 mg h^{-1}, (b) 240 mg h^{-1}, (c) 156 mg h^{-1}. Cost: (a) 340 € g^{-1}, (b) 83 € g^{-1}, (c) 43 € g^{-1}.

21.3 OVERLOADING EFFECTS

In preparative HPLC overloading effects are common. They can be divided into three categories:

(a) volume overload;
(b) mass overload (also called concentration overload);
(c) detector overload.

The effects are shown in Figure 21.3

In the *analytical chromatogram* the injected sample volume is so small that the peak widths are not influenced. Moreover, the sample mass is too low to give overloading.[6] The maximum allowed injection volume has been calculated in Section 19.2 and the

---

[6] The separation occurs in the linear part of the adsorption isotherm.

Concentration

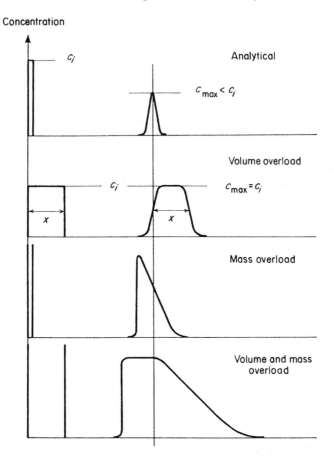

Figure 21.3 Overloading effects in HPLC.

sample mass that can be injected without overloading the system has been discussed in Section 2.7.

With *volume overload* the volume injected is so high that the peak width is affected (the sample concentration being low enough to avoid mass overload). As can be seen from Figure 21.3, the peak width at half-height, x, at the column outlet is identical with the inlet peak width. Also, the peak maximum concentration remains unchanged (provided that the peak is not diluted markedly at high k values). It is evident that the peaks becomes *rectangular* with volume overload.

In principle, volume overload may be increased until the peaks begin to touch each other. This maximum injection volume, v_L, can be calculated from the data given by the analytical chromatogram:[7]

$$V_L = V_0 \left[(\alpha-1)k_A - \frac{2}{\sqrt{N}} (2 + k_A + \alpha k_A) \right]$$

where V_0 is the void volume (eluent flow rate times breakthrough time), α is the separation factor of the peaks in question, k_A is the retention factor of the first peak of interest and N is the number of theoretical plates in the column. It is only possible to inject V_L if no mass overload occurs, i.e. if the sample solution is diluted. Otherwise, the injection volume needs to be lower.

Problem 41

A diluted solution contains component A, which is required to be isolated by preparative HPLC amongst accompanying impurities. The separation factor of A to the nearest adjacent peak is 1.5 and its retention factor is 4. The analytical column used has a theoretical plate number of 6400 and a void volume of 2 ml. Calculate the maximum allowed injection volume at which peak A actually touches its nearest neighbour.

Solution

$$V_L = 2 \left\{ (1.5-1)4 - \frac{2}{\sqrt{6400}} [2 + 4 + (1.5 \times 4)] \right\} = 3.4 \, \text{ml}$$

If the sample mass (calculated as injection volume times concentration) is higher than a certain value, the local concentration of the sample in the column is too high to allow a true equilibrium to be adjusted: this is called *mass overload*. The resulting peaks are *triangular* in shape. In most cases tailing will be observed together with decreased retention times, as can be seen in Figures 21.3 and 2.23. Sometimes mass overload is manifested as fronting together with increased retention times.

The user of preparative HPLC in general wants to obtain as much of a pure compound per unit time as possible. Therefore, it is necessary to work under conditions of overload.[8] If sample solutions are diluted, volume overload will

---

[7] This relationship is only valid if the sample solvent has the same elution strength as the mobile phase. If the solvent is markedly weaker, then even more can be injected (see Section 19.2).

[8] G. Cretier and J.L. Rocca, *Chromatographia*, **21**, 143 (1986); A.F. Mann, *Int. Biochem. Lab.*, **4**(2), 28 (1986); J.H. Knox and H.M. Pyper, *J. Chromatogr.*, **363**, 1 (1986); G. Guiochon and H. Colin, *Chromatogr. Forum*, **1**, 21 (1988); G.B. Cox and L.R. Synder, *LC GC Int.*, **1**(6) (1988); S. Golshan-Shirazi and G. Guiochon, *Anal. Chem.*, **61**, 1368 (1989).

preferentially occur whereas mass overload is common with concentrated samples. Often both effects are present and the peaks become truncated, as can be seen at the bottom of Figure 21.3 (with increasing retention the plateau is lost and the peaks become triangular). The maximum possible injected amount of a concentrated solution is determined empirically: the injection volume is increased until the peaks touch each other. Nondiluted samples are not suitable.

The sample does not have to be dissolved in a stronger solvent than the mobile phase. As large volumes generally are injected in preparative work, in this case the sample solvent should disturb the column equilibrium and reproducibility would no longer be guaranteed. The sample needs to be freely soluble in the eluent, otherwise the column becomes clogged.

Detector overload is indicated when the signal can no longer be recorded as a whole but appears as a rectangle, even at the highest attenuation. This gives the false impression of column overload. As already mentioned in Section 21.2, an insensitive detector is recommended in preparative work; in the case of a UV detector this means a small optical path length.

Fraction purity is guaranteed to a certain extent by the refractive index detector. If the peak of a UV-absorbing compound is partly overlapped by that of a nonabsorbing compound, the UV detector is unable to register this and the fraction obtained will not be pure.

21.4 FRACTION COLLECTION

The line between the detector and fraction collector must have a low volume in comparison with the mobile phase flow rate.

Problem 42

The line (PFTE tube) between the detector cell and collecting flask is 20 cm × 0.3 mm i.d. Not more than 1 s should elapse between the recorded signal and outflow of the sample zone, thus ensuring that fraction collection can be made correctly. What should the minimum mobile phase flow rate be?

Solution

$$\text{Tube volume} = \frac{d^2\pi}{4}l = \frac{0.3 \times 0.3 \times \pi \times 200}{4} = 14\,\text{mm}^3$$

$$\text{Minimum flow rate} = 14\,\text{mm}^3\text{s}^{-1} = 840\,\text{mm}^3\text{min}^{-1} = 0.84\,\text{ml min}^{-1}$$

For occasional preparative work, fractions may be collected in flasks or glass reaction vessels at the detector outlet. If a fraction collector is used, this should be

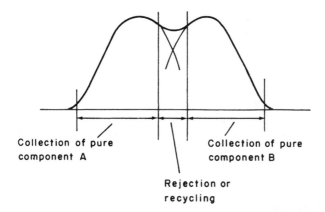

Figure 21.4 Preparative recovery of two poorly resolved peaks.

directly controlled by the recorder or monitor signal if possible. If large amounts of preparative product are often required, a fully automated system is a great advantage. Fraction collection must be program controlled by both time and signal, so the peaks can each be guided into a separate vessel without being intersected. If no peak appears, then the eluate should run back into the solvent reservoir. The next automatic injection should follow on as soon as the chromatogram is completed.

The purity of the fraction obtained decreases if the peaks are insufficiently resolved (Figure 21.4). This situation can be improved by rejecting the middle fraction, which can then be recycled if necessary (see Section 21.5).

The figures in Section 2.4 give more information on the purity of poorly resolved peaks. However, in the case of mass overload these drawings are no longer valid because the peaks are triangular and in fact influence each other. The profiles of touching peaks can be unfavourable so that the second peak cannot be obtained as a pure fraction, in contrast to the first one.[9]

As already mentioned, it is a prerequisite for preparative HPLC that the mobile phase is of the highest purity in order to guarantee the purity of the collected fractions after removal of the solvent.

Problem 43

With regard to accompanying compounds, a product can be obtained in 99.5% purity by preparative HPLC. Its concentration in the eluate is $5\,mg\,ml^{-1}$. The content of

---

[9] S. Ghodbane and G. Guiochon, *J. Chromatogr.*, **444**, 275 (1988).

nonvolatile impurities in the mobile phase is 0.002% (mass/volume). Calculate the purity of the isolated product after the removal of the solvent.

Solution

1 ml of eluate contains 99.5% of 5 mg = 4.975 mg of product as well as 0.025 mg of impurities from accompanying compounds and 0.020 mg of impurities from the solvent. This gives a sum of 0.045 mg of impurities and 4.975 mg of product. The purity of the product has dropped to 99.1%.

21.5 RECYCLING

If the resolution of components is inadequate, then the individual peaks cease to be distinguishable as the injected volume and sample mass are increased. In this instance, the resolution can be improved by circulating the eluate through the column again, based on one of the following techniques:

(a) The eluate is fed back to the pump after passing through the detector and pumped through the column once more. The pump must have a very small internal volume and must also work pulse-free (a pulse damper would involve too much extra-column volume). A relatively large column is essential (e.g. 25 cm × 2 mm i.d. would be unsuitable). However, despite all these measures there is a high degree of band broadening after every cycle. Up to ten cycles are possible, depending on the problem in question.
(b) An arrangement with two columns through which the mobile phase passes in the following sequence:

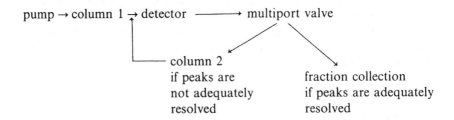

The eluate may be passed alternately through the two columns up to 20 times. The advantage of this system is that extra-column volumes are smaller than with the first method and the pump volume can be chosen at random. However, the detector cell must be pressure-tight.

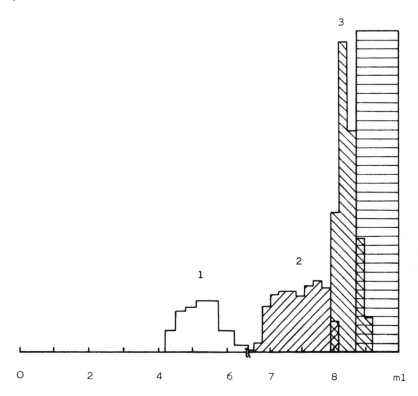

Figure 21.5 Separation of peptides by displacement chromatography [reproduced with permission from G. Subramanian, M.W. Phillips and S.M. Cramer, *J. Chromatogr.*, **439**, 431 (1988)]. Conditions: column, 25 cm × 4.6 mm i.d.; stationary phase, Zorbax ODS 5 μm; mobile phase, 1 ml min^{-1} phosphate buffer 50 mM, pH 2.2–methanol (60 : 40); temperature, 45 °C. Peaks: 1 = 9.6 mg *N*-benzoyl-L-arginine; 2 = 14 mg *N*-carboxybenzoxy-L-alanyl-L-glycyl-L-glycine; 3 = 15 mg *N*-carboxybenzoxy-L-alanyl-L-alanine. Displacer (horizontally shaded): 30 mg ml^{-1} of 2-(2-butoxy-ethoxy)ethanol in the mobile phase. The composition of the 150 μl fractions was analysed by reversed-phase HPLC.

21.6 DISPLACEMENT CHROMATOGRAPHY[10]

Even with overloaded conditions the ratio between sample mass and the amount of stationary phase required is low and uneconomical, as in most cases it will not be

---

[10] J. Frenz and C. Horváth, in *HPLC Advances and Perspectives*, vol. 5, C. Horváth, ed., Academic Press, New York, 1988, pp. 211–314; J. Frenz, *LC GC Int.*, **5**(12), 18 (1992).

higher than $10 \, mg \, g^{-1}$. A totally different approach for preparative liquid chromatography is displacement chromatography (Figure 21.5). In this method, comparatively enormous sample loads are brought on to the chromatographic bed. As the different compounds of the sample mixture show different affinities towards the stationary phase, they displace each other from the 'adsoptive' (in a broad sense) sites of this phase. If the separation performance of the column is high enough, pure but immediately consecutive fractions will be eluted.

To establish a real displacement train it is necessary to feed the column with a compound with an even higher affinity to the stationary phase after the sample mixture has been injected. This compound is the so-called displacer. The optimum displacer has to be found empirically. For silica as the stationary phase, quaternary amines are often suitable (if the sample components are not acids); for reversed phases, alcohols can be used. It is necessary to regenerate the column after each separation.

A drawback of displacement chromatography is the fact that optimum conditions cannot be found by 'scaling-up' of analytical methods.[11] However, after a successful but more or less time-consuming optimization of the displacement system, no other method allows this high throughput per unit time with this low consumption of the mobile phase.

---

[11] Analytical HPLC occurs in the linear part and displacement chromatography in the nonlinear part of the adsorption isotherm.

22 Separation of Enantiomers[1]

22.1 INTRODUCTION

Enantiomers are molecules that are non superimposable on their mirror images. Two mirror-image forms exist, the chemical and physical properties of which are identical, except for the sense of rotation of the plane of vibration of linear polarized light. They cannot be separated by any of the separation methods described so far.

Enantiomers can be separated by chromatography, provided that the system used is asymmetric, i.e. chiral. This can be achieved by various means:

(a) The mobile phase is chiral, the stationary phase nonchiral (i.e. a usual HPLC phase). Only a small amount of chiral compound need be added to any solvent used.
(b) The liquid stationary phase is chiral, the mobile phase nonchiral. The stationary phase is coated on a solid carrier.
(c) The solid stationary phase is chiral, the mobile phase nonchiral. This method is very easy to perform but the stationary phase is expensive.

In all the above cases, diastereomeric complexes are formed between the sample molecules and the asymmetric species in the chromatographic system and these will migrate with different velocities through the column. As an example, amino

---

[1] G. Gübitz and M.G. Schmid, eds., *Chiral Separations: Methods and Protocols*, Humana Press, Totowa, 2004; G. Subramanian, ed., *Chiral Separation Techniques*, Wiley-VCH, Weinheim, 3rd edn, 2007.

Practical High-Performance Liquid Chromatography, Fifth edition   Veronika R. Meyer
© 2010 John Wiley & Sons, Ltd

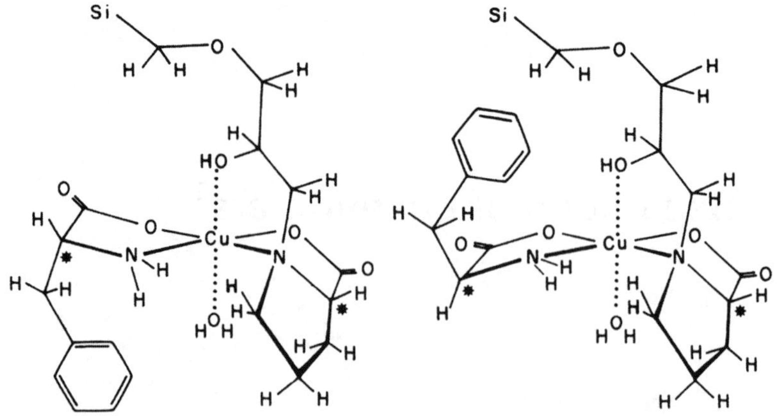

Figure 22.1 Model of diasteromeric complexes of (left) D- and (right) L-phenylalanine with a stationary phase of copper-loaded L-proline on silica [reproduced with permission from G. Gübitz, W. Jellenz and W. Santi, *J. Chromatogr.*, **203**, 377 (1981)].

acid–copper compounds give diastereomeric complexes of well known structures: two copper bonding sites may be occupied by the sample molecules (Figure 22.1). Amino acid–copper compounds may be bonded to silica, as shown in Figure 22.1, or they may be added to the mobile phase.

Enantiomers can be separated by traditional chromatographic methods, provided they have been previously derivatized with a chiral compound to produce diastereomers. This method of indirect separation of enantiomers is explained in Section 22.5.

The separation factor, α, is an important parameter to characterize chiral separation systems:

$$\alpha = \frac{k_2}{k_1} = \frac{t_{R_2} - t_0}{t_{R_1} - t_0}$$

where $k_2 > k_1$. As already mentioned in Section 2.5, the number of theoretical plates needed for the specific resolution of a pair of peaks decreases as the separation factor becomes greater; α values smaller than 1.05 cannot really be made use of in HPLC (in contrast to capillary gas chromatography). Values of α in excess of 3 are not really desirable in practice, as the corresponding peak pairs are then too far away from each other and are not readily recognized as enantiomers.

A polarimeter may be used as a detector if the conditions allow (high concentration and specific rotation).[2] It is also possible to use circular dichroism for detection and even to obtain CD spectra. The determination of the absolute configuration is possible with various methods.[3]

Enantiomer separation is of special importance in the pharamaceutical and clinical fields as many drugs are made of asymmetric molecules. Both forms as obtained by common (nonchiral) synthesis often produce different effects in the body and the pharmacokinetics may also differ.

22.2 CHIRAL MOBILE PHASES[4]

If a chiral reagent capable of forming a complex (in the strictest sense), an ion pair or any other adduct with the enantiomers in the sample is added to the mobile phase, there is a chance that the distribution coefficients of the diastereomers formed between the mobile and stationary phases will be different and, therefore, these can be separated on an HPLC column. An example is the separation of N-(1-pheny-lethyl)phthalamic acid enantiomers using quinine as a chiral ion-pair reagent (Figure 22.2).

The advantage of this method is that the user can select the compound that appears most likely to produce a separation from the wide range of chiral reagents listed in chemical catalogues. Luckily, not all optically active products are expensive and only small amounts are used in preliminary tests. There are no restrictions on choice of stationary phase and experience in nonchiral chromatography will doubtless form the basis for eluent selection (the main option being between aqueous and nonaqueous varieties). The chiral reagent does not necessarily need to be optically pure[5] (of course, there is no separation if it is racemic!) and, if available, the choice of the suitable antipode allows the favoured elution order to be obtained.

Enantiomer separation with chiral mobile phases is based on chemical equilibria which perhaps may fluctuate as a result of change in concentration, temperature or pH. It is worth varying these parameters systematically to optimize a separation. If the system is highly sensitive to changes in any one parameter, then this must be kept strictly constant once its optimum level has been found.

---

[2] D.R. Bobbitt and S.W. Linder, *Trends Anal. Chem.*, **20**, 111 (2001); G.W. Yanik, in: Subramanian (first footnote of this chapter), p. 561.

[3] C. Roussel *et al.*, *J. Chromatogr. A*, **1037**, 311 (2004).

[4] B.J. Clark, in: *A Practical Approach to Chiral Separation by Liquid Chromatography*, G. Subramanian, ed., VCH, Weinheim, 1994, p. 311–328.

[5] C. Pettersson, A. Karlsson and A. Gioeli, *J. Chromatogr.*, **407**, 217 (1987).

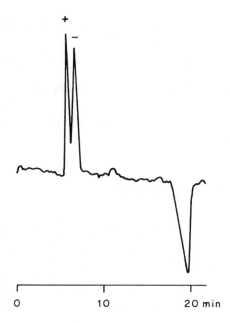

Figure 22.2 Separation of *N*-(1-phenylethyl)phthalamic acid enantiomers with a chiral ion-pair reagent [reproduced with permission from C. Pettersson, *J. Chromatogr.*, **316**, 553 (1984)]. Conditions: stationary phase, LiChrosorb SI 100, 5 μm; mobile phase, dichloromethane–butane-1,2-diol (99 : 1) containing 0.35 mM each of quinine and acetic acid; UV detector. The negative peak at ca. 19 min is a system peak.

One disadvantage of the method is that the enantiomer exist as diastereomeric associates after separation, whose dissociation may be impossible. Preparative separation is impossible under these circumstances.

22.3 CHIRAL LIQUID STATIONARY PHASES

If a chiral liquid is coated on the surface of the column packing particles, a chiral liquid–liquid partition system is obtained. The mobile phase must be saturated with stationary phase. A thermostat must be provided in order to keep the equilibrium of the solution constant. An example of this method is shown in Figure 22.3 with (+)-dibutyl tartrate as the stationary phase, with which several β-amino alcohols can be separated.

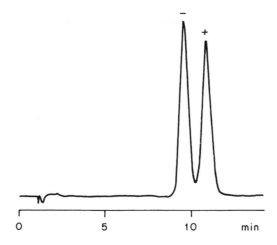

Figure 22.3 Separation of norephedrine enantiomers with (+)-dibutyl tartrate as the liquid stationary phase [reproduced with permission from C. Pettersson and H.W. Stuurman, *J. Chromatogr. Sci.*, **22**, 441 (1984)]. Conditions: column, 15 cm × 4.6 mm i.d.; stationary phase, (+)-dibutyl tartrate on 5 μm Phenyl Hypersil; mobile phase, phosphate buffer (pH 6) containing hexafluorophosphate, saturated with dibutyl tatrate; UV detector, 254 nm.

22.4   CHIRAL SOLID STATIONARY PHASES[6]

Enantiomer separation becomes very easy when a column with a solid chiral stationary phase (CSP) able to separate a given enantiomer pair is available. Unfortunately, there is no 'universal phase' for solving all separation problems and it is impossible that there ever will be. However, an immense number of CSPs have been developed and many of them are commercially available. A good deal of literature has also been published, from which a promising phase can usually be selected. Sometimes, the sample can be adapted to the phase by nonchiral derivatization. The most important CSPs are listed in Table 22.1. They are based on very different separation principles and can be grouped as follows:

(a) 'brush-type' CSPs where small molecules, usually with π-active groups, are bonded to silica;
(b) helical polymers, mainly cellulose and its derivatives;

---

[6] R. Däppen, H. Arm and V.R. Meyer, *J. Chromatogr.*, **373**, 1 (1986); D.W. Armstrong, *LC GC Int. Suppl.*, April 1998, p. 22.

TABLE 22.1   Some chiral stationary phases for HPLC[8]

'Brush-type' phases

Dinitrobenzoylphenylglycine $-Si-O-Si \cdots\cdots CH_2-NH-\overset{\overset{\displaystyle *}{|}}{\underset{\overset{\|}{O}}{C}}-CH-NH-C$

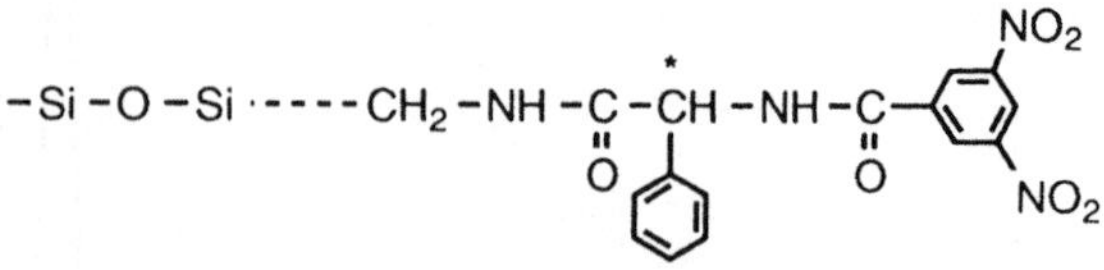

Dinitrobenzoylleucine $-Si-O-Si \cdots\cdots CH_2-NH-C-CH-NH-C$

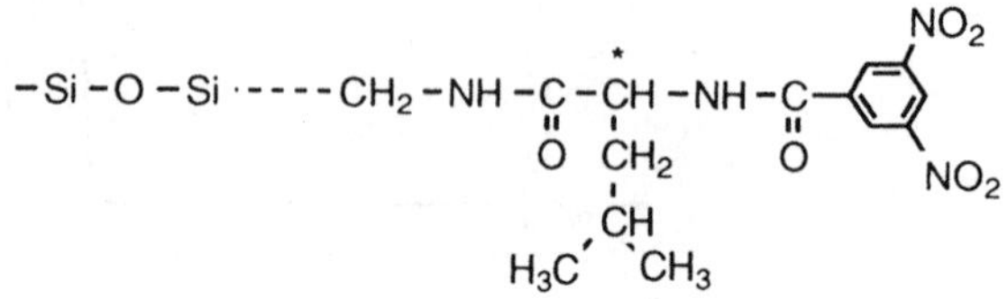

Naphthylalanine $-Si-O-Si \cdots\cdots CH_2-O-C-CH-NH$

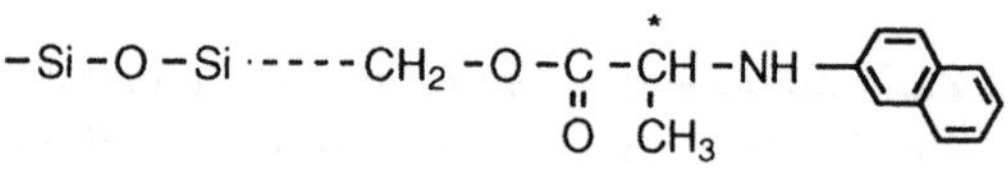

Naphthylleucine $-Si-O-Si \cdots\cdots CH_2-O-C-CH-NH$

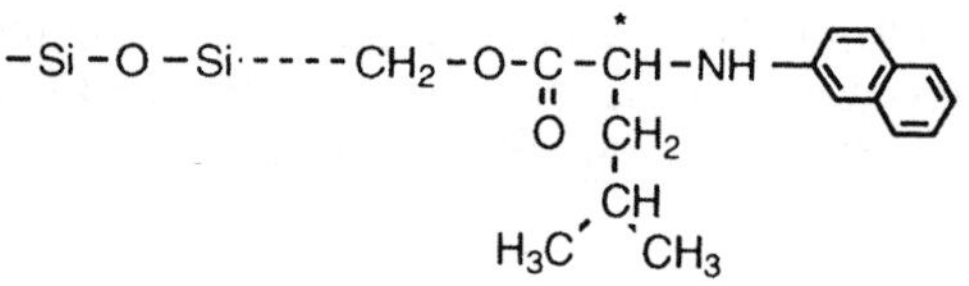

Chrysanthemoyl-phenylglycine $-Si-O-Si \cdots\cdots CH_2-NH-C-CH-NH-C-CH-CH-CH$

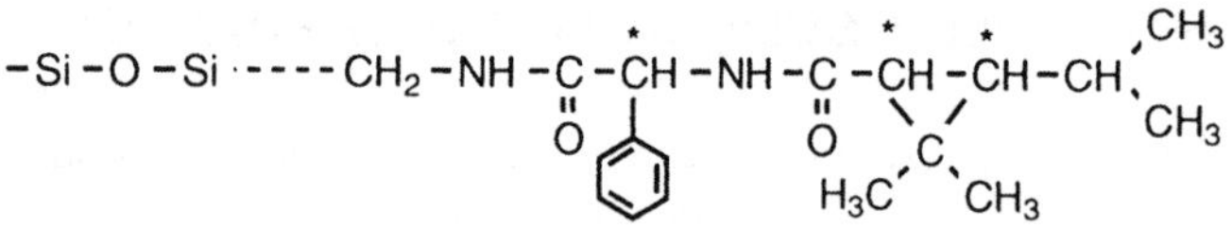

Naphthylethylurea $-Si-O-Si \cdots\cdots CH_2-NH-C-NH-CH$

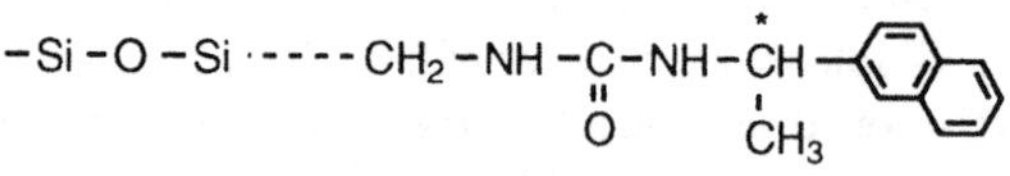

Helical polymer phases

Cellulose triacetate

Cellulose tribenzoate

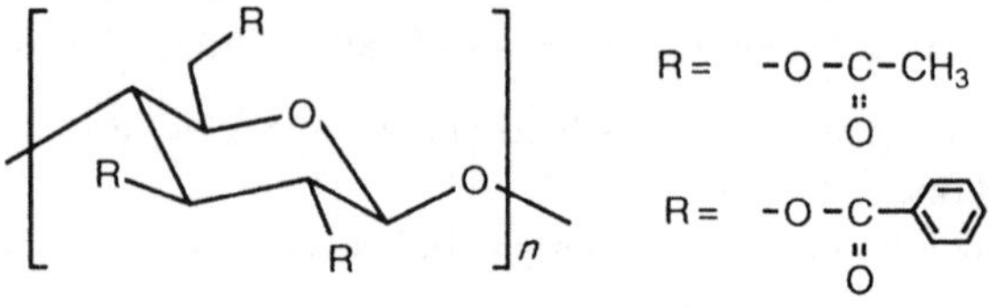

$R = -O-\underset{\underset{O}{\|}}{C}-CH_3$

$R = -O-\underset{\underset{O}{\|}}{C}-$

(continued)

---

[8]Most of these phases are chemically bonded to silica; - - - - means a spacer group, e.g. a CH_2 chain. The chirality of the phases is not specified here; some of the 'brush-type' phases are available in both enantiomeric forms. The chirality is given for cellulose, amylose, cyclodextrin, peptides, proteins.

TABLE 22.1 (*Continued*)

Amylose trisdimethyl-
phenylcarbamate

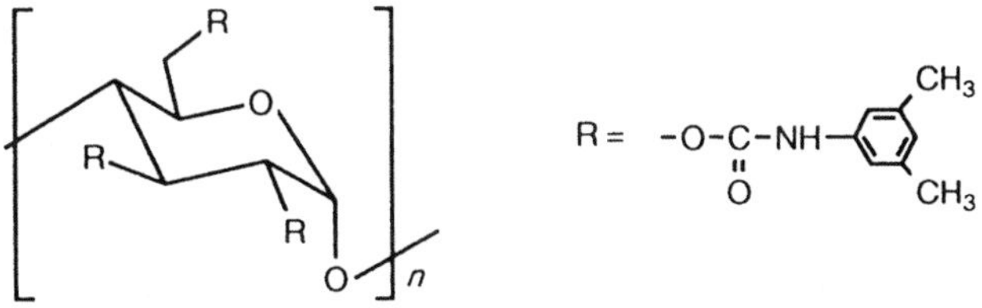

Poly(triphenylmethyl methacrylate)

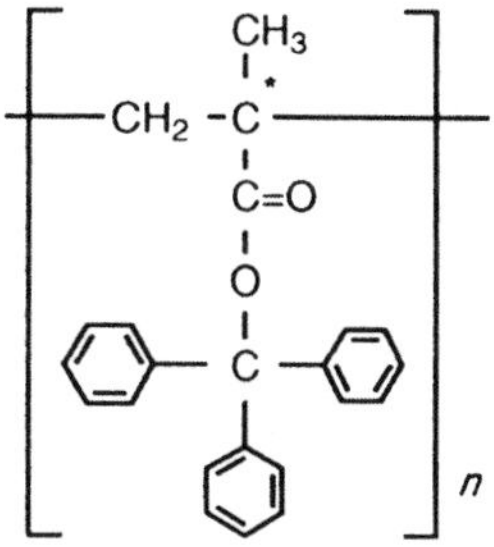

Cavity phases

β-Cyclodextrin

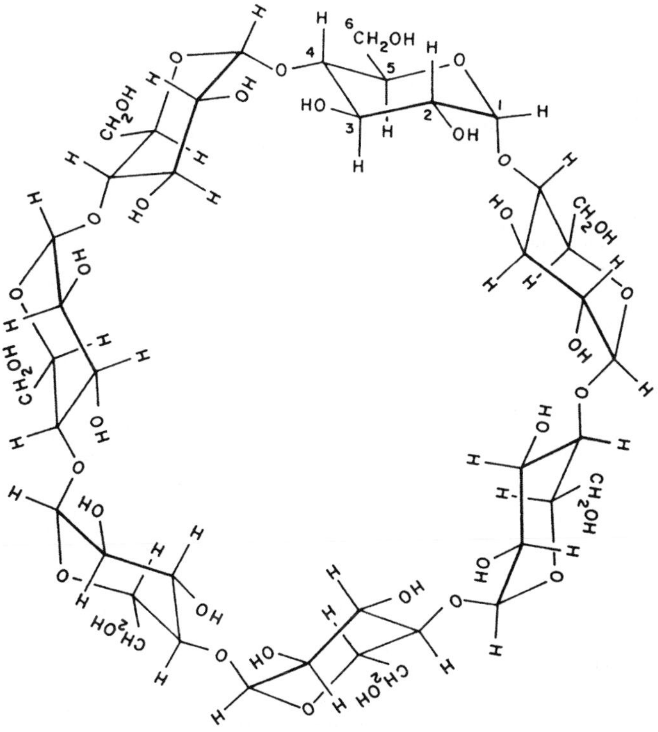

(*continued*)

TABLE 22.1   (*Continued*)

Crown ether

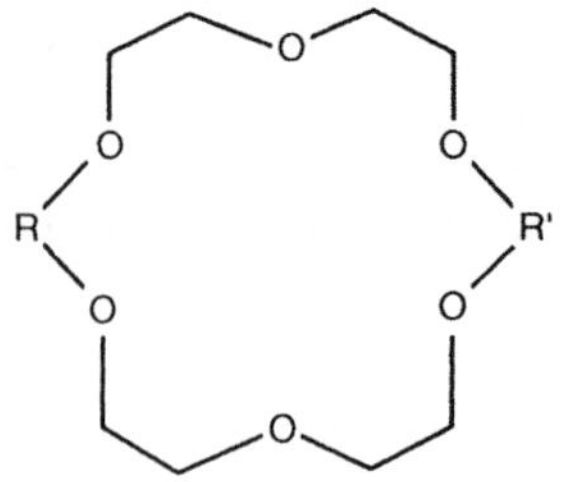

Vancomycin

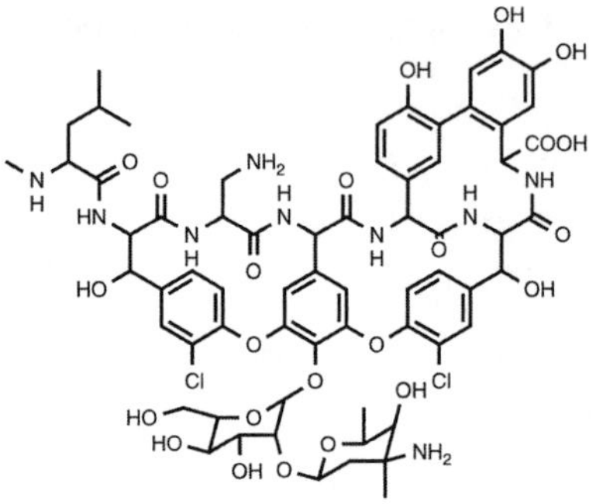

Protein phases
     Bovine serum albumin
     Human serum albumin
     α_1-Acid glycoprotein
     Ovomucoid
     Avidin
     Cellobiohydrolase I
     Pepsin

Ligand-exchange phases

Proline-copper

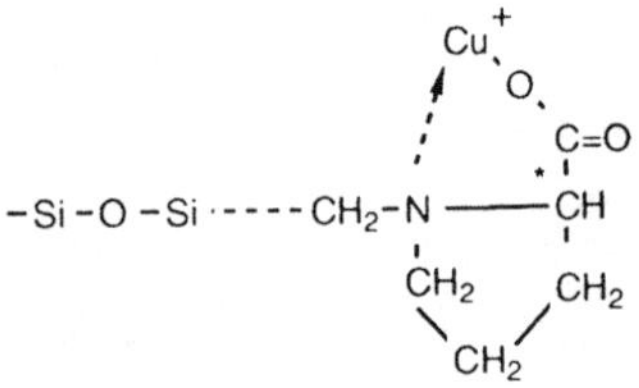

Valine-copper

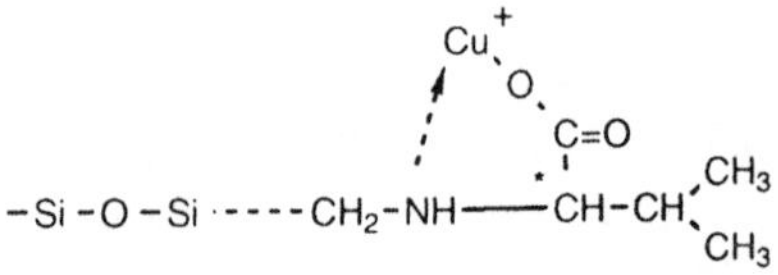

(c) cavity phases such as cyclodextrins, crown ethers and macrocyclic glycopeptide antibiotics;
(d) protein phases;
(e) ligand-exchange phases.

'Brush-Type' CSPs[7]

Section 7.5 outlined how silica can be derivatized with almost any functional group; the resulting monomer structures are known as 'brushes'. The first broadly used and still very important CSP is the 'brush-type' dinitrobenzoylphenylglycine (DNBPG), the first one shown in Table 22.1. According to its inventor, William H. Pirkle, it is often called 'Pirkle-phase', although a more correct name is 'Pirkle I' because it is not the only one of his phases that is on the market.

The DNBPG phase has a number of features which are typical for almost all of the 'brush-type' CSPs. It has two amide groups which are rigid (planar); therefore the whole chiral moiety will prefer a limited (i.e. not an unlimited) number of conformations which is important for chiral recognition. The amide groups can undergo dipole-dipole interaction as well as hydrogen bonds with suitable sample molecules. The dinitrobenzoyl group is a π-acceptor (the ring has a slightly positive charge) and will preferably interact with π-donors such as anilines, phenols, chlorobenzenes and naphthalenes. This interaction is assumed to be the most important one and there is no practical chance of obtaining a separation of enantiomers that do not bear this type of group. Therefore nonchiral deivatization is popular: amines and amino acids are converted to naphthamides, alcohols to naphthyl carbamates and carboxylic acids to anilides. Usually the derivatization is simple and improves detectability because the groups brought into the molecule have excellent UV absorption. The DNBPG phase is rather cheap and is available in both enantiomeric forms. This can be important for trace analysis where the small peak should be eluted in front of the neighbouring large peak, or in preparative applications where the first eluted peak can usually be easily obtained in pure form whereas the second peak is contaminated by its forerunner. Figure 22.4 provides a good example of the successful application of DNBPG silica for the separation of (D,L)-propranolol (as the oxazolidone derivative) in blood.

---

[7] C.J. Welch, *Advances in Chromatography*, P.R. Brown and E. Grushka, eds, **35**, 171 (1995); F. Gasparrini, D. Misiti and C. Villani, *J. Chromatogr. A*, **906,** 35 (2001); M.H. Hyun and Y.J. Cho, in: Gübitz and Schmid (first footnote of this chapter), p. 197.

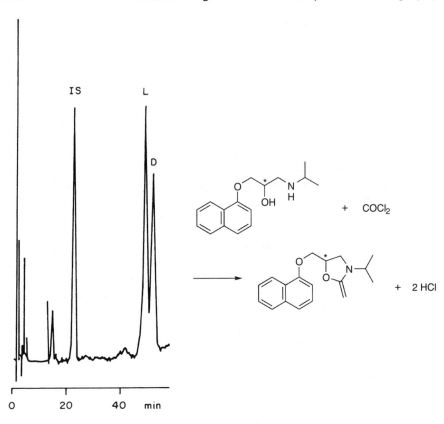

Figure 22.4 Separation of racemic propranolol as oxazolidone derivative with a dinitrobenzoylphenylglycine phase [reproduced with permission from I.W. Wainer, T.D. Doyle, K.H. Donn and J.R. Powell, *J. Chromatogr.*, **306**, 405 (1984)]. Conditions: sample, whole-blood extract, 2.5 h after administration of an 80 mg dose of propranolol racemate, derivatized to oxazolidone with phosgene; column, 25 cm × 4.6 mm i.d.; stationary phase, 3,5-DNB-phenylglycine, 5 μm; mobile phase, hexane–isopropanol–acetonitrile (97 : 3 : 1), 2 ml min⁻¹; fluorescence detector, 290–335 nm; IS, internal standard (oxazolidone derivative of pronethalol).

In a similar manner other 'brush-type' CSPs are used (see Table 22.1): dinitro-benzoylleucine, naphthylalanine, naphthylleucine, chrysanthemoyl-phenylglycine, naphthylethylurea, to name but a few. They all have their specificities for certain classes of compounds, although this is sometimes difficult to predict. To gain some idea of their individual abilities it is best to consult the literature or the excellent brochures provided by the manufacturers. 'Brush-Type' CSPs are robust and allow high sample loads.

Helical Polymers[9]

Underivatized microcrystralline cellulose is not an effective agent for the HPLC separation of enantiomers but its derivatives have turned out to be a most interesting and versatile family of CSPs. Triacetyl cellulose is available as a bulk material (in this form it is rather cheap) as well as coated on silica. Other derivatives are cellulose tribenzoate (also in Table 22.1), trisphenylcarbamate, trisdimethylphenylcarbamate, trischlorophenylcarbamate, tristoluyate and tricinnamate. They are expensive and perhaps less robust than the 'brush-type' CSPs but the range of enantiomers that can be separated by them (as a family) is the broadest and most impressive one of all the CSP groups. Their range of application is expanded by poly(triphenyl-methacrylate), its derivative where one of the phenyl groups is replaced by a pyridyl moiety, and by amylose derivatives. The retention and separation mechanisms are complex and not yet really known. Helical polymers are suited in an excellent manner for the resolution of enantiomers of 'twisted' molecules with D_{2d} symmetry (although a vast number of 'flat' compounds can be resolved as well).

Cavity Phases

There are three classes of cyclic chiral selectors in use as CSPs: cyclodextrins, crown ethers and macrocyclic glycopepetide antibiotics. They can undergo host-guest complexes with small molecules if these can fit into the ring structure; for enantio-differentiation this fit needs to be stereochemically controlled.

Cyclodextrins[10] are oligoglucoses with six, seven or eight units. Table 22.1 shows β-cyclodextrin, which is the heptamer and which is used most frequently; α-cyclodextrin has six glucose units and γ-cyclodextrin has eight. Glucose itself is chiral and in the cyclodextrin molecule, which in fact is a truncated cone, the obtained superstructure of directed primary (at the smaller rim of the cone) and secondary (at the wider rim) hydroxy groups yields chiral binding points which seem to be essential for enantioselectivity. These groups can also be derivatized, e.g. acetylated. Cyclodextrins are used like reversed phases, i.e. with very polar mobile phases. They can separate a wide range of chiral samples but it is difficult to predict their suitability for a given compound.

Chiral crown ethers[11] of the 18-crown-6 type can resolve amino acids as well as primary amines with a close neighbourhood of amino function and the centre of

---

[9] E. Yashima, *J. Chromatogr. A*, **906**, 105 (2001); C. Yamamoto and Y. Okamato, in: Gübitz and Schmid (first footnote of this chapter), p. 173; H.Y. Aboul-Enein and I. Ali, in: Gübitz and Schmid, p.183; I. Ali and H.Y. Aboul-Enein, in: Subramanian (first footnote of this chapter), p. 29.

[10] T. Cserháti and E. Forgács, *Cyclodextrins in Chromatography*, Royal Society of Chemistry, Cambridge, 2003, p. 51; C.R. Mitchell and D.W. Armstrong, in: Gübitz and Schmid (first footnote of this chapter), p. 61.

[11] M.H. Hyun, in: Subramanian (first footnote of this chapter), p. 275.

asymmetry. The interaction occurs between the amino protons and the crown ether oxygens. The group R and R' of the crown ether need to be large and rigid, e.g. binaphthyls, in order to force the small guest molecules into a well defined interaction with the host.

Macrocyclic antibiotics[12] such as vancomycin or teicoplanin are large molecules with several peptide-type ring structures (besides numerous phenyl rings); moreover they are glycosylated. They can be used in the normal-phase and reversed-phase mode as well.

Proteins[13]

It is well known in biochemistry that many proteins, especially enzymes but also transport proteins such as albumin, show high enantioselectivities in their inter-actions with small chiral molecules. It is possible to bind proteins to silica and to obtain a valuable class of CSPs that is mainly suited for the separation of chiral drugs. Several protein phases are commercially available: albumins, α-acid glycoprotein, ovomucoid, avidin, cellobiohydrolase I, and pepsin. They differ in their chromatographic and enantioselective properties which is not a surprise because their biological functions and their size, shape or isoelectric point are not identical at all.

Protein phases are expensive and delicate in handling and their performances (as plate numbers) and loadabilities are low. For many applications this is largely outweighed by their excellent enantioselectivity.

Ligand-Exchange Phases[14]

Amino acids bonded to silica and loaded with Cu^{2+} ions can interact in a steroselective manner with amino acids in aqueous solution. The copper ion forms a complex with both the bound and the sample amino acids, as shown in Figure 22.1. Ligand-exchange phases are suited for the separation of amino acids as well as of some β-amino alcohols and similar molecules because these compounds bear two polar functional groups in adequate spacing. This approach has found limited interest because the column efficiencies are rather low, the detectability of the nonderivatized sample compounds can be a problem and the mobile phase needs to contain copper.

---

[12] T.L. Xiao and D.W. Armstrong, in: Gübitz and Schmid (first footnote of this chapter), p. 113; T.E. Beesley and J.T. Lee, in: Subramanian (first footnote of this chapter), p. 1.

[13] J. Haginaka, *J. Chromatogr. A*, **906**, 253 (2001).

[14] A. Kurganov, *J. Chromatogr. A*, **906**, 51 (2001); V.A. Davankov, *J. Chromatogr. A*, **1000**, 891 (2003); V.A. Davankov, in: Gübitz and Schmid (first footnote of this chapter), p. 207.

If a separation problem cannot be solved with any of the commercially available phases, a possible way out is perhaps the laboratory synthesis of a specific, chiral *molecular imprinted polymer* (MIP).[15]

22.5   INDIRECT SEPARATION OF ENANTIOMERS[16]

If enantiomers are derivatized with a chiral, optically pure reagent a *pair of diastereomers* is formed. Diastereomers are molecules with more than one centre of asymmetry, which therefore differ in their physical properties. From the scheme in Figure 22.5 it is clear that they are not mirror images. Diasetereomers can be separated with a nonchiral chromatographic system, but in any case the derivatization reagent must be chosen very carefully.

For the reaction all hints for precolumn derivatization, as given in Section 19.8, need to be considered. Moreover, it is of the utmost importance that the reagent is of highest optical purity. Otherwise four isomers will be formed (two pairs of enantiomers), but the enatiomers cannot be separated in a nonchiral system and the result will be erroneous (see Table 22.2).

Another important point to consider is the prevention of racemization during derivatization. It is also a problem that the reagent needs to be present in excess during the reaction; if this excess cannot be removed prior to injection an interfering peak may occur in the chromatogram.

It is favourable for the functional group to be derivatized to be situated close to the chiral centre of the molecule. Too large a distance from the centre of asymmetry can lead to the impossibility to resolve the diastereomers. If possible one should try to form amides, carbamates or ureas. All these classes of compounds have a relatively rigid structure (in comparison with, for example, esters) which seems to facilitate the separation. If a choice is possible, one of the reagent's isomers should be taken that allows the minor compound of the pair of enantiomers to be eluted first; then the small peak will not be lost in the tailing of the leading large one.

Diastereomers *may* have differing detector properties. For quantitative analysis it is *necessary* to determine the calibration curve.

---

[15] E. Turiel and A. Martin-Esteban, *Anal. Bioanal. Chem.*, **378**, 1876 (2004); P. Spégel, L.I. Andersson and S. Nilsson, in: Gübitz and Schmid (first footnote of this chapter), p. 217; B. Sellergren, in: Subramanian (first footnote of this chapter), p. 399.
[16] N. Srinivas and L.N. Igwemezie, *Biomed. Chromatogr.*, **6**, 163 (1992); X.X. Sun, L.Z. Sun and H.Y. Aboul-Enein, *Biomed. Chromatogr.*, **15**, 2001 (116).

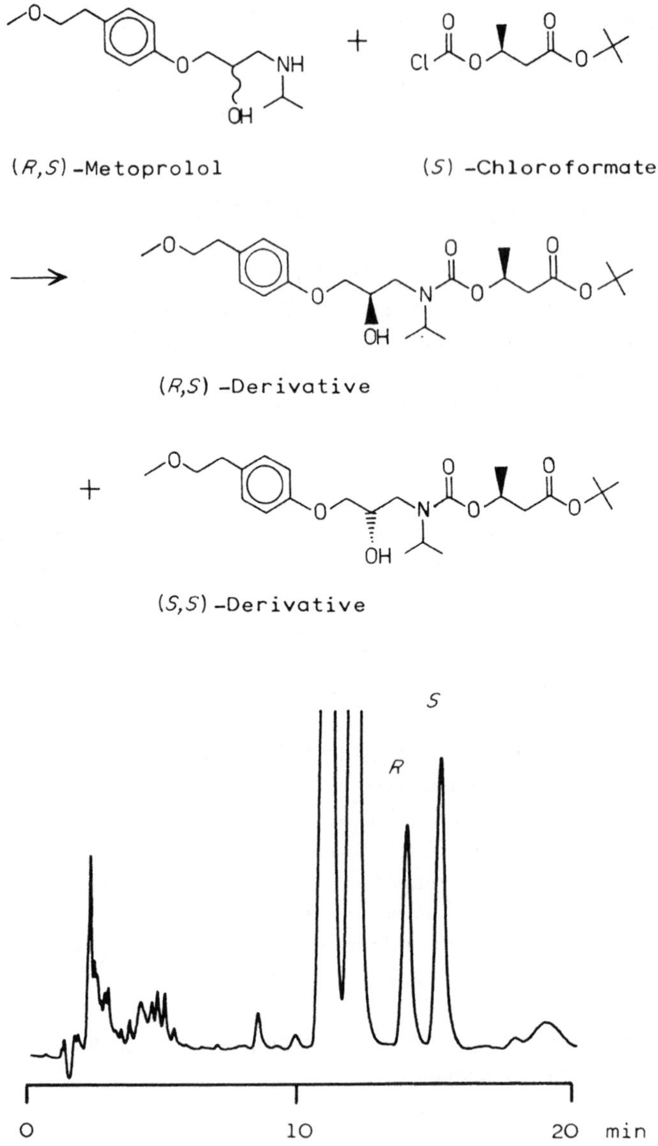

Figure 22.5 Reaction of (*R,S*)-metoprolol with (*S*)-*tert*-butyl 3-(chloroformoxy)-butyrate and determination of the ratio of enantiomers in human plasma (reproduced with permission from A. Green, S. Chen, U. Skantze, I. Grundevik and M. Ahnoff, poster at the 13th International Symposium on *Column Liquid Chromatography*, Stockholm, 1989.) Conditions: stationary phase, octadecyl silica; mobile phase, phosphate buffer–acetonitrile (1 : 1); fluorescence detector 272/312 nm.

TABLE 22.2   Purity of reagent and obtained purity of sample

Purity of optical reagent	Maximum detectable optical purity of sample
99.95%	99.9%
99.5%	99.0%
98%	96%

Indirect separation is often favourable with samples of biological origin. It is simpler than with the direct methods to obtain compounds of interest to be separated from interfering peaks. The analysis is done on silica (with nonaqueous samples) or on a C_{18} reversed phase (with biological samples), as shown in Figure 22.5.

23 Special Possibilities

23.1 MICRO, CAPILLARY AND CHIP HPLC[1]

Most columns used for HPLC separations nowadays have an internal diameter of 4.6 mm, although they use double the amount of solvent and the performance is not better than with the 3.2 mm i.d. columns. There are many good reasons why the diameter should, in fact, be reduced even further:

(a) Solvent consumption decreases with the square of the column diameter.
(b) As already outlined in Section 19.2, microcolumns are essential for trace analysis if the amount of sample is limited.
(c) Some LC-MS coupling techniques call for low eluent flow rates.
(d) Small and very small diameter columns are needed to achieve highest theoretical plate numbers.

Columns fall into three different categories:

(a) *Open capillaries.* A capillary of i.d. 50 μm or less has a chemically modified glass surface or liquid film as the stationary phase, the latter being either normal or reversed.
(b) *Packed capillaries.* It is possible to pack capillaries of, e.g., 15 cm × 75 μm i.d. with conventional HPLC phases or with a monolith.
(c) *Microcolumns.* These are similar to the usual HPLC columns but have an i.d. no greater than 1 mm.

---

[1] J.P.C. Vissers, *J. Chromatogr. A*, **856**, 117 (1999).

Open capillaries[2] of 5 μm i.d. and with 10^6 theoretical plates are rather easy to prepare. Theory shows that a capillary of i.d. <10 μm scores over a packed column as regards peak capacity, resolution and analysis time. The equipment poses a problem that is not insoluble (injection volume 50 pl, flow rate 2 nl min^{-1}).

Packed capillaries[3] offer a larger capacity, i.e. the amount of sample may be greater than with open capillaries. Lower permeability is a disadvantage.

Microcolumns[4] of i.d. 1 mm (steel tubes with an external diameter of 1/16 in) are available commercially and offer the advantage of increased sensitivity and savings on solvent. The apparatus required is as follows:

pump   25−250 μl min^{-1}
sample volume   0.2−1 μl (maximum sample load 10 μg)
detector volume   0.5−1 μl

Additionally there are fused-silica columns with ca. 0.2 mm i.d. Representing all other small-bore techniques, Figure 23.1 illustrates the performance of micro HPLC using this kind of column. Fifteen bile acids which are vital to any assessment of possible liver disease were identified in body fluids. The total solvent consumption of 210 μl, the ability to mix a gradient in this small volume and reproduce it and the refined reaction detector already described in Section 19.8 are special features of this particular system. The following abbreviations are used: UDC = ursodeoxycholic acid; C = cholic acid; CDC = chenodeoxycholic acid; DC = deoxycholic acid; LC = lithocholic acid; G (prefix) = glycine conjugate; T (prefix) = taurine conjugate.

A completely different development are the chips with integrated HPLC separation path. One possibility is a rectangular channel, e.g. 75 μm × 50 μm and a few centimetres long, which is packed with a common 5 μm stationary phase.[5] A chip module of the size of a credit card, put into a matching interface with liquid entrance and outlet, can be used as electrospray inlet for a mass spectrometer. However, true chip technology devices are only those which are completely prepared from silicon by lithography and etching, as known from semiconductor technology.[6] A possible design consists of regularly arranged cylinders

[2] R. Swart, J.C. Kraak and H. Poppe, *Trends Anal. Chem.*, **16**, 332 (1997).
[3] J. Hernández-Borges *et al.*, *J. Sep. Sci.*, **30**, 1589 (2007).
[4] R.P.W. Scott, *J. Chromatogr. Sci.*, **23**, 233 (1985); F.M. Rabel, *J. Chromatogr. Sci.*, **23**, 247 (1985).
[5] H. Yin and K. Killeen, *J. Sep. Sci.*, **30**, 1427 (2007).
[6] J. Eijkel, *Lab Chip*, **7**, 815 (2007); W. de Malsche *et al.*, *Anal. Chem.*, **79**, 5915 (2007).

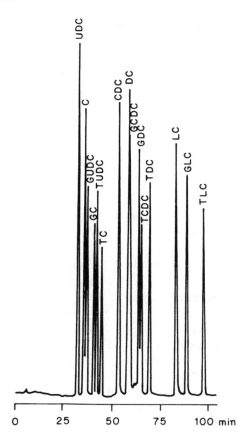

Figure 23.1 Separation of bile acids using micro HPLC [reproduced with permission from D. Ishii, S. Murata and T. Takeuchi, *J. Chromatogr.*, **282**, 569 (1983)]. Conditions: sample, 11 nl of solution containing 20 ng each of the acids; guard column, 5 cm × 0.2 mm i.d. fused silica; column, 20 cm × 0.26 mm fused silica; stationary phase, silica ODS CS-01, 5 μm; mobile phase, 60 mM phosphate buffer containing NAD and acetonitrile, gradient with increasing acetonitrile content, 2.1 μl min⁻¹; detector, derivatization with immobilized enzyme, then fluorescence, 365/470 nm. For abbreviations, see text.

of 5 μm width which rise vertically to the direction of flow. Their properties can be identical to normal-phase materials (as oxidized silicon) or to reversed phases (after derivatization). Such chips are able to perform true chromatographic separations but for the present they are only an interesting experimental development.

23.2 HIGH-SPEED AND SUPER-SPEED HPLC[7]

Routine analysis requires a short analysis time, especially when sample preparation is simple. In fact, the instrumental requirements are demanding but commercially available. Three categories can be distinguished:

$$\begin{array}{lll}
\text{conventional HPLC} & \text{breakthrough time } t_0 \approx 1 \text{ min} \\
\text{high-speed HPLC} & \approx 10 \text{ s} \\
\text{super-speed HPLC} & \approx 1 \text{ s}
\end{array}$$

For high- and super-speed[8] HPLC some conditions need to be fulfilled:

(a) The diffusion coefficient of the sample molecules in the mobile phase must be high; only mobile phases with very low viscosity can be considered. Separations at elevated temperature can be very attractive. The method is unsuitable for samples of higher molecular mass.

(b) As Figures 2.29 and 2.30 clearly show, fast separation is dependent on small particle size. Theory then dictates that the minimum possible retention time is proportional to the square of the particle diameter.

(c) Higher pressure than usual (UHPLC) is an interesting possibility, see Section 23.3.

(d) The column must be as short as possible.

(e) As stationary phases, the perfusion particles and monolithic materials mentioned in Section 7.3 are most favourable.

(f) The extra-column volume of the chromatograph must be minimal. An arrangement in which the column is mounted free from extra volumes between the injector and detector cell, as shown in Figure 23.2, is the best solution.

(g) The pump must be able to supply the required volumetric flow rate and pressure. The linear flow rate is the greater the smaller is the internal diameter of the column, for a given volumetric flow rate.

(h) The autosampler must work with a cycle time of 10 s or less.

(i) The time constant and the cell volume of the detector must be low enough not to affect separation in any way, i.e. they should have no effect on peak shape or resolution. Corresponding data are shown in Figures 23.3 and 23.4.

(j) Data processing, i.e. signal acquisition and integration, must be fast.

Obviously the column is operated far from its van Deemter minimum for both high- and super-speed HPLC and so cannot perform at its maximum performance. Although super-speed HPLC is more suitable for isocratic separation, super-speed gradients are also feasible.

---

[7] H. Chen and C. Horvárth, *J. Chromatogr. A*, **705**, 3 (1995); D. Blanco, N. Sánchez and M.D. Gutiérrez, *J. Liquid Chrom. Rel. Techn.*, **29**, 931 (2006); A.C. Mann, *LC GC Int.*, **20**, 290 (2007).

[8] J.J. Kirkland, *J. Chromatogr. Sci.*, **38**, 535 (2000).

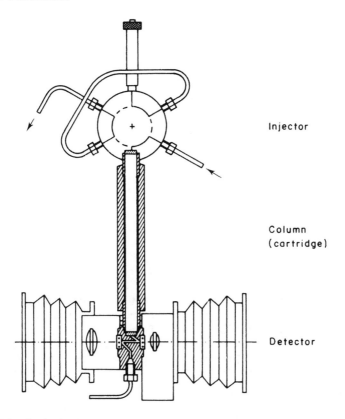

Figure 23.2 Optimized set-up of injector, column and detector (reproduced with permission of Kontron Analytik).

An example of a high- to super-speed separation of a pharmaceutical product is shown in Figure 23.3. The method is used routinely by the manufacturer for determining dissolution rates and content uniformity, with an arrangement as depicted in Figure 23.2.

Even faster is the separation depicted in Figure 23.4. A test mixture of five components is separated in ca. 3 s with a relative standard deviation during quantitative analysis of below 1.5%.

23.3 FAST SEPARATIONS AT 1000 BAR: UHPLC[9]

Figures 2.29–2.31 show at a glance that it is possible to get higher theoretical plate numbers or peak capacities if the working pressure is increased. On the other hand

[9] J.R. Mazzeo *et al.*, *Anal. Chem.*, **77**, 460 A (2005); M.E. Swartz, *J. Liq. Chromatogr. Rel. Tech.*, **28**, 1253 (2005); N. Wu and A.M. Clausen, *J. Sep. Sci.*, **30**, 1167 (2007).

Figure 23.3 High-speed separation of an antihypertensive drug (reproduced with permission of F. Erni, Novartis). Conditions: sample 10 μl of a solution of a capsule in artificial gastric juice; column, 10 cm × 2.1 mm i.d.; stationary phase, Spherisorb RP-18, 5 μm; mobile phase, 3 ml min^{-1} acetonitrile–0.4 M phosphoric acid (85 : 15); UV detector, 260 nm; cell volume, 2.8 μl; optical path length, 10 mm; time constant, 0.1 s (98%). Peaks: 1 = clopamide; 2 = dihydroergocristin; 3 = reserpine.

it is possible to perform faster separations with identical separation performance at higher pressure. (However, if the flow rate is increased without any consideration of the column dimensions the separation will be slightly poorer due to the van Deemter relationship. To work at increased flow alone is not a measure for optimization.)

It is not trivial to build pumps for use at 1000 bar which fulfill the common requirements such as pulse-free flow which is independent of the pressure drop. Nevertheless it is possible; the demand for high peak capacity (e.g. needed in proteomics) or for fast method development in industry led to the appearance of such instruments on the market. The technique is called ultra high pressure liquid chromatography (UHPLC).

It makes no sense to buy a UHPLC pump and to use it with conventional instrumentation. All the components of the system must match the increased requirements. In addition, special UHPLC columns with UHPLC phases are recommended in order to generate satisfactory results. Figure 23.5 shows the metabolic profile of a plasma sample which was obtained with UHPLC-MS.

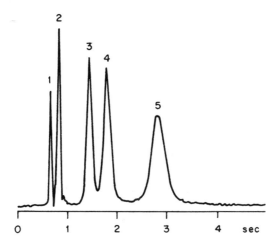

Figure 23.4 Super-speed separation of a five-component mixture [reproduced with permission from E. Katz and R.P.W. Scott, *J. Chromatogr.*, **253**, 159 (1982)]. Conditions: sample, 0.2 µl; column, 2.5 cm × 2.6 mm i.d.; stationary phase, Hypersil, 3.4 µm; mobile phase, 13 ml min^{-1} pentane containing 2.2% methyl acetate ($u = 3.3$ cm s^{-1}), 360 bar; UV detector, 254 nm; cell volume, 1.4 µl; time constant 6 ms; data acquisition, 100 points s^{-1}. Peaks: 1 = *p*-xylene; 2 = anisole; 3 = nitrobenzene; 4 = acetophenone; 5 = dipropyl phthalate.

The theoretical background for separations up to 3000 bar was discussed comprehensively by Martin and Guiochon.[10] It seems that various phenomena such as the compressibility of the mobile phase or thermal effects occur at pressures above 1000 bar, making it difficult to develop and optimize separations in this region.

23.4 HPLC WITH SUPERCRITICAL MOBILE PHASES[11]

A pure compound may be in the state of a gas, solid or liquid (or even multiphase) depending on the pressure and temperature, the interrelationship being shown in Figure 23.6. Following the vapour pressure curve which separates the gas and liquid states in the direction of increased pressure and temperature leads to an area in which the densities of both phases are identical. A phase that is neither gas nor liquid follows on from the critical point *P* (shaded area) and this is known as *the fluid* or *supercritical*

[10] M. Martin and G. Guiochon, *J. Chromatogr. A*, **1090**, 16 (2005).
[11] R.M. Smith, *J. Chromatogr. A*, **856**, 83 (1999).

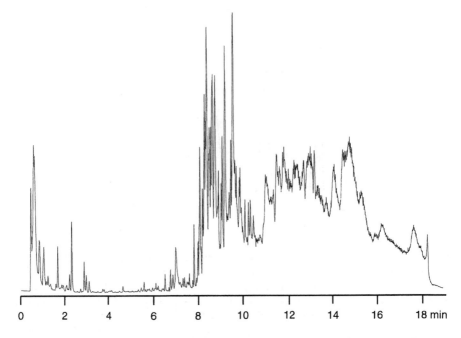

Figure 23.5 Separation with UHPLC [after T. Moritz, Nestlé Research Center, Lausanne, with permission; see also S.J. Bruce *et al.*, *Anal. Biochem.*, **372**, 237 (2008)]. Conditions: sample, extract from human plasma; column, 10 cm 2.1 mm i.d.; stationary phase, C8 UPLC 1.7 µm; mobile phase, 0.5 ml min^{-1} water–acetonitrile with 0.1% formic acid, linear gradient; pressure, 600 bar; detector, TOF-MS with positive ESI. The first peaks are amino acids and low-mass metabolites, the ones between 8 and 11 min are phospholipids (besides background peaks).

area. The fluid can be used as a mobile phase for chromatography which is then referred to as supercritical fluid chromatography (SFC).

Supercritical mobile phases open up an area that is related to both gas and liquid chromatography but is also subject to its own set of laws. Some of the physical phenomena are both interesting and unusual:

(a) The sample diffusion coefficients are midway between those in gases (comparatively large, as high flow rates are essential to GC) and values recorded in liquids (comparatively small, as low flow rates are a feature of HPLC).
(b) The mobile phase viscosity is greater than in gases but much less than in liquids.
(c) The solubility of sample molecules in the mobile phase increases as the pressure rises. Gradients, i.e. effects on retention time, can be produced not only by a

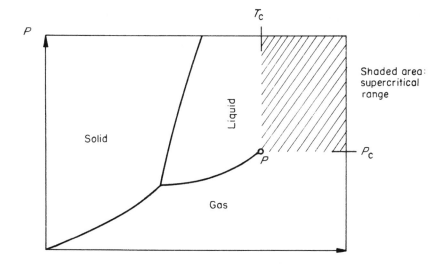

Figure 23.6   Phase diagram of a pure compound.

change in mobile phase composition but also by a pressure change. Strongly retained compounds are eluted faster by an increase of pressure.

The vapour pressure of volatile samples has a great influence on retention behaviour, this being one of the analogues with GC. Compounds that are potential mobile phases are listed in Table 23.1. The first two are obviously unsuitable for temperature-sensitive analytes. By far, carbon dioxide is used most frequently. Since it is very nonpolar a common B solvent is methanol.

Accurate pressure- and temperature-control facilities are required for supercritical chromatography. The mobile phase must be heated to the correct temperature in a spiral before the injection valve. The spiral, valve, column and detector should all be placed in an oven. A restrictor must be placed behind the detector so that the whole system can be maintained at a sufficiently high pressure. As columns it is possible to use either open capillaries, which allows us to obtain very high plate numbers, as well

TABLE 23.1   Potential mobile phases for supercritical HPLC

Compound	p_c (bar)	T_c (°C)	d (g cm^{-3})
n-Pentane	33.3	196.6	0.232
Isopropanol	47.0	253.3	0.273
Carbon dioxide	72.9	31.3	0.448
Sulfur hexafluoride	37.1	45.6	0.752

as packed columns, as was the case in the separation shown in Figure 23.7. Typical SFC detectors are UV, polarimeter (for the separation of enantiomers), MS, MS^2 and the flame ionization detector as in GC.

Polar analytes are less suited for SFC. The sample solvent can be methanol or everything which is less polar.

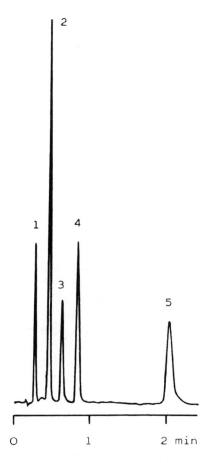

Figure 23.7 Separation of opium alkaloids with supercritical mobile phase [reproduced with permission from J.L. Janicot, M. Caude and R. Rosset, *J. Chromatogr.*, **437**, 351 (1988)]. Conditions: column, 12 cm × 4 mm i.d.; stationary phase, Spherisorb NH_2, 3 μm; mobile phase, carbon dioxide–methanol–triethylamine–water (87.62:11.80:0.36:0.22), 8 ml min^{-1}; pressure, 220 bar; temperature, 40.7 °C; UV detector, 280 nm. Peaks: 1 = narcotine; 2 = thebaine; 3 = codeine; 4 = cryptopine; 5 = morphine.

Preparative solutions are possible, even in industrial size[12] (unsaturated fatty acids, cyclosporine, phytol). Of special interest are also separations of enantiomers on chiral stationary phases which can be used in the normal-phase mode.[13]

23.5   HPLC WITH SUPERHEATED WATER[14]

A superheated liquid shows interesting properties even below its critical point. Superheated water at temperatures above 100 °C is especially attractive. Figure 23.8 shows its vapour pressure curve. To give an example, at 200 °C the pressure must be almost 16 bar in order to prevent boiling of the water. Higher temperatures are rarely used although the critical point of water is at 374 °C and 220 bar. The chromatographic conditions can be freely chosen as long as the pressure is at least as high as shown by the vapour pressure curve; it is well possible to perform separations in the region above the curve.

With increasing temperature the polarity of water decreases markedly; at 200 °C its polarity is almost the one of methanol. Therefore it is possible to run reversed-phase gradients with pure water alone by changing the temperature (Figure 23.9). The detector cell needs to be pressure-tight at the corresponding conditions. A capillary of adequate length and inner diameter, connected to the detector outlet, acts as a restrictor.

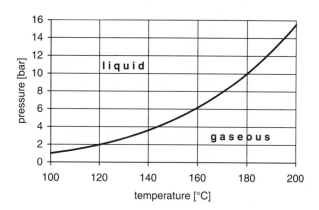

Figure 23.8   Boiling point curve of water between 100 and 200 °C.

[12] W. Majewski, E. Valery and O. Ludemann-Hombourger, *J. Liq. Chromatogr. Rel. Tech.*, **28**, 1233 (2005).
[13] D. Mangelings and Y. Vander Heyden, *J. Sep. Sci.*, **31**, 1252 (2008).
[14] R.M. Smith, *Anal. Bioanal. Chem.*, **385**, 419 (2006); R.M. Smith, *J. Chromatogr. A*, **1184**, 441 (2008); K. Hartonen and M.L. Riekkola, *Trends Anal. Chem.*, **27**, 1 (2008).

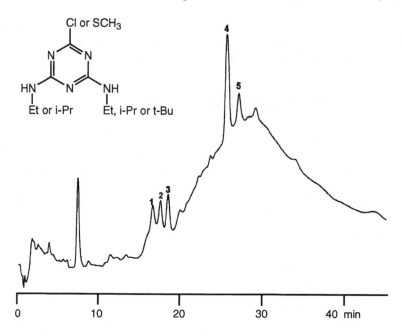

Figure 23.9 Separation of triazine herbicides with superheated water and temperature gradient [reproduced with permission from R. Tajuddin and R.M. Smith, *J. Chromatogr. A*, **1084**, 194 (2005)]. Conditions: sample, extract from ericaceous compost, spiked with triazine herbicides; column, 10 cm × 2.1 mm i.d.; stationary phase, Hypercarb 5 µm (porous graphitic carbon); mobile phase, 1 ml min^{-1} water; composed temperature gradient from 130 to 220 °C; pressure, 35 bar; UV detector, 222 nm. Peaks: 1 = propazine; 2 = atrazine; 3 = simazine; 4 = ametryn; 5 = terbutryn. Compounds 1–3 with chlorine at R_1, 4 and 5 with thioether.

The advantages of superheated water are striking: it is cheap, available at any place, without any harm to the environment, has excellent UV transparency and is compatible with the nonspecific flame ionization detector. Increasing the temperature leads to decreasing viscosity, thus enabling fast mass transfer. Most analytes are stable even at the high temperatures needed, at least for the duration of the separation. Traditional reversed phases are not suited but more advanced materials with special bonding chemistry must be used. Polystyrenes, zirconia-based phases or porous graphitic carbon are suitable as well, however, note the individual temperature stability of all these phases. Separations with superheated water are perfomed with packed capillaries as well as with common-diameter HPLC columns.

23.6 ELECTROCHROMATOGRAPHY[15]

It is not only possible to force the mobile phase through a capillary or column by means of a pump but also by electroendosmosis. Thereby one utilizes the fact that an electric double layer occurs on all boundary layers. Silica or quartz glass are surfaces covered with fixed negative excess charges and a solution in contact with it forms positive boundary charges. If a potential gradient of approximately $50\,kV\,m^{-1}$ is applied the solution flows in the direction of the negative electrode.

The distinct advantages of this technique are as follows:

(a) The flow profile is not parabolic, as with pump transport, but rectangular. The band-broadening component caused by flow distribution (A term) falls off.

(b) For this reason it is not necessary to use narrow inner diameters if open capillaries are used.[16] The separation performance obtained is independent of capillary diameter. Its length can be chosen in accordance with the demands, as flow

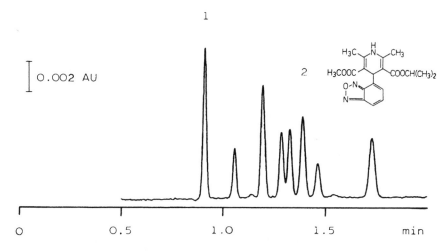

Figure 23.10 Separation with electrochromatography [reproduced with permission from H. Yamamoto, J. Baumann and F. Erni, *J. Chromatogr.*, **593**, 313 (1992)]. Conditions: capillary, 14.3/32.4 cm × 50 μm i.d.; stationary phase, ODS Hypersil 3 μm; mobile phase, sodium tetraborate 4 mM–acetonitrile (20:80), 2.6 mm s⁻¹; voltage, 30 kV; UV detector, 220 nm. Peaks: 1 = thiourea; 2 = isradipin; other peaks are byproducts of isradipin.

---

[15] J. Jiskra, H.A. Claessens and C.A. Cramers, *J. Sep. Sci.*, **26**, 1305 (2003); U. Pyell, ed., *Electrokinetic Chromatography*, John Wiley & Sons, Ltd, Chichester, 2006.

[16] L.A. Colón *et al.*, in: *Advances in Chromatography*, vol. 42, P.R. Brown and E. Grushka, eds, Dekker, New York, 2003, pp. 43–106.

resistance does not need to be considered. Separation performance is as high as in capillary gas chromatography.

(c) Since flow resistance is unimportant it is possible to use packed capillaries with very small particles (≤ 1 μm). The reduced plate height h is less than 2, i.e. better than in conventional HPLC.

The mobile phase needs to be conductive; buffer solutions between 0.001 and 0.1 M in strength are used. The linear flow rates are in the 1 mm s^{-1} region. A drawback is the strong heating of the mobile phase as it flows; therefore efficient cooling is necessary and capillary diameters are limited. The transition to capillary electrophoresis, where sample molecules *need* to be charged (in contrast to electrochromatography), is not well defined.[17] An excellent and fast separation of a drug and its byproducts is presented in Figure 23.10.

---

[17] F.M. Everaerts, A.A.A.M. van de Goor, T.P.E.M. Verheggen and J.L. Beckers, *J. HRC*, **12**, 26 (1989).

24 Appendix 1: Applied HPLC Theory

The principles and results of HPLC theory that are useful in everyday work have been presented in Chapters 2 and 8. The following is a comprehensive repetition and summary of important equations. Derivations and references are not presented here; both can be found in the two chapters mentioned and in many sections of the book (see Index); references can also be found in the paper on which this appendix is based.[1]

The basis of the considerations is an HPLC column of typical dimensions, 25 cm in length L_c and 4.6 mm inner diameter d_c, which is packed with a stationary phase of 5 μm particle diameter d_p. Let us investigate what can be calculated from these and some more data.

Plate Number N

By means of the reduced plate height h it is possible to calculate N. With the assumption that the column is packed very well (not excellently), i.e. that $h = 2.5$, we obtain:

$$N = \frac{L_c}{hd_p} = \frac{250\,000}{2.5 \times 5} = 20\,000 \tag{1}$$

---

[1] V.R. Meyer, *J. Chromatogr.*, **334**, 197 (1985); see also: M.A. Stone, Mathematics of optimization and scaling for the practical chromotographer, *J. Liq. Chromatogr. Rel. Tech.*, **30**, 605 (2007), including the optimization of gradients.

Practical High-Performance Liquid Chromatography, Fifth edition   Veronika R. Meyer
© 2010 John Wiley & Sons, Ltd

Comment

The value for N of 20 000 is rather high and in fact is seldom reached by 25 cm HPLC columns ($d = 5$ μm) used routinely. A value of 10 000 or 15 000 is, unfortunately, more realistic.

Void Volume V_0

The total porosity, ε_{tot}, is about 0.8 for many column packings of totally porous particles without chemical derivatization. Thus V_0 can be obtained from:

$$V_0 = \frac{d_c^2 \pi}{4} L_c \varepsilon_{tot} = \frac{4.6^2 \pi}{4} 250 \times 0.8 \, \text{mm}^3 = 3320 \, \text{mm}^3 = 3.3 \, \text{ml} \qquad (2)$$

Comment

It should be mentioned that the value of V_0 would be reduced by half if a column of 3.2 mm i.d. were to be used instead of one of 4.6 mm i.d. since V_0 is proportional to d_c^2. From this point of view it is not clear why the 4.6 mm column is so popular; one of 3.2 mm would enable a saving in solvent consumption of 50%!

Linear Flow Velocity u and Flow Rate F

The column is regarded as being used at its optimum flow which corresponds to $v_{opt} = 3$. For the calculation of the matching flow velocity of the mobile phase, a mean diffusion coefficient, D_m, of the solute molecules in the mobile phase of 1×10^{-9} m^2s^{-1} is assumed. Thus u and F are given by:

$$u = \frac{v D_m}{d_p} = \frac{3 \times 1 \times 10^{-9}}{5 \times 10^{-6}} \text{m s}^{-1} = 0.6 \times 10^{-3} \text{m s}^{-1} = 0.6 \, \text{mm s}^{-1} \qquad (3)$$

$$F = \frac{u d_c^2 \pi \varepsilon_{tot}}{4} = \frac{0.6 \times 4.6^2 \times \pi \times 0.8}{4} \text{mm}^3\text{s}^{-1} = 8.0 \, \text{mm}^3\text{s}^{-1}$$
$$= 0.48 \, \text{ml min}^{-1} \qquad (4)$$

Comment

The value of the diffusion coefficient is often higher in adsorption chromatography due to the use of nonviscous organic solvents, i.e. the optimum flow velocity is higher. In contrast, the diffusion coefficients of solutes in the methanol–water mixtures often used for reversed-phase chromatography are smaller due to the rather high viscosity of these eluents. Although this effect is of minor practical relevance it should be kept in mind. Moreover, it is of great importance in the chromatography of macromolecules that have very small diffusion coefficients which result from their large molar volumes. This means that large molecules must be chromatographed at low flow velocities.

Chromatography at higher flow rates than the optimum one results in a decrease in the separation ability of the column. For wellpacked microparticulate HPLC columns this effect is nearly negligible if the reduced velocity is less than 10.

Retention Time t_R

The retention time of an unretained solute, t_0, is:

$$t_0 = \frac{L_c}{u} = \frac{250}{0.6}\,s = 417\,s \approx 7\,min \tag{5}$$

For two retained solutes having retention factors, $k = 1$ and $k = 10$:

$$t_R = k \times t_0 + t_0 = 1 \times 7 + 7\,min = 14\,min$$
$$t_R = 10 \times 7 + 7\,min = 77\,min \tag{6}$$

Comment

The retention times that result from chromatography at the optimum flow rate are rather high. Note that all retention times are independent of the inner diameter of the column if the analysis is performed at a given reduced velocity. All values obtained by use of the following equations, except (9) and (12), are independent of the flow rate and the retention time.

Retention Volume V_R

The volume of mobile phase, V_R, necessary to elute peaks of either $k = 1$ or $k = 10$ are:

$$V_R = F \times t_R = 0.48 \times 14\,ml = 6.7\,ml$$
$$V_R = 0.48 \times 77\,ml = 37\,ml \tag{7}$$

Comment

As already mentioned in the discussion of equation (2), the retention volume V_R, of a peak is proportional to the square of the column diameter! If the first eluted peak has

$k = 0$, i.e. it is a nonretained solute, its retention volume is half the value calculated for $k = 1$: only 3.3 ml (the void volume). In this case the values obtained from equations (10), (11), (13) and (14) would be decreased!

Peak Capacity n

If $k = 10$ is not exceeded the peak capacity is:

$$n = 1 + \frac{\sqrt{N}}{4} \ln(1 + k_{\max}) = 1 + \frac{\sqrt{20\,000}}{4} \ln(1 + 10) = 86 \tag{8}$$

Peak Width 4σ and Peak Volume V_p

If the first peak has $k = 1$, its base width and its volume are:

$$4\sigma = 4\frac{t_R}{\sqrt{N}} = 4\frac{14 \times 60}{\sqrt{20\,000}} \,\text{s} = 23.8\,\text{s} \tag{9}$$

$$V_p = 4\sigma \times F = 23.8 \times 8\,\mu l = 190\,\mu l \tag{9a}$$

Comment

The peak width in seconds (equation 9) is influenced by the mobile phase flow rate, whilst the peak width in microlitres (the peak volume, equation 9a) is not, if we assume that the decrease in plate number by a flow rate other than the optimum one is negligible. Yet the peak width in seconds is an important parameter for the calculation of the detector time constant (equation 12)!

Injection Volume V_i

The maximum allowed injection volume, V_i, to avoid an excessive broadening of the first peak is defined as:

$$V_i = \theta V_R \frac{K}{\sqrt{N}} \tag{10}$$

where K is a parameter characteristic of the quality of injection and is assumed to be equal to 2 and θ^2 defines the fraction of peak broadening. Thus at 1% peak broadening, i.e. $\theta^2 = 0.01$ and $\theta = 0.1$:

$$V_i = 0.1 \times 6700 \frac{2}{\sqrt{20\,000}}\,\mu l = 9.5\,\mu l$$

At 9% peak broadening, i.e. $\theta^2 = 0.09$ and $\theta = 0.3$:

$$V_i = 0.3 \times 6700 \frac{2}{\sqrt{20\,000}} \mu l = 28 \, \mu l$$

Comment

Usually the injection quality can be assumed to correspond with $K = 2$ but for excellent injection it is lower, thus allowing larger sample volumes. However, it is more important that the maximum injection volume depends on the retention volume of the solute; therefore, for capillary columns with their extremely small retention volumes (since V_R decreases with the square of the capillary diameter), the allowed injection volume has values in the nanolitre range.

There is one way to avoid the limitations in injection volume. If the sample solvent is markedly weaker than the mobile phase for the HPLC analysis, i.e. if its elution strength is low, the solutes are concentrated at the top of the column. In this case the injection volume may be unusually high, in the range of millilitres or even litres. However, a prerequisite is that the amount of solute is small enough to prevent the adsorption isotherm from becoming nonlinear.

At this point the attention of the reader is drawn to a fact whose importance will be apparent later in the discussion of the detection limit (equations 14–16). If for the calculation of the injection volume in equation (10) the retention volume (37 ml) of the second peak with $k = 10$ is used, V_i is found to be much higher: 157 μl may be injected without broadening of the second peak by more than 9%. Of course, the first peak would be broadened dramatically but one can imagine situations where the early eluted peaks are not of interest.

Detector Cell Volume V_d

The maximum allowed volume of the detector cell, V_d, is defined as:

$$V_d = \frac{\theta V_R}{\sqrt{N}} = \frac{V_i}{2} \tag{11}$$

Thus, at 1% peak broadening, $V_d = 9.5/2 \, \mu l \approx 5 \, \mu l$ and at 9% peak broadening, $V_d = 28/2 \, \mu l = 14 \, \mu l$.

Comment

It is obvious that a standard 8 μl detector cell causes a peak broadening of more than 1% for the early eluted peaks on a 25 cm × 4.6 mm column.

Detector Time Constant τ

The maximum allowed time constant of the detector, τ, is defined as:

$$\tau = \theta \frac{t_R}{\sqrt{N}} \qquad (12)$$

At 1% broadening of the first peak:

$$\tau = 0.1 \frac{14 \times 60}{\sqrt{20\,000}}\, s = 0.6\, s$$

At 9% peak broadening:

$$\tau = 0.3 \frac{14 \times 60}{\sqrt{20\,000}}\, s = 1.8\, s$$

Comment

The time constant of a detector should not exceed 0.5 s if used with columns of 25 cm in length, a requirement that is satisfied by all modern HPLC detectors. However, if the columns are as short as 10 or 5 cm, a time constant of 0.5 s is too high. The time constant must be decreased if the velocity of the mobile phase is increased.

Capillary Tube Length l_{cap}

The maximum allowed length of a connecting capillary between the injector, column and detector is defined as:

$$l_{cap} = \frac{384\theta^2 D_m V_R^2}{\pi N F d_{cap}^4} \qquad (13)$$

At 1% peak broadening and a capillary inner diameter d_{cap} of 0.25 mm:

$$l_{cap} = \frac{384 \times 0.01 \times 10^{-5} \times 6.7^2}{\pi \times 20\,000 \times 8 \times 10^{-3} \times 0.025^4}\, cm = 8.8\, cm$$

At 9% peak broadening:

$$l_{cap} = \frac{384 \times 0.09 \times 10^{-5} \times 6.7^2}{\pi \times 20\,000 \times 8 \times 10^{-3} \times 0.025^4}\, cm = 79\, cm$$

Comment

It is important to realize that the capillary length depends on the square of the retention volume of the peak of interest and on the inverse fourth power of the inner diameter of the capillary. In practice, the inner diameter of connecting tubes should not exceed 0.25 mm; they should be as short as possible.

Dilution Factor c_i/c_p

Due to the chromatographic process, the injected solutes become diluted, i.e. the concentration of the solute in the peak maximum, c_p, is lower than that in the injected solution, c_i. The dilution factor is defined as:

$$\frac{c_i}{c_p} = \frac{V_R}{V_i}\sqrt{\frac{2\pi}{N}} \qquad (14)$$

Thus, if 9% peak broadening is allowed, for $k = 1$:

$$\frac{c_i}{c_p} = \frac{6700}{28}\sqrt{\frac{2\pi}{20\,000}} = 4.2$$

and for $k = 10$:

$$\frac{c_i}{c_p} = \frac{37\,000}{28}\sqrt{\frac{2\pi}{20\,000}} = 23$$

Concentration at Peak Maximum c_p

If the concentration of each of the solutes in the sample is 1 ppm (10^{-6} g ml^{-1}), their concentrations at the respective peak maxima are, for $k = 1$:

$$c_p = c_i\frac{c_p}{c_i} = 10^{-6}\frac{1}{4.2}\text{g ml}^{-1} = 0.24 \times 10^{-6}\text{g ml}^{-1} \qquad (15)$$

and for $k = 10$:

$$c_p = 10^{-6}\frac{1}{23}\text{g ml}^{-1} = 0.043 \times 10^{-6}\text{g ml}^{-1}$$

Comment to Equations (14) and (15)

Of course the later eluted peak is diluted more than the early eluted one. However, if only the peak with $k = 10$ is important, the injection volume can be increased to 157 µl as shown in the discussion of equation (10). In this case the dilution factor of this

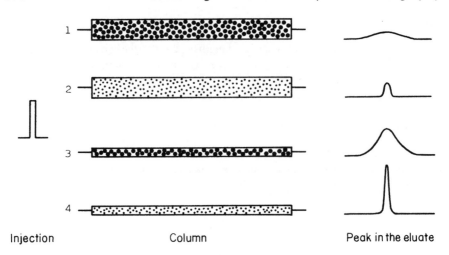

Figure 24.1 Peak shapes as obtained with different column diameters and separation performances (as a function of particle diameter at a given column length). Columns 1 and 3 are packed with a coarse stationary phase, columns 2 and 4 with a fine one. The packing quality, defined as reduced plate height h, is the same in all four cases. Separation performance is independent of column inner diameter; therefore peaks 1 and 3 as well as 2 and 4, respectively, are of the same width. The peak height in the eluate can be calculated from equation (15): it is higher when the column is thinner and the separation performance is better. Peak areas cannot be compared although the same amount is injected in any case if a concentration-sensitive detector is used: optimum flow rate depends on particle size and hence also the residence time in the detector.

second peak is as small as that for the first peak with an injection volume of 28 µl, namely 4.2. Therefore, the peak maximum concentrations will be the same in both cases, $0.24 \times 10^{-6} \, g \, ml^{-1}$. Note that this is only true if the respective maximum allowed injection volume is used for both peaks. In this case, c_p is independent of the peak retention volume (i.e. of the column inner diameter or of the k value) or of the plate number of the column.

In contrast to these considerations, Figure 24.1 shows the influence of column diameter and separation performance (or particle size of the stationary phase) on peak shape if a constant volume of a solution with a given concentration is injected.

UV Detector Signal A

The second component is nitrobenzene with a molar absorption coefficient, $\varepsilon = 10^4$ (254 nm), and a molar mass of $123 \, g \, mol^{-1}$. The signal, A, which will result in the UV detector can be calculated using the Lambert–Beer law, $A = \varepsilon \, cd$, where c is the concentration in $mol \, l^{-1}$ and d the path length in cm. Thus, if $c_p = 0.043 \times$

$10^{-6}\,\mathrm{g\,ml}^{-1} = 0.35 \times 10^{-6}\,\mathrm{mol\,l}^{-1}$ and $d = 1\,\mathrm{cm}$ (as is the case for many UV detectors), we obtain:

$$A = 10^4 \times 0.35 \times 10^{-6} \times 1 = 3.5 \times 10^{-3}\ \text{absorption units (a.u.)} \qquad (16)$$

Detection Limit

A common definition of the detection limit is a signal of four times (sometimes also twice) the noise level of the detector. The typical noise level of a UV detector is 10^{-4} a.u. This means a detection limit of 4×10^{-4} a.u. which is about one order of magnitude lower than the signal calculated in equation (16). Since the injected solution had a concentration of 1 ppm, the detection limit is about 0.1 ppm or, when related to the injection volume of 28 µl, 2.8 ng.

Comment to Equation (16) and Detection Limit

In the case of the detection limit we have to distinguish between the minimum detectable concentration and the minimum detectable mass. The minimum detectable concentration of a solute in the sample solution depends only on the detector properties and on the optical properties of the solute, i.e. its absorbance, if the maximum tolerable sample volume with respect to the retention volume of this solute is injected. The minimum detectable concentration is independent if the column dimensions, plate number or capacity factor.

In contrast to the concentration, the minimum detectable mass depends on the retention factor of the solute. The earlier a peak is eluted, the smaller is the maximum injection volume and, with the concentration of the sample solution being constant, the smaller is the absolute mass of solute injected. The same is true if the column inner diameter is reduced.

For trace analysis, this means that small bore columns should be used and low retention factors are advantageous if the sample volume is limited, as is often the case in clinical or forensic chemistry. If enough sample is available, e.g. in food analysis, it is not necessary to use small-bore columns and low retention factors. However, the analysis time, solvent consumption and column overload by accompanying substances (not discussed here) need to be kept in mind.

Pressure Drop Δp

The pressure drop can be calculated by use of the dimensionless flow resistance parameter, ϕ:

$$\Delta p = \frac{4\phi L_c \eta F}{d_c^2 \pi d_p^2} = \frac{\phi \times L_c(\mathrm{mm}) \times \eta(\mathrm{mPa\,s}) \times F(\mathrm{ml\,min}^{-1})}{4.7 \times d_c^2(\mathrm{mm}^2) \times d_p^2(\mu\mathrm{m}^2)}\ \mathrm{bar} \qquad (17)$$

With $\phi = 1000$ for slurry-packed particulate stationary phases and viscosity $\eta = 1$ mPa s (water or water–acetonitrile mixture):

$$\Delta p = \frac{1000 \times 250 \times 1 \times 0.48}{4.7 \times 4.6^2 \times 5^2} \text{bar} = 48 \text{ bar}$$

Comment

The pressure drop is markedly lower if low-viscosity organic solvents are used, e.g. hexane. In contrast to this, it is higher with methanol-water mixtures.

Conclusions

It is obvious that the calculations described are interesting and moreover are useful for routine laboratory work. They allow one to recognize some of the instrumental and chromatographic limitations and benefits of HPLC. However, the basic problem of chromatography is not solved with these considerations. What physicochemical conditions are needed to make possible the separation of the different types of molecules present in the sample mixture?

The different parameters that allow such separation are to be found in the resolution equation, which is the basic equation of liquid chromatography:

$$R_S = \frac{1}{4}\left(\frac{\alpha - 1}{\alpha}\right)\sqrt{N}\left(\frac{k}{1+k}\right) \tag{18}$$

In this equation there are only three parameters. Two of them, the plate number, N, and the retention factor, k, have been met in the calculations described. In principle, they can be altered at will. The plate number can always be adjusted by using a shorter column or by coupling two or more columns; the retention factor can be adjusted by using a mobile phase of greater or lower elution strength. In contrast, the separation factor, α, which represents the selectivity parameter, depends only on the physico-chemical properties of the separation system used, such as adsorption, hydrophobic or ion-exchange equilibria. Often these phenomena are not easy to understand and the chemist has to search for the best separation system by trial and error. Finding systems of high selectivity is the fine art of chromatography and the use of calculations such as those presented here is no more than an aid to this end.

25 Appendix 2: How to Perform the Instrument Test

Bruno E. Lendi and Diego Bilgerig
OmniLab Ltd, CH-8932, Mettmenstetten, Switzerland

25.1 INTRODUCTION

The present operating procedure describes the procedure and the documentation for the performance qualification (PQ) of HPLC systems.[1] It can be used for both isocratic and gradient systems with UV detection and it is independent of the instrument manufacturer. The procedure includes the tests of pump, autosampler, UV detector and column oven. It can be put into practice immediately.

All the proposed setpoint values are well proven recommended data and can be adapted to the individual requirements.

For laboratories working in accordance to GLP or GMP it is recommended to perform the test after the installation of a new HPLC system, later at least once a year and in any case after a change of location. If a component (e.g. the pump) needs to be repaired the relevant part of the test should be performed again.

25.2 TEST SEQUENCE

The following sequence is recommended for an efficient test procedure:

(a) Preparation of the mobile phases including degassing. Priming of the system, preferably the day before the test.

---

[1] L. Kaminski *et al.*, *J. Pharm. Biomed. Anal.*, **51**, 557 (2010).

Practical High-Performance Liquid Chromatography, Fifth edition   Veronika R. Meyer
© 2010 John Wiley & Sons, Ltd

(b) If the system is not computer-controlled it is necessary to write down the important instrument parameters and set-ups. These data will be needed after the test in order to re-establish the original conditions.

(c) Set-up of the working conditions for isocratic and gradient profile tests (see Section 25.3), thorough rinsing of both channels. Take care that the absorbance (baseline) is not higher than 1 AU at 100% B.

(d) Stabilization of the system under working conditions.

(e) Documentation of the system and instrument types, etc.

(f) Visual inspection of all components of the system.

(g) Pump test (Section 25.4). After a successful test both the pressure and flow meter must be removed. If the expected performance fails the work must be stopped and the necessary service or repair needs to be planned.

(h) Test for linearity and repeatability (test sequence of Section 25.3). The required number of vials are put into the autosampler, the test sequence is programmed and the test is started. It is absolutely necessary to work with both adequate time constant of the detector and integration parameters.

(i) If the column is not located within the oven during the test sequence the column oven test (Section 25.7.) can be performed simultaneously.

(j) Test for detector noise (Section 25.5.) immediately after the test sequence.

(k) Determination of the wavelength accuracy of the detector (Section 25.5).

(l) Registration of the gradient profile (Section 25.4).

(m) Calculations and documentation.

(n) All printouts are dated and signed, and the whole documentation is signed.

25.3 PREPARATIONS

The testing of a gradient system needs approximately a full working day.

Prepare all the mobile phases, test samples, and measuring equipment as well as the column needed for the instrument test.

System Data

Note the following data prior to the tests:

(a) system name;

(b) type and serial number of the individual components of the system;

(c) injection volume;

(d) optical path length of the detector cell;

(e) manufacturer and batch number of the column;

(f) manufacturers, batch numbers, expiration data, and weighing data of the compounds used for the test solutions.

Measuring Equipment

For the measurement of pressure, flow, and temperature it is necessary to use certified instruments only. The current certificates are copied and attached to the documentation.

Mobile Phases

For all HPLC sytems mobile phase 1 (MP 1) is needed: 600 ml of methanol (HPLC grade) and 400 ml of water are mixed, equilibrated to room temperature, and degassed. For gradient systems mobile phase 2 (MP 2) is also needed: 1000 ml of MP 1 and 2 ml of acetone are mixed and degassed for a short time only (no vacuum!).

Test Solutions

Note all the compound data as mentioned above. All solutions are prepared quantitatively with measuring flasks and pipettes.

Stock solution: for the tests for repeatability/linearity/carry over: 199.5–200.5 mg benzophenone (puriss. p.a.) is dissolved to 100 ml of methanol.

100% test solution: Since the absorptivity of the 100% test solution needs to be between 0.7 and 0.9 AU the stock solution is to be diluted with MP 1 according to the injection volume and the detector flow cell path length. The matching test solution is prepared as follows:

Test solution number	Optical path length of cell	Injection volume	System-specific 100% test solution
1	10 mm	1 µl	Stock solution (undiluted)
2	5 mm	5 µl	40 ml stock solution to 100 ml with MP 1
3	10 mm	5 µl	20 ml stock solution to 100 ml with MP 1
4	10 mm	20 µl	5 ml stock solution to 100 ml with MP 1
5	8 mm	20 µl	6 ml stock solution to 100 ml with MP 1
6	6 mm	20 µl	15 ml stock solution to 100 ml with MP 1
7	6 mm	50 µl	6 ml stock solution to 100 ml with MP 1
8	8 mm	100 µl	3 ml stock solution to 100 ml with MP 1

(Only one of these test solutions needs to be prepared.)

50% test solution: 25 ml of the selected 100 % test solution is diluted to 50 ml with MP 1.
1% test solution: 5 ml of the 50% test solution is diluted to 250 ml with MP 1.
0.1% test solution: 5 ml of the 1% test solution is diluted to 50 ml with MP 1.

Anthracene solution: for the test of the wavelength accuracy: 9.5–10.5 mg anthracene p.a. is dissolved to 100 ml. 1 ml of this solution is dissolved to 100 ml with MP 1, giving a concentration of approx. $1\,\mu g\,ml^{-1}$.

If the detector has a built-in holmium oxide filter the wavelength accuracy is tested with this filter in accordance to the manufacturer's test procedure.

Isocratic Working Conditions

Mobile phase: MP 1
Flow rate: $1.5\,ml\,min^{-1}$
Wavelength: 254 nm
Run length: 8 min
Detector time constant (or response time, rise time): $\leq 1\,s$
Column: LiChrospher 100 RP 8, 5 µm, steel, 12.5 cm × 4 mm i.d.
Temperature: ambient
Test sequence:

Injection	Name	Description
1	Blank	Mobile phase 1
2	0.1%/1	0.1% test solution
3	0.1%/2	0.1% test solution
4	1%/1	1% test solution
5	1%/2	1% test solution
6	50%/1	50% test solution
7	50%/2	50% test solution
8	100%/1	100% test solution
9	100%/2	100% test solution
10	100%/3	100% test solution
11	100%/4	100% test solution
12	100%/5	100% test solution
13	100%/6	100% test solution
14	Blank	Mobile phase 1 (carry over)

Gradient Profile Conditions

Mobile phase: MP 1/MP 2
Flow rate: $2\,ml\,min^{-1}$
Wavelength: 280 nm
Run length: 70 min

Detector time constant (or response time, rise time): ≤ 1 s
Column: LiChrospher 100 RP 8, 5 µm, steel, 12.5 cm × 4 mm i.d.
Temperature: ambient
Gradient profile:

Time [min]	MP 1 [%]	MP 2 [%]	Description
0	100	0	Step 0
5.0	100	0	
5.1	95	5	Step 1
10.0	95	5	
10.1	75	25	Step 2
15.0	75	25	
15.1	50	50	Step 3
20.0	50	50	
20.1	25	75	Step 4
25.0	25	75	
25.1	5	95	Step 5
30.0	5	95	
30.1	0	100	Step 6
40.0	0	100	
40.1	100	0	Step 7
50.0	100	0	
50.1	50	50	Step 8
55.0	50	50	
55.1	0	100	Step 9
60.0	0	100	
60.1	50	50	Step 10
65.0	50	50	
65.1	100	0	
70.0	100	0	

25.4 PUMP TEST

Visual Inspection

The pump is switched on (without flow) and the instrument (pump head, displays, etc.) as well as the connecting fittings and tubes are checked visually. In the case of complaint the necessary measures are planned and documented.

Pressure Display at Zero Flow

The drain valve is opened and the reading of the pressure display is noted.

Setpoint value:

☑ Pressure display ±1 bar.

In the case of a deviation the display is adjusted (if possible); the facts are documented.

Pressure Accuracy, Flow Accuracy, Flow Precision

The measuring instruments for flow and pressure are connected and the pump is switched on according to the isocratic working conditions. The system is run for 10–15 min. Afterwards the displays of the instruments as well as the pressure displayed at the pump are documented during 6 min at 1 min intervals.

Setpoint values:

☑ Pressure accuracy: ±10 bar (pump display minus pressure gauge display).
☑ Flow rate accuracy: max. $0.1\ ml\ min^{-1}$ (set value minus flowmeter display).
☑ Flow precision: max. $0.02\ ml\ min^{-1}$ (highest minus lowest measured value).

Gradient Profile[2]

Gradient systems are tested for the accuracy and precision of the gradient profile (Figure 25.1), the delay volume, and the mixing volume with channels A and B.

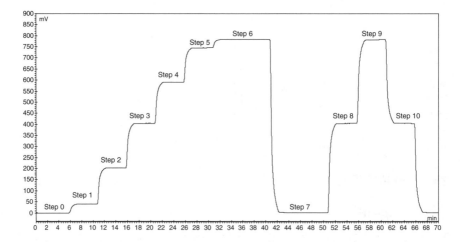

Figure 25.1 The gradient profile as generated with this instrument test. The *y* axis can be in mAU instead of mV.

---

[2] S. Marten, A. Knöfel and P. Földi, *LC GC Eur.*, **21**, 372 (2008).

Ternary and quaternary systems are also tested with A to C and C to D, respectively, if necessary. The gradient profile conditions (Section 25.3) are programmed, the system is rinsed thoroughly, and the gradient is run and registered. The profile is evaluated as shown in Figure 25.2.

Remarks

If necessary, the duration of the individual steps can be prolonged to obtain a stable baseline during the last 2 min.

If the test needs to be repeated later on it is necessary to use the original system configuration because the delay volume of the system is calculated from the gradient profile.

Gradient Accuracy

The difference between experimental and programmed %B values of steps 1–5 is calculated for the judgement of the gradient accuracy.

Setpoint value:

☑ Gradient accuracy: ±2 percentage points.

Gradient Precision

The evaluated data (in mV or AU) of steps 3, 8, and 10 is used for the judgement of the gradient precision. Of the three results, the largest difference between the set and experimental value is the relevant test datum.

Setpoint value:

☑ Gradient precision: ±2 percentage points.

Determination of Delay Volume and Mixing Volume

Delay time (t_{delay}) and mixing time (t_{mix}) are determined as shown in Figure 25.3. For the calculation of the respective volumes the times are multiplied with the flow rate of the pump. The data are for information only and are not judged.

25.5 UV DETECTOR TEST

Visual Inspection

The detector is switched on and the displays and connections are checked visually. In the case of complaint the necessary measures are planned and documented.

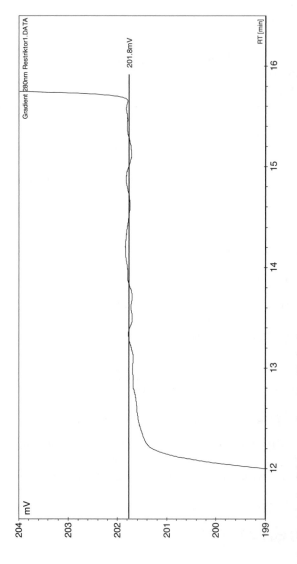

Figure 25.2 Evaluation of the gradient profile (as an example, only one step is shown). For the graphical determination it is necessary to enlarge every step of the chromatogram individually and to print it out. The height of the steps is determined graphically in mV or mAU. The particular proportion of channel B (in %) is calculated from these step heights (Section 25.8).

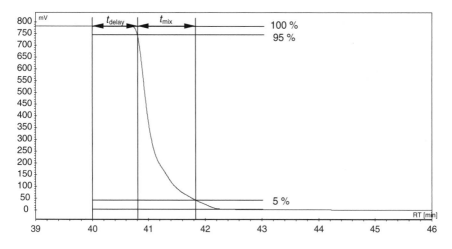

Figure 25.3 Determination of delay time and mixing time.

Noise

The whole chromatographic system is run for at least 30 min under the isocratic working conditions. For the determination of the dynamic noise the detector signal is registered during 10 min or longer. The noise is determined as shown in Figure 25.4 by

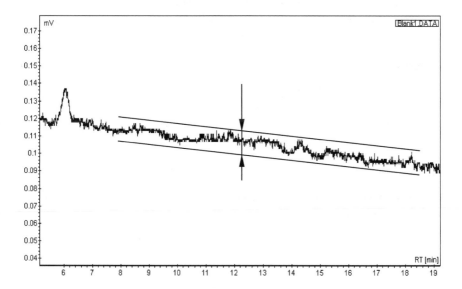

Figure 25.4 Graphical determination of the noise.

using a suitable part of the signal of at least 10 min duration. Disturbing or ghost peaks must be absent in this part of the baseline.

The noise in mm, determined from the printout, is calculated as AU and documented.

Setpoint value:

☑ Noise: 1×10^{-4} AU or less.

Wavelength Accuracy

The detector cell is filled with the anthracene solution. The measuring wavelength is varied between 248 and 254 nm in 1 nm steps. The resulting absorbance is documented and its maximum is determined. As an alternative approach if a scanning detector is used, the UV spectrum can be registered and attached to the documentation.

Setpoint value:

☑ Wavelength accuracy: 251 nm $\pm$ 2 nm.

If the detector is equipped with a built-in holmium oxide filter, the wavelength accuracy is tested with this filter in accordance to the manufacturer's testing procedure.

Remark: If the automated zeroing after a wavelength change cannot be switched off it is necessary to inject (into the detector cell) first the mobile phase, then to adjust zero, then to inject the anthracene solution, and finally to read off the absorbance.

Linearity

For the calculation of the linearity the injections 2–9 of the test sequence are evaluated. The correlation coefficient of the regression function is determined from the peak areas.

Setpoint value:

☑ Correlation coefficient: 0.99 or higher.

The mean values of the peak areas of the 0.1%, 1%, 50%, and 100% test solutions are calculated, respectively. The following ratios are calculated from the means and documented.

Setpoint values:

☑ Ratio 1% to 0.1%: 8.0–12.0
☑ Ratio 100% to 1%: 97–103

25.6 AUTOSAMPLER TEST

Visual Inspection

The autosampler is switched on and the displays and connections are checked visually. In the case of complaint the necessary measures are planned and documented.

Repeatability

For the determination of repeatability injections 8–13 of the test sequence are evaluated. The precision (in %) is calculated from the peak areas (Section 25.8).

Setpoint value:

☑ Precision: ± 0.8% or less.

Carry Over

The carry over of the autosampler is calculated from injections 13 and 14 of the test sequence.

Setpoint values:

☑ Carry over: 0.1% or less for autosamplers with flushing, 0.2 % or less for autosamplers without flushing.

Remark: The injection volume and the number of flushing cycles is to be noted in the documentation.

25.7 COLUMN OVEN TEST

Visual Inspection

The column oven is switched on and the displays and connections are checked visually. In the case of complaint the necessary measures are planned and documented.

Accuracy and Stability of the Temperature

The test is performed at $20\,°C$ and $40\,°C$. If the oven is not designed for cooling the $20\,°C$ test is renounced. In addition, the oven can also be tested at a higher temperature, namely the one which is used on a regular base.

The temperature in the column compartment can be very different depending on the oven type, therefore the temperature probe is attached as follows:

(a) In forced air circulation ovens the probe is mounted as close to the column as possible.

(b) In ovens without air circulation the probe is mounted directly on the metal block (heat exchanger).

The location of the temperature probe must be documented (e.g. 'measured at the column').

The oven is stabilized at the setpoint temperature for at least 30 min before the measurement is performed. For the test the experimental temperature, measured by the temperature probe, and the oven display are registered and documented for 30 min in intervals of 10 min.

Setpoint values:

☑ Accuracy of the display: max. $\pm 1\,°C$ (mean of the oven display minus setpoint value).

☑ Accuracy of the temperature: max. $\pm 3\,°C$ (mean of the probe measurements minus setpoint value).

☑ Temperature stability: max. $1\,°C$ (highest minus lowest measured value).

25.8 EQUATIONS AND CALCULATIONS

Calculation of the percentage amount of channel B in the gradient test:

$$\text{step } x[\%] = \frac{\text{height}_{\text{step } x} - \text{height}_{\text{step } 0}}{\text{height}_{\text{step } 6} - \text{height}_{\text{step } 0}} \cdot 100$$

Calculation of the autosampler precision:

$$\text{precision } [\%] = \frac{\text{standard deviation}}{\text{mean of measured values}} \cdot 100$$

$$\text{standard deviation} = \sqrt{\sum_{i=1}^{n} \frac{(\text{value}_i - \text{mean of measured values})^2}{n-1}}$$

Calculation of the autosampler carry over:

$$\text{carry over } [\%] = \frac{\text{area}_{\text{injection } 14}}{\text{area}_{\text{injection } 13}} \cdot 100$$

Correlation coefficient for detector linearity:

$$r = \frac{n(\sum xy) - (\sum x) - (\sum y)}{\sqrt{[n \sum x^2 - (\sum x)^2] \cdot [n \sum y^2 - (\sum y)^2]}}$$

25.9 DOCUMENTATION

All measured values, calculations and findings are noted in the documentation. The documentation must be signed and dated by hand by the technician. Computer printouts must be dated and signed, too. Tests which do not pass must be documented, even if the respective components can be adjusted or repaired on the spot. Repeated tests must be marked as such.

26 Appendix 3: Troubleshooting

Bruno E. Lendi

OmniLab Ltd, CH-8932 Mettmenstetten, Switzerland

26.1 PRESSURE PROBLEMS

Pressure Too High

Soiled mobile phase (and, as a consequence, also soiled frits, capillaries, valve and/or column):
→ Filter the mobile phase; dirty buffers are best thrown away and prepared freshly. Pure water needs to be replaced more frequently than organic solvents due to the formation of algae and bacteria.

Soiled pump outlet frit:
→ Replace the frit. If black particles are visible on the frit (rather common) it is necessary to replace the high-pressure seal and perhaps also the piston. In addition, check if the seal is resistant to the solvent in use. (However, today's seal materials should be resistant to all common solvents, including water.)

Clogged capillaries:
→ Replace the capillaries or install them in reverse direction. Caution: do not overtighten the fittings of peek capillaries.

Wrong fitting (or only wrong ferrule) installed:
→ Use the correct fitting and ferrule or use fingertight fittings. Do not mount old capillaries and fittings from other systems, as this can lead to high backpressure or extra-column volumes.

Practical High-Performance Liquid Chromatography, Fifth edition   Veronika R. Meyer
© 2010 John Wiley & Sons, Ltd

High-pressure injection valve clogged:
→ Replace rotor seal and/or stator. If the surface of the stator is scratched it must be replaced in any case. Otherwise the new rotor seal will be ruined within a short time.

Column too old or contaminated:
→ Replace the column or flush it, perhaps with reversed direction of flow (without connection to the detector).

Precipitation of the sample:
→ Check the miscibility of sample, sample solvent, and mobile phase. It is always best to dissolve the sample in the mobile phase (for gradient separations in the mixture which is present at the time of injection).

Wrong composition of the mobile phase:
→ Check the mixture or gradient; often it is best to prepare fresh mobile phase as requested by the standard operating procedure.

Temperature fluctuation:
→ A temperature drop of 10 °C of the mobile phase can result in a backpressure rise of up to 40 bar.

Inaccurate pump flow:
→ Check, if necessary call the manufacturer service.

Pressure Too Low

Worn high pressure seals of the pump:
→ Replace the seals. Important: do not use water to break in the new seals. 2-Propanol or methanol are recommended.

Leaky pump valves:
→ Clean by a high flow rate or in an ultrasonic bath, or replace the valves.

Leak after the pump:
Check systematically and repair.

Wrong composition of the mobile phase:
→ Check the mixture or gradient; often it is best to prepare fresh mobile phase as requested by the standard operating procedure.

Temperature fluctuation:
→ A temperature rise of 10 °C of the mobile phase can result in a backpressure drop of up to 40 bar.

Inaccurate pump flow:
→ Check, if necessary call the manufacturer service.

Pressure Fluctuations

Damaged pump valves and/or worn seals:
→ Replace valves or seals.

Air or gas bubbles in the pump head:
→ Flush the pump with open drain valve at high flow rate. Degas the mobile phase or use an on-line degasser. (This is absolutely necessary with low-pressure gradient systems.)

Clogged inlet filter of the pump:
→ Check and replace if necessary. Aqueous mobile phases are a higher risk than organic solvents.

26.2 LEAK IN THE PUMP SYSTEM

Leak of the piston flushing seals or of the high pressure seals:
→ Replace the seals, check the pistons for damage and replace if necessary.

Rise of the piston flushing solvent level:
This is a definite sign that the high pressure seals must be replaced.

26.3 DEVIATING RETENTION TIMES

Usually the problem comes from the pump or the gradient unit:
→ Check the flow accuracy with measuring flask and stop-watch. It may be necessary to replace the seals or valves.

Leak before the column:
→ Check and repair systematically.

Temperature fluctuations (especially in summertime):
→ Use a column oven with heating and cooling opportunities.

The more volatile component evaporates from the mobile phase reservoir:
→ Cool the reservoir. Cover it with an air-tight lid with valve (thus preventing evaporation, resulting in the advantageous side-effect that the laboratory air is free of solvent vapors). Mix the solvents within the HPLC system even in the case of isocratic separations.

Gradient profile or solvent composition deviates from the written method:
→ Check the pump valves for leaks. A common cause are crystallized salts in the valves. The remedy is simple: flush with salt-free solvents before the pump is switched off! Caution: different HPLC systems have different mixing and dwell volumes. With gradient separations, check the equilibration time between injections and prolong it if necessary. Check the gradient method.

Generally, it should be found out if the dead time t_0 has changed (this indicates a pump or leak problem) or if the k values are different (this indicates a solvent or temperature problem).

26.4 INJECTION PROBLEMS

Sample Carry Over

Usually the problem comes from the injection system:
→ Replace worn parts of the autosampler, i.e. needle, capillary between needle and valve, rotor seal and/or stator.

The less common but tricky reasons:
→ Look for poor capillary joints and replace them. Replace the injector loop. Replace capillaries with rough inner surface. Check the column and replace it if necessary. Overfilled sample vials. Wrong flushing solvent of the autosampler without cleaning effect.

Poor Injection Reproducibility

Usually the problem comes from the injection system:
→ Check and replace the worn parts as described above.

In addition, almost every part of the HPLC system is suspicious (!):
→ Poor performance of the pump. Nonreproducible gradient. The sample is too cold. The mobile phase is not clean. The flushing solvent reservoir of the autosampler is empty. Part of the sample remains in the injection system. Sample is precipitating or crystallizing in the injection system. The sample positioning in the autosampler is no longer adjusted. Poorly suited septa are used. A vacuum is built up in the sample vials. Too rapid transfer of the sample from vial to loop. Leaky injection syringe in the autosampler; replace the piston or the whole syringe.

Check if the autosampler injects the programmed volume: the sample vial is weighed before and after the injection. If the sampling is done without loss of solution the mass difference must match the volume (with consideration of the sample solution density). If the autosampler draws more solution in order to reduce the carry over this extra volume must be considered in the calculation.

26.5 BASELINE PROBLEMS

Noisy Baseline

Low energy of the UV lamp:
→ Replace and adjust if necessary.

Low energy at low UV wavelengths:
→ Replace the lamp even if the energy at e.g. 254 nm is still acceptable. Replace the flow cell windows. Check if the mobile phase is UV transparent and clean. Is the solvent degassed properly? Use a degasser.

Noise is still too high after UV lamp replacement:
→ Check the correct adjustment of the lamp. Clean the flow cell windows or replace them.

The pump does not work properly:
→ Check and repair as described above.

Air or gas bubble in the detector cell:
→ Flush the cell with high flow rate. If the cell is pressure-resistant, bubble formation can be prevented by the installation of a long capillary tube (e.g. 0.25 mm × 10 m) at the detector outlet.

Electrical signals from the environment:
→ Use an electronic filter, install the HPLC instrument at another place.

Air draught:
→ Switch off the air-conditioning system, install the instrument at another place.

Baseline Drift

Mobile phase:
→ The reasons can be warming-up, incomplete mixing or incomplete degassing. It is recommendable to use a degasser and a column oven.

Soiled column:
→ Often the problem disappears by being more patient with regard to flushing. Compounds from earlier injections can still be within the column which need to be eluted with suitable mobile phases.

Strong pressure fluctuations of the pump:
→ See the proposals given above under 'Pressure Fluctuations'. If the drift occurs only in the low UV, it is well possible that the pump and/or detector are not suited. Use a pump with low pulsation and a detector with a flow cell which is insensitive to refractive index changes and is suitable for gradient separations.

Soiled detector flow cell:
→ Clean the windows or replace them.

Replace the UV lamp?
→ No, this may only help when old instruments and single-beam diode array detectors are in use. With modern detectors, the common cause comes from the electronics which can only be repaired by the manufacturer.

When should the UV lamp be replaced?
→ This is a simple question with a less simple answer. Many manufacturers of modern lamps and detectors recommend a replacement after 5000–7000 hours of use. If low-UV detection is performed an earlier replacement is necessary. In this region, the emitted energy drops earlier. As a general rule it can be stated that too early a replacement is to the pleasure of the manufacturer and that too late a replacement gives imprecise or even wrong analyses.

26.6 PEAK SHAPE PROBLEMS

Negative Peaks (see also Sections 6.9 and 13.4)

Unsuitable mobile phase:
→ The sample was dissolved or diluted in a solvent with higher transmittance than the mobile phase at the chosen wavelength. The solvent is not clean or of poor quality.

The sample is excited by the monochromatic light and emits fluorescence light:
→ Check the method and change it.

Tailing, Peaks Too Broad, Double Peaks

Old or damaged column:
→ Especially in the case of double peaks, the problem usually comes from a poor column. It must be replaced.

Extra-column volumes in the system:
→ The main causes are worn rotor seals, poor capillary fittings and capillaries with too large an inner diameter.

Detector cell with too large an internal volume:
→ This circumstance needs special attention with low flow rates and small columns. It is necessary to use a detector cell with low internal volume.

Injection volume too large:
→ This can lead to nonlinearity, peak broadening, tailing, and even a damaged column. Check the method.

Ghost Peaks (Section 19.9)

Peaks from the previous injection are showing up:
→ Too little time between consecutive injections. Increase the time interval. Unsuitable gradient profile. Especially at the end of the gradient, the profile must be calculated in such a way that all compounds, also the noninteresting and highly retained ones, are eluted.

An unknown peak always appears at the same retention time:
→ If the peak becomes smaller with every blind injection there is a high probability of sample carry over. Measures as explained above under 'Sample Carry Over'. If the peak appears with more or less constant size it may be an injection peak (system peak). Measures can be difficult or impossible. The mobile phase must not be detector-active, i.e. the UV cutoff wavelength must be taken into consideration. Perhaps the peak disappears when a mobile phase from another batch or another manufacturer is used, especially in low UV and with acetonitrile or trifluoroacetic acid. Injected air can produce a peak.[1]

Peaks Too Small (Peaks Too Large)

Wrong sample or sample diluted too much (too little):
→ Check the sample solution.

Erroneous injection:
→ Check the autosampler, check if the correct injection volume is requested by the control unit, or check the manual injection. See also the sample volume check at 'Poor Injection Reproducibility' above.

Wrong detector range, signal output (in mV) is not taken over correctly by the control unit:
→ Check the detector or the set-ups at the data system.

Wrong detection wavelength:
→ Check the detector and/or the method.

26.7 PROBLEMS WITH LIGHT-SCATTERING DETECTORS

The large and nonspecific sensitivity of light-scattering detectors (ELSD) has its pitfalls. The most common ones are the following.

Background signal too large:
→ The mobile phase must be selected with great care. Especially water, but also the organic solvents, must be of high purity. Unfortunately, the quality 'Gradient Grade' is not always a guarantee for the purity required by this detection method. Even glass containers can be the reason for an increased background due to sodium ions. Salts should not be used at all; if they are indispensable their concentration must be as low as possible.

---

[1] J.W. Dolan, D.H. Marchand and S.A. Cahill, *LC GC Int.*, **10**, 274 (1997) or *LC GC Mag.*, **15**, 328 (1997) (with printing errors: ml instead of μl).

The linearity is poorer than with UV detection:
→ Note that the linear range of an ELSD is comparably small. If the concentrations of the various analytes differ markedly it can be necessary to perform two or more injections with different volumes thus allowing the quantitation of an analyte at its best suited amount. Usually the different sensitivity levels of the detector are not suitable attenuators. If possible, do not change the attenuation during an analysis.

Known and expected peaks are not visible:
→ Note the different volatility of the analytes. If the compounds are volatile or semi-volatile it is often necessary to lower the temperature of the evaporating tube. Depending on the instrument it can be even possible to work with an aqueous mobile phase at 30 °C without an excessive background rise.

Irreproducible analytical results:
→ Many reasons can be a possibility. If other causes, such as a malfunction of the autosampler, can be excluded the nebulizer tube could be contaminated or clogged.
→ In many cases the problem is solved after a cleaning of the tube for a couple of minutes in an ultrasonic bath. Sometimes the gas pressure for the nebulizer needs to be increased or decreased.

Carry over:
→ The two main causes are the evaporator tube heating coil and/or the nebulizing chamber. The chamber can usually be cleaned with a bottle brush. The tube can be heated to 100 °C and with some luck the contamination disappears within a few minutes. After cooling the tube must be rinsed thoroughly with a suitable mobile phase.

26.8 OTHER CAUSES

Chromatography's enemy is hurry:
→ Long equilibration times previous to the first injection help to save time. Flush long enough after the last injection, especially if salt solutions have been used.

Salts in the mobile phase:
→ It is best to avoid them. Unfortunately, this is not possible in many cases. Possible measures: Before shutdown, replace the solvents with salts by pure solvents, then flush the whole system (including also the degasser) for a long enough time. If this is not possible it is necessary to keep a constant minimum flow rate of at least 100 µl/min.

Prevention by maintenance:
→ Every system runs as well as it is maintained. Some approximate figures which should not be exceeded: The pump seals should be replaced after one or two years. The autosampler rotor seal should be replaced not later than after 7000 injections

(with salt solutions after 3500 injections). Replace the syringe piston annually. Replace the UV detector lamp (the lamps) after 4000–7000 hours. Tungsten-halogen lamps for the visible range should be replaced after no more than 2000–3000 hours, otherwise a complete breakdown (with unwanted consequences) can be the result. Clean or replace the cell windows every two years, depending on the sample types even earlier.

26.9 INSTRUMENT TEST

$\rightarrow$ Even if the laboratory does not work under GLP guidelines it is recommendable to check the whole HPLC system with a validated standard operating procedure (SOP) in regular time intervals. The results should be documented. The guideline can be found in Chapter 25.

27 Appendix 4: Column Packing[1]

Column preparation is not the only critical factor, but nevertheless represents one of the most important aspects affecting the quality of the chromatographic system.

Particles below 20 μm in diameter cannot be packed in dry form as they tend to become lumpy, thus precluding the preparation of high-performance columns. The material must be suspended in a liquid to give a slurry. A high-pressure pump then conveys the slurry into the column at great speed. There are almost as many methods for wet packing as there are column manufacturers. Proponents are convinced that their own ways and means are the best, making it difficult for the uninitiated to choose between them. Slurry production itself is based on a few essential principles:

(a) The dry product may be suspended in a fluid of the same density as the packing material, thus preventing sedimentation. This is known as the *balanced-density method*.[2] Sedimentation separates the stationary phase according to size; hence the larger particles would sink to the bottom and all the smaller ones would be at the top, thus imparting the separating performance.

As the density of silica is about 2.2 g em^{-3}, halogenated hydrocarbons such as dibromomethane (CH_2Br_2), tetrachloroethylene (C_2Cl_4), carbon tetrachloride (CCl_4) and diiodomethane (CH_2I_2) must be used for the slurry.

---

[1] Dry packing suitable for particles larger than 20 μm is not covered here. However, the following sources give useful information on this aspect: L.R. Snyder and J.J. Kirkland, *Introduction to Modern Liquid Chromatography*, John Wiley & Sons, Inc., New York, 2nd ed., 1979, p. 207; J. Klawiter, M. Kaminski and J.S. Kowalczyk, *J. Chromatogr.*, **243**, 207 (1982). Suggestion: add material in batches and tap the column vertically.

[2] Determination of the density of granular solids was described by F. Patat and K. Kirchner, *Praktikum der technischen Chemie*, Walter de Gruyter, Berlin, 4th ed., 1986, pp. 46, 47 (pycnometer method). Guidelines can also be found on the internet. Possible problem: the pores of porous materials must be filled with liquid!

Practical High-Performance Liquid Chromatography, Fifth edition   Veronika R. Meyer
© 2010 John Wiley & Sons, Ltd

    Slurry preparation method: 78 volume parts of dibromomethane ($d =$ 2.49 g ml^{-1}) and 22 volume parts of dioxane ($d = 1.03$ g ml^{-1}) give a mixture of density 2.17 g ml^{-1}.

(b) A high-viscosity liquid such as paraffin oil or cyclohexanol is used for the slurry. Sedimentation is relatively slow. The disadvantage of this method is that the slurry gives a large pressure drop and cannot be pumped into the column at such a fast rate. Hence, the partly packed column should be heated, e.g. by gradual immersion in a water-bath, to reduce the viscosity of the suspending fluid in the packing and reduce the flow resistance accordingly.

(c) Any solvent can be used for slurry preparation without bothering about density or viscosity. Good columns can be obtained by working quickly. Tetrachloroethylene with its high density of 1.62 g ml^{-1} is a good compromise as no mixture is required (a silica bead 5 µm in diameter takes only 0.4 mm min^{-1} to sink). A 0.01 M solution of ammonia in water can also be used for slurry preparation, the electrostatic effect of the ammonia preventing agglomeration.

(d) A slurry vessel equipped with a magnetic stirrer which moves continuously as filling proceeds has also been proposed.[3]

The slurry is homogenized in an ultrasonic bath as about a 5% mixture. A higher concentration seems to reduce the attainable plate number. The column is bolted to a precolumn, which is also packed but is not used for chromatography, and to a steel slurry reservoir generally of the form shown in Figure 27.1.

    The slurry is poured in and topped with solvent. No air should be allowed into any part of the column–guard column–slurry reservoir system. The lid is screwed on tightly and a connection made to the pump. The pump must work at maximum stroke or pressure as soon as it is activated. The column must be filled as quickly as possible under a pressure many times greater than the subsequent operating level (with the exception of materials that are not absolutely pressure resistant such as styrene–divinylbenzene, in which case the manufacturer's maximum pressure guidelines should be followed). The column may be filled from top to bottom or vice versa[4] (slurry reservoir below, column above, as shown in Figure 27.2) and tests should be made to establish which arrangement provides the greatest number of theoretical plates in each individual case. For security reasons the arrangement must always be vertical!

    Subsequent collapse of the packing can be prevented by pumping about 0.5 l of solvent through at this maximum pressure level, that used later for chromatography being perfectly adequate. A silica column should be equilibrated with an eluotropic series of solvents (e.g. methanol → ethyl acetate → *tert*-butyl-methyl ether → hexane) if the slurry has been topped up with water. Finally, the column

---

[3] H.R. Linder, H.P. Keller and R.W. Frei, *J. Chromatogr. Sci.*, **14**, 234 (1976).

[4] P.A. Bristow, P.N. Brittain, C.M. Riley and B.R. Williamson, *J. Chromatogr.*, **131**, 57 (1977).

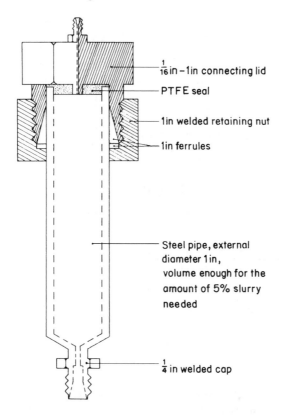

$\frac{1}{16}$ in – 1in connecting lid

PTFE seal

1in welded retaining nut

1in ferrules

Steel pipe, external
diameter 1 in,
volume enough for the
amount of 5% slurry
needed

$\frac{1}{4}$ in welded cap

Figure 27.1 Slurry reservoir for HPLC column packing.

is separated from the guard column or slurry reservoir and connected to the chromatograph.

Method for Packing Silica-Based Columns

(a) Stationary phase[5] and suspension liquid (approximately 5% mass/volume) are homogenized in an ultrasonic bath.

(b) A frit and end fitting are fixed to the column, which is then stoppered with a 'plug' and connected to the precolumn. Both are filled to the top with suspension liquid by means of a large syringe and the pump is switched on briefly to ensure that the tubing is also filled with liquid.

---

[5] If the stationary phase has a poor particle size distribution or if regenerated material is packed again it is recommended that the smallest particles be removed by sedimentation. Suitable liquids therefore are dioxane–chloroform (9 : 1) for silica or dioxane–ethanol (9 : 1) for reversed-phase materials.

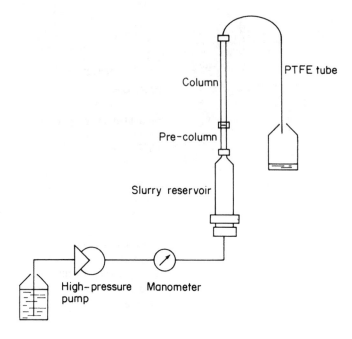

Figure 27.2 Equipment for packing HPLC columns.

(c) The slurry reservoir is connected to the precolumn, slurry is poured in and suspension liquid used to top it up; then the lid is tightened.

(d) The pump is connected to the slurry reservoir, the plug is removed and a PTFE tube connected. The pump is switched on at maximum capacity.

(e) 0.5 l of solvent is pumped through at maximum pressure. If silica is to be packed, then the solvent used subsequently as the mobile phase should also be used at this point. Methanol is recommended for reversed-phase columns.

(f) The pump is switched off, the column is detached from the precolumn and closed with an end fitting with frit; finally, it is built into the liquid chromatograph, but not yet connected with the detector. It is conditioned with mobile phase.

Index of Separations

Practical High-Performance Liquid Chromatography, Fifth edition   Veronika R. Meyer
© 2010 John Wiley & Sons, Ltd

Subject Index

α value, *see* Separation factor

A solvent, 83
Absorbance, 97, 370
Absorption unit, 97
Accreditation, 318
Accuracy, 311, 313
Acetonitrile for HPLC, 177
Acidity of solvents, 83
Activated gel, 251
Additives to mobile phase, 44, 66, 86, 167, 179, 212, 243
Adsorption chromatography, 159
  applications, 168
  mobile phase, 162
  test mixture, 147
Adsorption isotherm, 162
Affinity chromatography, 249
  applications, 252
  elution possibilities, 250
Agarose, 134
Alumina, 133
Amino phase, 129, 197
Amperometric detector, 103
  pulsed, 104
Analysis, 285
  qualitative, 285
  quantitative, 291

trace, 287, 371
Antioxidant, 66
APCI, 112
APESI, 112
Apparatus test, 373
Asymmetry of peaks, 41, 142, 392
Atmospheric pressure chemical ionization, 112
Atmospheric pressure electrospray ionization, 112
AU, AUFS, 97
Autosampler, 76
  problems, 390
  test, 383
Axially compressed columns, 117

B solvent, 83
  and peak height, 300
Balanced-density method, 397
Band broadening, 19
Band width, spectral, 98
Baseline problems, 93, 304, 390
Basicity of solvents, 83, 166
BHT, 66
Bed reactor, 307
Bimodal columns, 242
Binary mixture, 88, 163, 175
Boiling point, table, 82

Practical High-Performance Liquid Chromatography, Fifth edition   Veronika R. Meyer
© 2010 John Wiley & Sons, Ltd